Selected Papers from 2020 IEEE International Conference on High Voltage Engineering (ICHVE 2020)

Selected Papers from 2020 IEEE International Conference on High Voltage Engineering (ICHVE 2020)

Editors

Issouf Fofana
Bo Zhang

MDPI • Basel • Beijing • Wuhan • Barcelona • Belgrade • Manchester • Tokyo • Cluj • Tianjin

Editors
Issouf Fofana
Université du Québec à Chicoutimi
Canada

Bo Zhang
Tsinghua University
China

Editorial Office
MDPI
St. Alban-Anlage 66
4052 Basel, Switzerland

This is a reprint of articles from the Special Issue published online in the open access journal *Energies* (ISSN 1996-1073) (available at: https://www.mdpi.com/journal/energies/special_issues/ICHVE2020).

For citation purposes, cite each article independently as indicated on the article page online and as indicated below:

LastName, A.A.; LastName, B.B.; LastName, C.C. Article Title. *Journal Name* **Year**, *Volume Number*, Page Range.

ISBN 978-3-0365-6175-2 (Hbk)
ISBN 978-3-0365-6176-9 (PDF)

Cover image courtesy of Issouf Fofana.

Contents

About the Editors vii

Issouf Fofana and Bo Zhang
High-Voltage Engineering and Applications in Our Modern Society
Reprinted from: *Energies* **2022**, *15*, 8341, doi:10.3390/en15228341 1

Hao Yang, Haotian Zhang, Wen Cao, Xuanxiang Zhao, Ran Wen, Junping Zhao, Shengwu Tan and Pengchao Wang
Optical Diagnostic Characterization of the Local Arc on Contaminated Insulation Surface at Low Pressure
Reprinted from: *Energies* **2021**, *14*, 6116, doi:10.3390/en14196116 5

Liang He, Jie Yang, Ziwei Zhang, Zongwu Li, Dengwei Ding, Minghu Yuan, Rong Li and Mao Chen
Research on Mechanical Defect Detection and Diagnosis Method for GIS Equipment Based on Vibration Signal †
Reprinted from: *Energies* **2021**, *14*, 5507, doi:10.3390/en14175507 17

Xiaoyan Han, Donghui Luo, Xin Zhang, Yongxing Cao, Yu Zhang and Gege Ban
Study on Evaluation Method of Impulse Dispersion Performance of Grounding Device Using X-ray Imagining System
Reprinted from: *Energies* **2021**, *14*, 5124, doi:10.3390/en14165124 33

Ziwei Zhang, Dengwei Ding, Liang He, Weidong Liu, Cuifen Bai and Junjun Liu
Transient Voltage UWB Online Monitoring System for Insulation Failure Analysis and Fault Location of GIL
Reprinted from: *Energies* **2021**, *14*, 4863, doi:10.3390/en14164863 49

Donghui Luo, Jialun Li, Yongxing Cao, Bo Tan, Wei Li and Hanyu Wang
Research on the Influence of Typical Soil Parameters on Critical Breakdown Field Strength and Residual Resistivity Based on Discharge Topography
Reprinted from: *Energies* **2021**, *14*, 4810, doi:10.3390/en14164810 61

Xuri Xu, Yu Gao, Jing Li, Zheng Song, Huicun Zhao and Tao Han
Numerical Simulation on Charge Transport and DC Breakdown in Polyethylene-Based Micro-h-BN/Nano-SiO_2 with Filler Orientation Dependent Trap Energy
Reprinted from: *Energies* **2021**, *14*, 4645, doi:10.3390/en14154645 75

Lixiao Mu, Xiaobing Xu, Zhanran Xia, Bin Yang, Haoran Guo, Wenjun Zhou and Chengke Zhou
Autonomous Analysis of Infrared Images for Condition Diagnosis of HV Cable Accessories
Reprinted from: *Energies* **2021**, *14*, 4316, doi:10.3390/ en14144316 89

Xia Zhao, Haibin Shen, Men Guo, Ziming He, Yupeng Li and Ran Wen
DC Aging Mechanism of Co_2O_3-Doped ZnO Varistors
Reprinted from: *Energies* **2021**, *14*, 4011, doi:10.3390/en14134011 105

Dongdong Zhang, Hong Xu, Jin Liu, Chengshun Yang, Xiaoning Huang, Zhijin Zhang and Xingliang Jiang
Research on the Non-Contact Pollution Monitoring Method of Composite Insulator Based on Space Electric Field
Reprinted from: *Energies* **2021**, *14*, 2116, doi:10.3390/en14082116 117

Chenmeng Zhang, Kailin Zhao, Shijun Xie, Can Hu, Yu Zhang and Nanxi Jiang
Research on the Time-Domain Dielectric Response of Multiple Impulse Voltage Aging Oil-Film Dielectrics
Reprinted from: *Energies* **2021**, *14*, 1948, doi:10.3390/en14071948 . . . . **133**

Bingbing Dong, Yu Gu, Changsheng Gao, Zhu Zhang, Tao Wen and Kejie Li
Three-Dimensional Electro-Thermal Analysis of a New Type Current Transformer Design for Power Distribution Networks
Reprinted from: *Energies* **2021**, *14*, 1792, doi:10.3390/en14061792 . . . . **149**

Hongtao Jiang, Xiaohong Zhang, Junguo Gao and Ning Guo
Dielectric and AC Breakdown Properties of SiO_2/MMT/LDPE Micro–Nano Composites
Reprinted from: *Energies* **2021**, *14*, 1235, doi:10.3390/en14051235 . . . . **163**

Mingzhen Li, Jialong Bu, Yupeng Song, Zhongyi Pu, Yuli Wang and Cheng Xie
A Novel Fault Location Method for Power Cables Based on an Unsupervised Learning Algorithm
Reprinted from: *Energies* **2021**, *14*, 1164, doi:10.3390/en14041164 . . . . **179**

Donghui Luo, Yongxing Cao, Yu Zhang, Shijun Xie, Chenmeng Zhang and Shuping Cao
Study on Structural Parameters and Analysis Method of Soil Successive Impulse Discharge Channel
Reprinted from: *Energies* **2021**, *14*, 877, doi:10.3390/en14040877 . . . . **199**

Fubao Jin, Shanjun Zhang and Yuanxiang Zhou
Pattern Recognition of Development Stage of Creepage Discharge of Oil–Paper Insulation under AC–DC Combined Voltage Based on OS-ELM
Reprinted from: *Energies* **2021**, , 552, doi:10.3390/en14030552 . . . . **217**

John Owens, Ang Xiao, Jason Bonk, Michael DeLorme and Agnes Zhang
Recent Development of Two Alternative Gases to SF_6 for High Voltage Electrical Power Applications
Reprinted from: *Energies* **2021**, *14*, 5051, doi:10.3390/en14165051 . . . . **229**

About the Editors

Issouf Fofana

Issouf Fofana, IET Fellow, has held the position of Canada Research Chair on insulating liquids and mixed dielectrics for electrotechnology (ISOLIME) from 2005 to 2015. At his university, he serves as Director of the International Research Centre on Atmospheric Icing and Power Network Engineering (CENGIVRE) and chairs the research chair on the Ageing of Power Network Infrastructure (ViAHT). Prof. Fofana is a member of a number of technical/scientific committees of international conferences (including IEEE ICDL, IEEE CEIDP, IEEE ICHVE, CATCON, and ISH, etc.) and of the IEEE DEIS Administrative Committee. He is also serving the scientific community as Subject Editor in the Transformers, Transmission Lines and Cables category in *IET Generation, Transmission & Distribution*; as a Guest/Academic Editor for *IEEE TDEI* and Energies; and as chair of the IEEE DEIS Technical Committee on Dielectric Liquids. He is also member of a few CIGRE and ASTM working groups. Prof. Fofana's research in the area of HV engineering has focused on insulation systems relevant to power equipment. His lifetime publication record includes more than 350 scientific publications and 3 patents.

Bo Zhang

Bo Zhang, IET Fellow, IEEE Senior Member, is a professor of the Department of Electrical Engineering of Tsinghua University. He is the convener of CIGRE WG C4.50 and member of several CIGRE working groups, vice-chair of IEEE WGs P2869 and P2970, and the secretary of the Asia-Pacific International Conference on Lightning. He is also guest/academic Editor for *CSEE Journal of Power and Energy Systems, Energies, and Electric Power Systems Research*. Prof. Zhang's research interests include intelligent monitoring and digital twins of power transmission and transformation equipment, electromagnetic environments, and the electromagnetic compatibility of power systems. He has published more than 100 peer reviewed papers in IEEE/IET journals and has co-authored two books published by IEEE/IET. He received the IEEE EMC Technical achievement award and the Scientific Committee Award from the International Conference on Lightning Protection.

energies

MDPI

Editorial

High-Voltage Engineering and Applications in Our Modern Society

Issouf Fofana [1,*] and Bo Zhang [2]

[1] Modelling and Diagnostic of Electrical Power Network Equipment Laboratory (MODELE), Department of Applied Sciences, Université du Québec à Chicoutimi, Chicoutimi, QC G7H 2B1, Canada
[2] Department Electrical Engineering, Tsinghua University, Beijing 100084, China
* Correspondence: ifofana@uqac.ca

1. Introduction

Electrical energy is polymorphic, with voltage levels varying between a few volts to MVs and frequencies from a few Hz to MHz. This variability offers flexibility of use. For engineering applications, the choices of electrical parameters are dictated by technical and/or economic criteria. For example:

- In aviation, frequencies from 400 Hz to a few kHz are used.
- Energies produced at power stations are increased to kV levels with the step-up transformers to reduce Joule losses in long-distance transportations [1,2]. Consequently, many AC–DC transmission line projects up to 1200 kV have been constructed or are under way in many countries [3–15]. The application of high voltage in electrical power transmission is the most common, but electrical engineers also use this know-how in many other fields (e.g., [15–21]). Table 1 lists some of the main applications.

Table 1. A few of the main fields of high-voltage applications.

Field	Applications
AC and DC transmission grids	Transmission lines, cables insulators, instrument transformers, distribution/power transformers, generators, reactors, circuit breakers, disconnectors, surge arresters, capacitors, rectifiers, gas-insulated switchgears (GIS), substations, groundings, electromagnetic compatibility, etc.
Lightning	Danger of explosion and fire, disturbances of sensitive electronics, lightning capture, protection against lightning (lightning rod, spark gaps, lightning arrester, ground wire), aviation, etc.
Geomagnetic disturbances	Protection of transmission lines, antenna protection, protection of electronic devices, armored cages, etc.
Electronics	Cathode ray tube, piezoelectric generators, electric ignition, electronic flash, discharge lamps, UV bacteriological filter, etc.
Physics	Electronic microscopy, particle accelerators, laser printers, electromagnetic induction, etc.
Medicine	Biological effects of electric fields, X-ray diagnosis, X-ray therapy, ozone therapy, dielectrophoresis, heaters, etc.
Agro-food	Elimination of bacteria using high-voltage pulses (electroporation), plant growth, etc.
Aerospace and defense applications	Taser, lighting, laser weapons, all-electric aircraft, unmanned aerial vehicles (UAVs), light tactical vehicles, etc.
Materials processing	Treatment of water, sludge, medical waste, cleaning of gas pipes, melting by inertial confinement, treatment of nuclear waste, etc.
Mining engineering	Selective fragmentation of minerals using high-voltage pulses, etc.
Static electrification	Electrostatic generators, electrostatic motors, electrostatic filters, xerocopy, electrostatic printers, electrostatic paint, etc.

Citation: Fofana, I.; Zhang, B. High-Voltage Engineering and Applications in Our Modern Society. *Energies* **2022**, *15*, 8341. https://doi.org/10.3390/en15228341

Received: 5 September 2022
Accepted: 18 October 2022
Published: 8 November 2022

This Special Issue, in its final form, focuses the theoretical and practical developments in high-voltage engineering and applications pertaining to electrical engineering.

2. An Outlook of the Special Issue

The 2020 IEEE International Conference on High-Voltage Engineering and Application (ICHVE 2020) was organized by the by Tsinghua University, Beijing (China), and endorsed by the IEEE Dielectrics and Electrical Insulation Society (DEIS). The conference, chaired by Prof. Jinliang He, was held in Beijing, China, from September 6–10, 2020. It was the seventh of a series of successful conferences after Chongqing, China, in 2008; New Orleans, USA, in 2010; Shanghai, China, in 2012; Poznan, Poland in 2014; Chengdu, China, in 2016, and Athens, Greece, in 2018. This conference attracted a great deal of attention from international researchers in the field of high-voltage engineering. It not only provided an excellent platform to share knowledge and experiences on high-voltage engineering, but it also provided the opportunity to present the latest achievements in power engineering, including topics of ultra-high voltage, smart grids, new insulation materials, and their dielectric properties. Due to the impact of the COVID-19 pandemic, this event was held online for the first time. However, it is worthwhile to mention that this conference was by far the largest international one on high-voltage engineering. The conference received 1132 abstracts. Since the pandemic affected some research activities, 730 full texts were received, and 695 papers were finally accepted after review processes by the conference. More than 5000 experts, scholars, and graduate students from all over the world from 20 different countries had the opportunity to attend 52 oral sessions (including 3 keynote speeches and 23 invited speeches), along with 13 open forums through the online platforms Zoom, Weibo, and Bilibili and the opportunity to become acquainted with the trends in high-voltage research. The main research areas covered by the conference papers are shown in Figure 1, which provides a graphical overview of the proportion of papers on each topic.

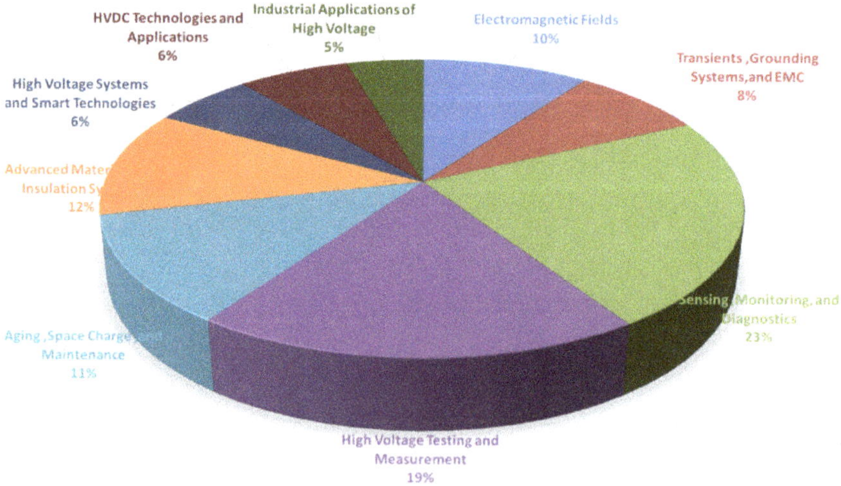

Figure 1. Proportion of papers by topics.

The conference also announced the two winners of the 2020 IEEE Sun Caixin–Stan Gezbosky Awards. Recipients are outstanding researchers recognized for their important contributions to the field of high-voltage engineering. Professor Xingliang Jiang from Chongqing University was the winner of the Lifetime Achievement Award, and Dr. Chuanyang Li from the University of Connecticut was the winner of Young-Professional Achievement Award. Both gave lectures. The conference also selected the 10 best student papers.

This Special Issue includes sixteen (16) high-quality papers presented during the 2020 IEEE International Conference on High-Voltage Engineering and Application (ICHVE 2020) spanning the above research areas. The papers were enriched with additional research outcomes, and the number of submitted papers increased in size by 50% compared to the original conference. The Special Issue was welcomed with great interest, as ICHVE attracts recent advancements in all fields of high-voltage engineering and applications. The editors would like to thank all contributors to this Special Issue.

3. Closing Remarks

The papers published in this Special Issue report on the progress made in various high-voltage applications. A mix of experimental and modelling/simulation investigations are reported. The outcomes will doubtlessly contribute to improving power system design and condition diagnosis/monitoring and, consequently, improve the reliability of these technologies. The articles published in this Special Issue also indicate that research in this field of engineering is very active and that the applications are quite diversified.

Faced with the continuous modernization of our society, more high-voltage applications are expected in various areas: "Things are getting smaller and more powerful". With the years to come, many exciting applications are therefore expected.

Author Contributions: Conceptualization, I.F. and B.Z.; writing—original draft, I.F. and B.Z.; writing—review and editing, I.F. and B.Z. All authors have read and agreed to the published version of the manuscript.

Funding: This research received no funding.

Acknowledgments: The guest editors are grateful for the invitation to guest edit this Special Issue. They are also indebted to the editorial staff of *Energies* for the kind co-operation, patience, and committed engagement. Both guest editors would also like to thank all the authors for contributing to this Special Issue. Thanks are also extended to the blind reviewers for helping improving the content of the accepted papers.

Conflicts of Interest: The authors declare no conflict of interest.

References

1. Papailiou, K.O. Overhead Lines. In *CIGRE Green Books*; CIGRE: Paris, France, 2017.
2. EPRI. *EPRI AC Transmission Line Reference Book—200 kV and Above*, 3rd ed.; EPRI: Palo Alto, CA, USA, 2005.
3. Long, R.; Zhang, J. Risk Assessment Method of UHV AC/DC Power System under Serious Disasters. *Energies* **2017**, *10*, 13. [CrossRef]
4. Hedtke, S.; Xu, P.; Pfeiffer, M.; Zhang, B.; He, J.; Franck, C.M. HVDC Corona Current Characteristics and Audible Noise During Wet Weather Transitions. *IEEE Trans. Power Deliv.* **2020**, *35*, 1038–1047. [CrossRef]
5. Trinh, N.G.; Maruvada, P.S.; Flamand, J.; Valotaire, J.R. A Study of the Corona Performance of Hydro-Quebec's 735-kV Lines. *IEEE Power Eng. Rev.* **1982**, *PAS-101*, 30. [CrossRef]
6. Asplund, G.; Astrom, U.; Lescale, V. 800 kV HVDC for Transmission of Large Amount of Power Over Very Long Distances. In Proceedings of the 2006 International Conference on Power System Technology, Chongqing, China, 22–26 October 2006; pp. 1–10. [CrossRef]
7. Swarup, S.K. Selection of design parameters of equipment for 800 kV transmission systems in India. In Proceedings of the International Conference on AC and DC Power Transmission, London, UK, 17–20 September 1991; pp. 235–240.
8. Åström, U.; Lescale, V.F.; Menzies, D.; Weimin, M.; Zehong, L. The Xiangjiaba-Shanghai 800 kV UHVDC project, status and special aspects. In Proceedings of the 2010 International Conference on Power System Technology, Hangzhou, China, 24–28 October 2010; pp. 1–6. [CrossRef]
9. Guo, X.; Fu, Y.; Yu, J.; Xu, Z. A Non-Uniform Transmission Line Model of the ±1100 kV UHV Tower. *Energies* **2019**, *12*, 445. [CrossRef]
10. Wang, X.; Tang, C.; Huang, B.; Hao, J.; Chen, G. Review of Research Progress on the Electrical Properties and Modification of Mineral Insulating Oils Used in Power Transformers. *Energies* **2018**, *11*, 487. [CrossRef]
11. Krylov, S.V.; Timashova, L.V. Experience of live-line maintenance on 500-1200 kV lines in Russia. In Proceedings of the ESMO '93. IEEE 6th International Conference on Transmission and Distribution Construction and Live-Line Maintenance, Atlanta, GA, USA, 6–9 June 1993; pp. 359–368. [CrossRef]

12. Maruvada, P.a.; Trinh, N.G.; Dallaire, R.D.; Rivest, N. Corona Studies for Biploar HVDC Transmission at Voltages between ±600 kV and ±1200 kV Part 1: Long-Term Bipolar Line Studies. *IEEE Trans. Power Appar. Syst.* **1981**, *PAS-100*, 1453–1461. [CrossRef]
13. Hammons, T.J.V.F.L.; Uecker, K.; Haeusler, M.; Retzmann, D.; Staschus, K.; Lepy, S. State of the Art in Ultrahigh-Voltage Transmission. *Proc. IEEE* **2012**, *100*, 360–390. [CrossRef]
14. Allard, S.; Mima, S.; Debusschere, V.; Quoc, T.T.; Criqui, P.; Hadjsaid, N. European transmission grid expansion as a flexibility option in a scenario of large scale variable renewable energies integration. *Energy Econ.* **2020**, *87*, 104733. [CrossRef]
15. Aguet, M.; Ianoz, M. *Haute Tension*; EPFL, Presses Polytechniques et Universitaires Romandes: Lausanne, Switzerland, 2001; p. 440, Volume XII.
16. Behjat, V.; Rezaei-Zare, A.; Fofana, I.; Naderian, A. Concept Design of a High-Voltage Electrostatic Sanitizer to Prevent Spread of COVID-19 Coronavirus. *Energies* **2021**, *14*, 7808. [CrossRef]
17. Vergura, S. Criticalities of the Outdoor Infrared Inspection of Photovoltaic Modules by Means of Drones. *Energies* **2022**, *15*, 5086. [CrossRef]
18. Huang, W.; Chen, Y. The application of high voltage pulses in the mineral processing industry—A review. *Powder Technol.* **2021**, *393*, 116–130. [CrossRef]
19. Vishnyakov, V.I. Pulsed high-voltage electrical discharges in water: The resource for hydrogen production and water purification. *Int. J. Hydrog. Energy* **2022**, *47*, 12500–12505. [CrossRef]
20. Gao, Y.; Tian, E.; Zhang, Y.; Mo, J. Utilizing electrostatic effect in fibrous filters for efficient airborne particles removal: Principles, fabrication, and material properties. *Appl. Mater. Today* **2022**, *26*, 101369. [CrossRef]
21. Koyama, S.; Tamura, Y.; Ishikawa, G.; Ishikawa, Y. Acceleration of germination and early growth of plant seeds by high frequency and low intensity alternating electric fields. *Eng. Agric. Environ. Food* **2019**, *14*, 95–101. [CrossRef]

energies

MDPI

Article

Optical Diagnostic Characterization of the Local Arc on Contaminated Insulation Surface at Low Pressure †

Hao Yang [1,*], Haotian Zhang [1], Wen Cao [1], Xuanxiang Zhao [1], Ran Wen [1], Junping Zhao [2], Shengwu Tan [3] and Pengchao Wang [3]

1 School of Electronics and Information, Xi'an Polytechnic University, Xi'an 710048, China; haotianzhang163@163.com (H.Z.); caowen@xpu.edu.cn (W.C.); xuanxiangzhao@163.com (X.Z.); wenran1994@126.com (R.W.)
2 School of Electrical Engineering, Xi'an Jiaotong University, Xi'an 710049, China; junping_zhao@mail.xjtu.edu.cn
3 Pinggao Group Company Limited, Pingdingshan 467000, China; shengwutan@163.com (S.T.); wangpengchao188@163.com (P.W.)
* Correspondence: yanghao@xpu.edu.cn

† This paper is an extended version of our paper published in 2020 IEEE International Conference on High Voltage Engineering and Application (ICHVE), Beijing, China, 6–10 September 2020; pp. 1–4.

Abstract: Flashover of contaminated insulators is a major problem for power systems at high altitude. Laboratory experiments have shown that the optical diagnostic method can provide extensive information on the physical process of contamination flashover. In this paper, a study of the local arc on a wet polluted surface under low pressure by using the optical diagnostic method is presented. The thickness of the continuous spectrum, spectral line intensity and the spectral composition varies significantly in different stages of the local arc development. Thermodynamic parameters of the local arc (including electron temperature, electron density and conductivity) are obtained by analyzing the spectra. Both the electron temperature and the conductivity increase with the increase in leakage current and air pressure. Although the electron density does not change significantly with an increase in leakage current, it increases significantly with an increase in air pressure. The findings of this work could be used as supplementary information for the investigation of local arc parameters, thus providing a reliable reference for the calculation of contamination flashover at high altitude.

Keywords: contamination flashover; the local arc; low pressure; optical diagnostic method

Citation: Yang, H.; Zhang, H.; Cao, W.; Zhao, X.; Wen, R.; Zhao, J.; Tan, S.; Wang, P. Optical Diagnostic Characterization of the Local Arc on Contaminated Insulation Surface at Low Pressure. *Energies* **2021**, *14*, 6116. https://doi.org/10.3390/en14196116

Academic Editors: Issouf Fofana and Bo Zhang

Received: 17 August 2021
Accepted: 21 September 2021
Published: 26 September 2021

1. Introduction

Line insulators are some of the most widely used components in transmission lines, and their performance directly affects the safe operation of power systems. When contamination that has accumulated on the surface of the insulators contains a conductive medium, it might cause a contamination flashover accident in the fog and drizzle [1–3]. The contamination status of insulators and the climate circumstance are very close in the same area [4,5]. When contamination flashover occurs on a single insulator string, other insulators in this area are also on the edge of flashover [6]. It is known that contamination flashover of insulators often causes massive blackouts, resulting in huge economic losses [7–9].

Contamination flashover of outdoor insulators remains a major problem for transmission lines [10,11]. Especially in high-altitude areas, the contamination flashover voltage of the line insulators is greatly reduced due to the lower air pressure [12,13]. Additionally, transmission lines inevitably have to pass through high-altitude areas [14,15]. Therefore, the study on the flashover process on the insulator surface under low pressure is of great significance for predicting the flashover voltage and preventing contamination flashover accidents.

Contamination flashover is the process in which the local arc develops on the surface of the insulator, until the high-voltage electrode and ground electrode are connected [16,17]. Hence, it is helpful to understand the physical process by obtaining the thermodynamic parameters of the local arc. Until now, former researchers mainly focused on the electrical parameters of the local arc. However, as a special kind of plasma, the local arc is mainly composed of free electrons and charged ions and has special properties quite different from resistance.

The thermodynamic parameters that characterize plasma mainly include electron temperature, excitation temperature, rotation temperature, electron density, thermal conductivity and electrical conductivity. It has been found that these parameters of the local arc play a very important role in the calculation of contamination flashover. M. Slama, from the Ampère Laboratory of the Central University of Technology, proposed a dynamic mathematical equation describing the relationship between the resistivity of the local arc channel and the arc temperature [18,19]. Based on the law of conservation of energy, W. Sima, from Chongqing University, obtained the mathematical expressions of the contamination flashover voltage and arc temperature [20]. It can be concluded that thermodynamic parameters of the local arc play a very important role in calculating the arc characteristics and predicting the flashover voltage.

However, there are relatively less experimental studies on the measurement of thermodynamic parameters of the local arc. A. Nekahi, from the University of Quebec, used fiber optic spectrometers to obtain the emission spectra of AC and DC of the local arcs on iced surfaces. By fitting the three molecular spectra of OH (A-X), NH (A-X) and N^{2+} (B-X), the relationship between the rotation temperature of the local arc and the leakage current was obtained, and the results show that the local arc rotation temperature increases with the increase in the leakage current amplitude [21,22]. S. Li of Tsinghua University also designed a spectral temperature measurement system to obtain the emission spectrum of the local arc in the critical flashover stage [23]. However, there is a lack of research on the thermodynamic parameters of the local arc on the wet contaminated insulation surface under low pressure.

In this paper, an optical diagnostic system is designed for investigating the characteristics of the local arc on the polluted surface under low pressure. A special tracing element is employed to obtain the spectrum of the local arc. Typical parameters including electron temperature, electron density and conductivity are obtained by analyzing the spectra of the local arc. This work can enrich the investigation of the local arc under low pressure and may provide useful parameters for the calculation of flashover voltage.

2. Experimental Arrangements and Methods

2.1. Test Facilities

In this paper, an optical diagnostic platform of the local arc under low pressure was established, as shown in Figure 1. The experiments were carried out in the QD660 artificial climate room of Xi'an Jiaotong University. The climate chamber had a cylindrical structure with a diameter of 4 m and a height of 5.2 m. Its size met the requirements of the artificial contamination flashover test for insulator strings of AC 110 kV and below. The artificial climate chamber could be used to achieve any environmental parameter between -50 °C + 60 °C and 0.05 MPa~0.1 MPa. The power was provided by a 160 kV/4 A, 600 kVA transformer with a short-circuit impedance less than 5%. The leakage current was measured by a power-frequency Rogowski coil.

The spraying system consisted of high-pressure water pumps, pipes and atomizing nozzles. The deionized water was pressurized by a high-pressure water pump, and sent through a pipeline to atomization nozzles, installed at multiple locations around the wall of the artificial climate chamber. By changing the size of the nozzle, the diameter of the droplets ranging from 15 to 2000 μm could be obtained. The polluted specimen was installed horizontally in the center of the chamber, and the spraying system was then used to wet the specimen before the test.

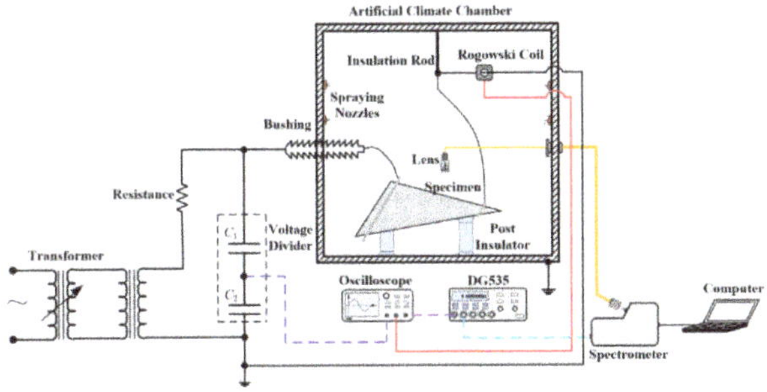

Figure 1. Spectrum diagnostic platform.

The spectrometer was used to record the spectra of the local arc. The wavelength range of the optical fiber spectrometer was 250–880 nm, which meets the spectral measurement requirements. In order to increase the light intensity coupled into the spectrometer, a fiber collimator with a diameter of 45 mm was used to convert the arc radiation light into parallel light, and then the parallel light was coupled into the fiber through a light collector, as shown in Figure 2. A four-channel digital delay pulse signal generator (DG535) was used to trigger the spectrometer to acquire waveform and spectra synchronously.

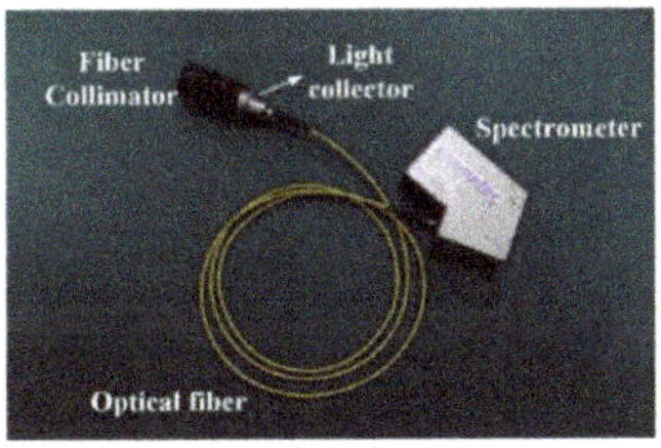

Figure 2. Spectral acquisition device.

2.2. Specimen and Test Methods

The development of the local arc over the insulator was random, resulting in low efficiency and poor repeatability of experimental measurement. Hence, a triangle glass plate was employed in our experiment, as shown in Figure 3a. The electric field intensity around the top corner of the triangular glass plate surface was higher than the rest of the region, as shown in Figure 3b, from our previous study. The local arc developed from the ground electrode and gradually developed to the high-voltage electrode. The glass triangular plate was polluted with a 2-cm narrow band for convenient measurement.

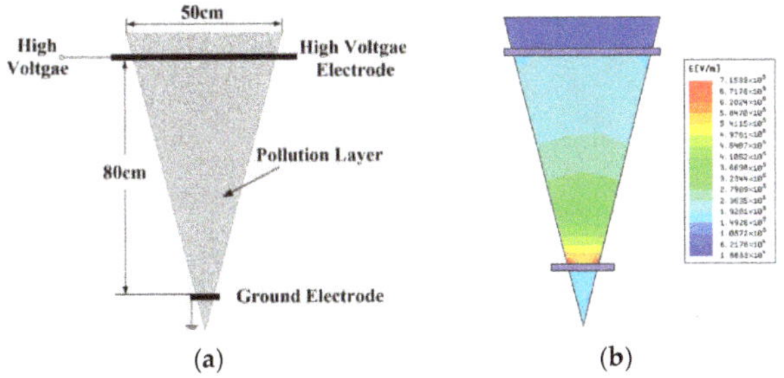

Figure 3. Triangular glass plate and its surface electric field distribution, (**a**) triangular glass plate used as test sample, (**b**) simulation results of the electric field distribution of the triangular glass surface.

The solid layer method was used to pollute the specimens artificially, according to IEC 60,805 [24]. The artificial contamination was a mixture of sodium chloride, kaolin and de-ionized water. The equivalent salt deposit density (ESDD) was chosen as 0.1 mg/cm^2, and the non-soluble deposit density (NSDD) was chosen as 0.1 mg/cm^2. In the initial stage of the local arc, only a single spectral line could be found for the same element, i.e., Na I and K I, which did not meet the requirement of the spectral line relative intensity method, which requires "at least two atomic spectrum lines of the same element". Therefore, under the premise of not affecting the characteristics of the arc discharge, a trace amount of lithium was added as a tracer element to the artificial pollution for the occurrence of the useful spectral lines.

2.3. Calculation of Plasma Parameters

(1) Electron temperature

The electron temperature (T_e) of the local arc can not only characterize the state of the particles in the plasma, but also be used to calculate the conductivity of the local arc. In this paper, T_e of the local arc was calculated based on the spectral line relative intensity method. Using the two different excited state parameters of the same atom, s and t, the formula for T_e is obtained:

$$T_e = \frac{-5040(E_s - E_t)}{\log(\lambda I/gA)_s - \log(\lambda I/gA)_t} \tag{1}$$

In the formula, E_s and E_t are the energy of two different excited states of s and t. Taking the atomic excitation energy as the independent variable and the formula $\log(\lambda I/gA)$ as the dependent variable, the Boltzmann diagram is obtained. The relationship between the slope k of the straight line in the Boltzmann diagram and T_e is:

$$T_e = -\frac{5040}{k} \tag{2}$$

In order to ensure high accuracy of the results, the energy level difference between the two spectral lines should be as large as possible, usually greater than 2 eV. The difference in excitation energy between the two spectral lines of Li I 610.4 nm and Li I 670.8 nm is 2.63 eV, which meets the requirements of the spectral line relative intensity method. Therefore, Li I 610.4 nm and Li I 670.8 nm were chosen as the characteristic spectral lines, and Boltzmann's diagram was employed to calculate the T_e of the local arc.

(2) Electron density

Electron density (n_e) measurement methods include the Stark broadening method, Saha equation method, continuous spectroscopy method, Inglis–Teller method, etc. This paper chose the Stark broadening method with its high accuracy and simple measurement method to measure n_e. The relationship between the broadening value of the spectral line $\Delta\lambda$ and n_e can be expressed as:

$$\Delta\lambda = wn_e + 1.75\alpha n_e^{1/4}(1 - 1.2\rho)w \tag{3}$$

where w is the Stark stretching factor; ρ is the Debye shielding factor; and α is the broadening factor. According to reference [25], the equation above can be changed into:

$$n_e = \frac{\Delta\lambda_{NaStarkFWHM}}{4.854} T^{0.126} \tag{4}$$

where $\Delta\lambda_{NaStarkFWHM}$ is the width at half maximum (FWHM) of the Na I 568.8 nm spectral line, broadened by the Stark effect in nanometres.

(3) Electric conductivity

Because there are a large number of electrons and ions in the arc plasma, it has a certain electrical conductivity (σ_s). The Spitzer formula is not suitable to calculate the conductivity of low-temperature plasma. R. Mohanti and J. Gilligan, from North Carolina State University, conducted theoretical research on the conductivity of low-temperature plasma and modified the Spitzer formula [26]. The conductivity σ_s can be calculated with T_e and n_e, as shown in our former research [27]:

$$\sigma_s = \frac{3.1 \times 10^{-2} {T_e}^{3/2}}{\ln\left(1 + 2.2 \times 10^{14} T_e^3 / n_e\right)} \tag{5}$$

3. Typical Emission Spectrograms of the Local Arc

The flashover development process can be divided into four stages: initiation, development, critical flashover and flashover. Assuming that the creepage distance is L, in the initial stage, the local arc length is shorter than $1/10\ L$ and the leakage current amplitude (i_m) is less than 50 mA. In the development stage, the local arc length is between 1/10 and $1/2\ L$ and i_m is between 50–150 mA. During the critical flashover stage, the local arc length is between 1/2 and $2/3\ L$ and i_m is between 200 and 300 mA. The optical fiber spectrometer was set to continuous acquisition mode to acquire the emission spectrum of the local arc from the initial stage to the flashover occurrence. The time-resolved emission spectrum of the local arc was obtained, as shown in Figure 4.

(1) The initial stage

In the initial stage of the local arc, i_m is less than 50 mA. In the spectrum shown in Figure 4a, no continuous spectral components can be observed; only the linear spectral lines of lithium, sodium and potassium atoms with a low relative intensity value can be found.

(2) Development stage

As i_m continues to increase, the magnitude of the radiation spectrum line increases significantly. The energy injected into the arc channel continues to increase, and the collision between electrons and atoms becomes more intense, leading to the appearance of the atomic line of nitrogen in the spectrum, as shown in Figure 4b.

(3) Critical flashover stage

In the critical flashover stage, the copper atoms participate in the arc discharge, and three copper atomic lines Cu 510.5 nm, 515.8 nm and 521.8 nm appear in the spectrum. The dramatic increment in injection energy in the arc channel broadens the continuous spectrum thickness in the radiation spectrum, which leads to the increase in the ionization degree of particles in the channel, especially the alkali metal particles. Therefore, the K II 397.3 nm line spectrum was detected in the line, as shown in Figure 4c.

(4) Flashover stage

After the occurrence of flashover, the arc channel penetrates between the high-voltage electrode and the ground electrode, resulting in the energy of the power supply injecting into the arc channel almost instantaneously. The enhancement of the collisions between electrons and atoms/ions in this channel accelerate the ionization process of oxygen and nitrogen molecules. Therefore, oxygen ion line O II 485.7 nm and nitrogen ion line N II began to appear in the spectrum, as shown in Figure 4d.

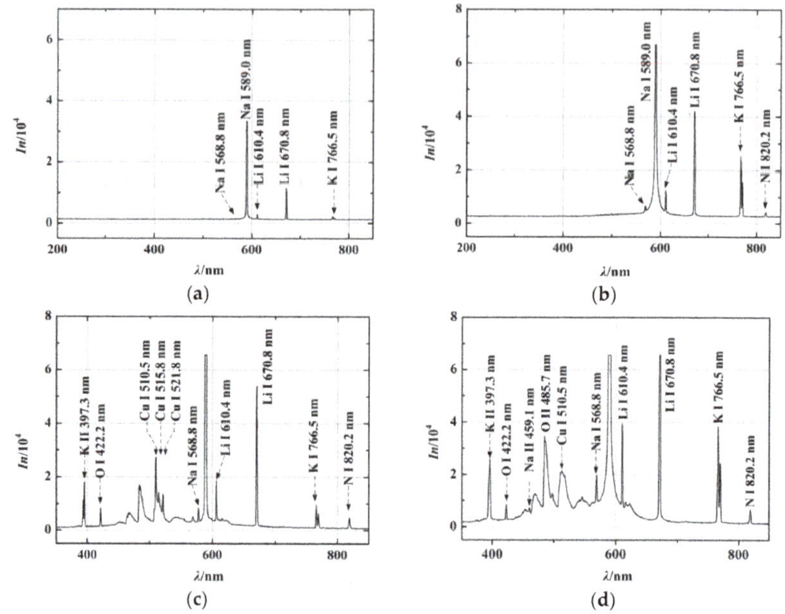

Figure 4. Typical spectrum of the local arc in different stages, (**a**) spectrum of local arc in initial stage, (**b**) spectrum of local arc in development stage, (**c**) spectrum of local arc in critical flashover stage, (**d**) spectrum of local arc in flashover stage.

4. Parameters of the Local Arc under Low Pressures

Spectral diagnosis of the local arc on a wet contaminated insulation surface under low pressure was carried out in this section. Several vital parameters, such as electron temperature, electron density and conductivity were obtained.

4.1. Electron Temperature of the Local Arc

According to the actual altitude of the transmission line, four typical low-pressure values of 0.06, 0.07, 0.08 and 0.09 MPa were selected for the experiment in this paper. The experimental results show that the emission spectra of the local arc under different pressures are almost the same, similar to Figure 4. However, there is a significant difference in the ratio of the relative intensity of the discrete spectral lines. Li I 610.4 nm and Li I 670.8 nm were chosen as the characteristic spectral lines. The Boltzmann diagram method, mentioned in Section 2.3, was used to calculate the local arc electron temperature. The Boltzmann plots of the local arc emission spectra under different air pressures at 0.1 A are shown in Figure 5. The slope k of the straight lines can be obtained for Equation (2) to calculate T_e.

Because the statics of T_e show a large dispersion, the mathematical expectation value of the data was used to represent T_e at a specific leakage current peak. T_e of the local arc under different pressure is shown in Figure 6.

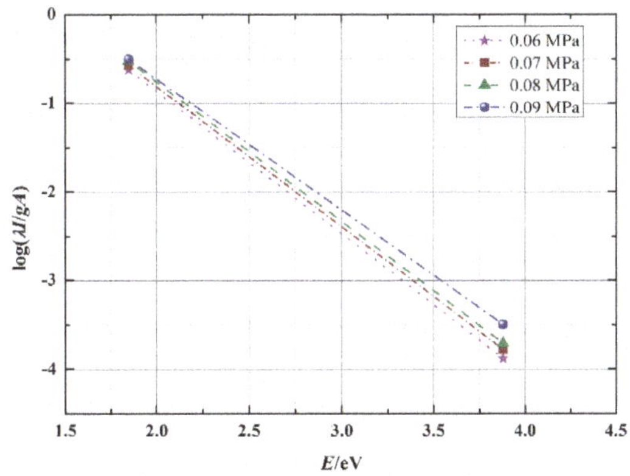

Figure 5. Boltzmann plots of the local arc emission spectra under different air pressures at 0.1 A.

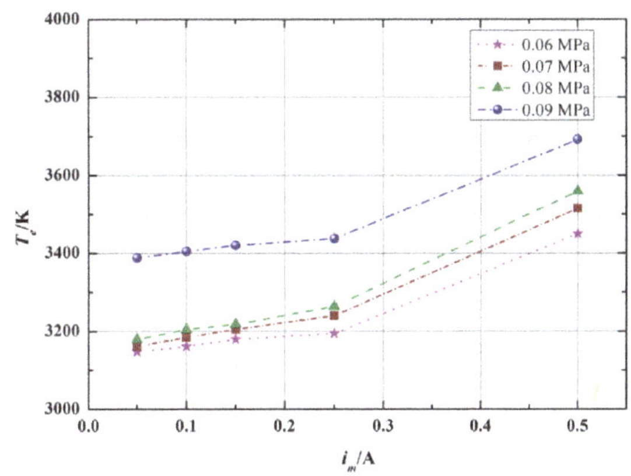

Figure 6. The value of T_e at different air pressures.

It can be seen from the figure that when the air pressure is 0.06 MPa, T_e is between 3150 K and 3450 K, and when the air pressure is 0.09 MPa, T_e rises to the range between 3400 K and 3700 K. It can be concluded that T_e increases with the increase in i_m at a certain pressure. When i_m increases, the injection energy of the arc channel also increases, resulting in the enhancement of collisions between electrons and atoms/ions. Hence, T_e increases rapidly with the increase in i_m.

Moreover, it can be concluded that T_e increases with the increase in air pressure. Our previous research results show that the voltage gradient of local arc (E) in a high-pressure environment is higher than that in a low-pressure environment when i_m is the same [28]. The injection energy per unit length of the local arc in a high-pressure environment is relatively higher. In such cases, the movement of electrons in the arc channel intensifies due to higher energy. In other words, the migration speed of electrons increases. With the increase in air pressure, the number density of air molecules increases, resulting in the enhancement of inelastic collisions of electrons. Therefore, excitation and ionization collision frequencies are higher at a higher pressure, resulting in an increase in T_e.

4.2. Electron Density of the Local Arc

As mentioned above, the Stark broadening method was adopted to calculate n_e. In the critical flashover and flashover stages, the local arc radiation spectrum contains strong continuum components. In order to obtain the broadening value, the Lorentz linear function was used to perform the Na I 568.8 nm fitting. The value of n_e under different air pressure is shown in Figure 7.

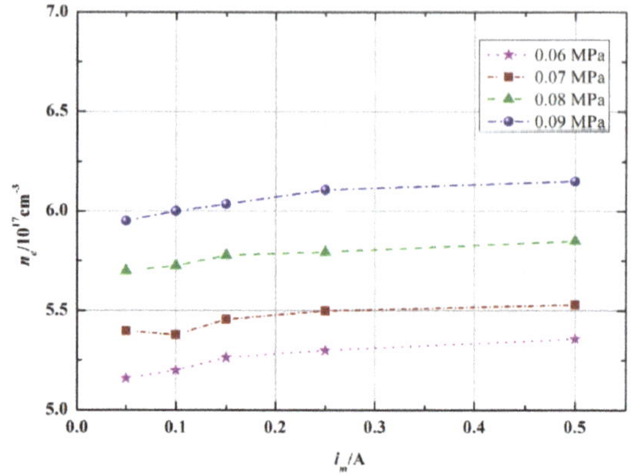

Figure 7. The value of n_e under different air pressures.

When i_m is between 50 mA and 500 mA, and the air pressure is between 0.06 MPa and 0.09 MPa, the n_e of the arc channel is between 5.2×10^{17} cm^{-3} and 6.3×10^{17} cm^{-3}. With the increase in i_m, the ionization degree of gas particles increases, resulting in the increase in free electrons. In addition, the radius and volume of the local arc increase correspondingly. Therefore, n_e does not change much during the development stage of the local arc. It can also be deduced that n_e increases with the increase in the air pressure. When the outside air pressure increases, the energy of the arc channel increases and the local arc channel shrinks, leading to the increase in n_e.

4.3. Conductivity of the Local Arc

Conductivity is one of the basic electrical parameters of the local arc, and it is also one of the critical parameters for calculating the voltage of the local arc. Compared with the "three-electrode" method [28], the non-contact spectroscopy method makes it much easier to obtain σ_s of the local arc propagating over the polluted surface. The σ_s of the arc channel can be calculated from Equation (5) by substituting T_e and n_e with the values in Figures 6 and 7. Figure 8 shows the relationship between σ_s and i_m under different pressure. Since the index value of T_e is larger than n_e in Equation (5), σ_s and T_e have the same changing trend.

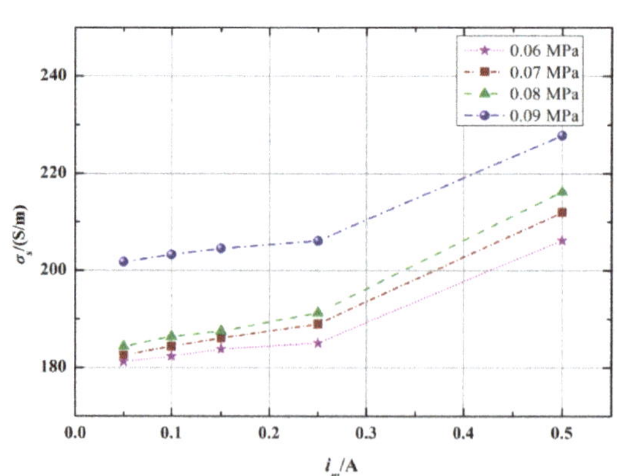

Figure 8. The value of σ_s at different air pressures.

Additionally, σ_s can also be obtained by the calculation of electrical parameters of the local arc. The calculated results are compared with the spectral diagnosis results in the following part. In Rizk's former study, the relationship between E and i_m under low pressure was obtained, as shown in the following equation [29]:

$$E = 716(p/p_0)^{0.77} i_m^{-0.56} \tag{6}$$

where p is the actual pressure, p_0 is the ambient pressure. The relationship between σ_s and E can be expressed as follows:

$$\sigma_s = \frac{i_m}{\pi E r^2} \tag{7}$$

where r is the arc boundary radius, and the relationship between r and p has been revealed by Rizk in Reference [29]:

$$r \propto p^{-0.465} \tag{8}$$

and the radius r_d of the arc channel in ambient pressure is also a function of i_m [30].

$$r_d = \sqrt{\frac{i_m}{1.45\pi}} \tag{9}$$

Thus, we obtain:

$$\sigma_s = \frac{i_m^{0.56}}{494(p/p_0)^{-0.16}} \tag{10}$$

By solving Equation (10), σ_s at a different pressure could be obtained, as shown in Figure 9. Compared with Figure 7, the calculated value of σ_s is smaller. Since the plasma inside the local arc is not completely uniformly distributed, the true value of r is not the arc boundary radius measured by actual experiments. The local arc channel is populated with high-density charged particles and the outer area of the channel is a heated and luminous area consisting of neutral atoms. It is generally believed that the maximum radial light intensity gradient of the channel is at the intersection of the ionization zone and the neutral atomic zone. That is, the radius corresponding to the maximum value is taken as the true radius value of the conductive area of the arc channel. Hence, the value of r in Equation (7) is larger than the true value, resulting in a smaller value of σ_s. However, the deviations of σ_s between the true value and the calculated value is trivial, within an order of magnitude. We can still consider our proposed approach of the spectral diagnostic method, which is relatively effective in measuring the local arc conductivity under low air pressure.

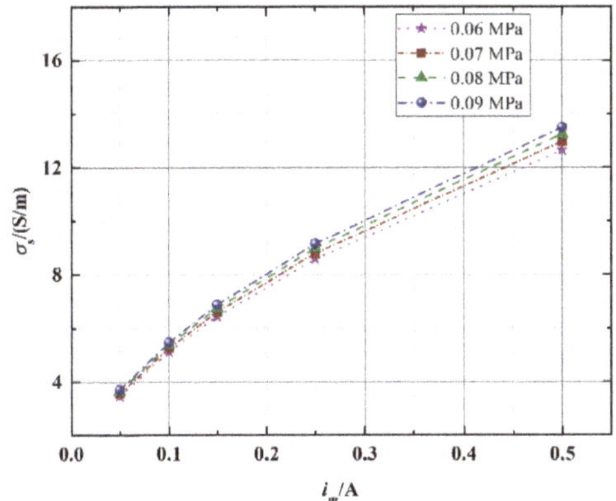

Figure 9. The calculated value of σ_s at different air pressures.

5. Conclusions

In this paper, we carried out the spectral diagnosis of the local arc on the wet contaminated insulation surface under low pressure. Key parameters of the local arc, including electron temperature, electron density and conductivity were calculated by spectroscopic measurements. The following conclusions were deduced:

(1) The spectra of the local arc are quite similar, even though the air pressure is different. The spectral characteristics of the different development stages of the local arc can be easily distinguished.
(2) Thermodynamic parameters of the local arc (including electron temperature and electron density) are obtained by analyzing the spectra. Electron temperature increases significantly with the increase in leakage current and air pressure, while, the electron density does not change much with the increase in leakage current, but it increases significantly with the increase in air pressure.
(3) The conductivity of the local arc is also calculated by employing the value of electron temperature and electron density in this paper. The conductivity increases with the increase in leakage current and air pressure. The deviation of conductivity between the true value and the calculated value is trivial.

Former researchers have used thermodynamic parameters of the local arc to predict the contamination flashover. In our further study, the obtained parameters of the local arc under low pressure will be used for the calculation of contamination flashover voltage of on-line insulators. This will be explored in our following paper.

Author Contributions: Conceptualization, H.Y.; Formal analysis, H.Y., W.C., X.Z., R.W. and J.Z.; Investigation, W.C., X.Z., R.W. and J.Z.; Writing—original draft, H.Z.; Writing—review and editing, S.T. and P.W. All authors have read and agreed to the published version of the manuscript.

Funding: This work was supported in part by the Science and Technology Project of State Grid Corporation of China (No. B491BG180016).

Institutional Review Board Statement: Not applicable.

Informed Consent Statement: Not applicable.

Data Availability Statement: Data sharing not applicable. No new data were created or analyzed in this study. Data sharing is not applicable to this article.

Conflicts of Interest: The authors declare no conflict of interest.

References

1. Tan, S.; Wang, P.; Zhao, J.; Yang, H.; Wen, R. Study on the Typical Optical Emission Spectrograms of AC Arc Discharge over the Polluted Insulation Surface. In Proceedings of the 2020 IEEE International Conference on High Voltage Engineering and Application (ICHVE), Beijing, China, 6–10 September 2020; pp. 1–4. [CrossRef]
2. Zhang, C.; Wang, L. Experimental Investigation on Pollution Flashover Performance of Multiple Parallel Suspension Insulators. *IEEE Trans. Dielectr. Electr. Insul.* **2016**, *23*, 2840–2849. [CrossRef]
3. Yang, Z.; Jiang, X.; Han, X.; Zhang, Z.; Hu, J. Influence of Pollution Chemical Components on Ac Flashover Performance of Various Types of Insulators. *High Volt.* **2019**, *4*, 105–112. [CrossRef]
4. He, J.; Gorur, R. Flashover of Insulators in a Wet Environment. *IEEE Trans. Dielectr. Electr. Insul.* **2017**, *24*, 1038–1044. [CrossRef]
5. Yang, L.; Zhang, F.; Wang, Z.; Jiang, X.; Hao, Y.; Li, L.; Liao, Y.; Zhang, F. Evaluation of Wetting Condition and Its Effects on contamination flashover Voltage of Aerodynamic Insulators. *IEEE Trans. Dielectr. Electr. Insul.* **2016**, *23*, 2875–2882. [CrossRef]
6. Shu, L.; Liu, Y.; Jiang, X.; Hu, Q.; He, G.; Yu, Z.; Xiao, L. Three-Dimensional Electric Field Simulation and Flashover Path Analysis of Ice-Covered Suspension Insulators. *High Volt.* **2020**, *5*, 327–333. [CrossRef]
7. Slama, M.E.A.; Albano, M.; Haddad, A.M.; Waters, R.T.; Cwikowski, O.; Iddrissu, I.; Knapper, J.; Scopes, O. Monitoring of Dry Bands and Discharge Activities at the Surface of Textured Insulators with AC Clean Fog Test Conditions. *Energies* **2021**, *14*, 2914. [CrossRef]
8. Shariatinasab, R.; Saghafi, S.; Khorashadizadeh, M.; Ghayedi, M. Probabilistic Assessment of Insulator Failure under Contaminated Conditions. *IET Sci. Meas. Technol.* **2020**, *14*, 557–563. [CrossRef]
9. Savadkoohi, E.M.; Mirzaie, M.; Seyyedbarzegar, S.; Mohammadi, M.; Khodsuz, M.; Pashakolae, M.G.; Ghadikolaei, M.B. Experimental Investigation on Composite Insulators AC Flashover Performance with Fan-Shaped Non-Uniform Pollution under Electro-Thermal Stress. *Int. J. Electr. Power Energy Syst.* **2020**, *121*, 106142. [CrossRef]

10. Lan, L.; Zhang, G.; Wang, Y.; Wen, X.; Wang, W.; Pei, H. The Influence of Natural Contamination on Pollution Flashover Voltage Waveform of Porcelain Insulators in Heavily Polluted Area. *IEEE Access* **2019**, *7*, 121395–121406. [CrossRef]
11. Zhang, Z.; Zhang, W.; You, J.; Jiang, X.; Zhang, D.; Bi, M.; Wu, B.; Wu, J. Influence Factors in Contamination Process of XP-160 Insulators based on Computational Fluid Mechanics. *IET Gener. Transm. Distrib.* **2016**, *10*, 4140–4148. [CrossRef]
12. Qiao, X.; Zhang, Z.; Jiang, X.; Liang, T. Influence of DC Electric Fields on Pollution of HVDC Composite Insulator Short Samples with Different Environmental Parameters. *Energies* **2019**, *12*, 2304. [CrossRef]
13. Zhang, C.; Wang, L.; Guan, Z.; Zhang, F. Pollution Flashover Performance of Full-Scale ±800 KV Converter Station Post Insulators at High Altitude Area. *IEEE Trans. Dielectr. Electr. Insul.* **2013**, *20*, 717–726. [CrossRef]
14. Huang, D.; Ruan, J.; Cai, W.; Li, T.; Wei, Y.; Liu, J. Flashover Prevention on High-Altitude H Transmission Line Insulator Strings. *IEEE Trans. Dielectr. Electr. Insul.* **2009**, *16*, 88–98. [CrossRef]
15. Rudakova, V.; Tikhodeev, N. Influence of Low Air Pressure on Flashover Voltages of Polluted Insulators: Test Data, Generalization Attempts and Some Recommendations. *IEEE Trans. Power Deliv.* **1989**, *4*, 607–613. [CrossRef]
16. Liao, Y.; Wang, Q.; Yang, L.; Kuang, Z.; Hao, Y.; Zhang, C. Discharge Behavior and Morphological Characteristics of Suspended Water-Drop on Shed Edge during Rain Flashover of Polluted Large-Diameter Post Insulator. *Energies* **2021**, *14*, 1652. [CrossRef]
17. Mei, H.; Guan, X.; Fu, X.; Zhao, C.; Wang, L. Influence of Tower Anticorrosion Coating as Contaminant on Operation Characteristics of Composite Insulator. *High Volt.* **2018**, *3*, 193–198. [CrossRef]
18. Michelarakis, M.; Widger, P.; Beroual, A.; Haddad, A. Electrical Detection of Creeping Discharges over Insulator Surfaces in Atmospheric Gases Under AC Voltage Application. *Energies* **2019**, *12*, 2970. [CrossRef]
19. Slama, M.E.A.; Beroual, A.; Hadi, H. Analytical Computation of Discharge Characteristic Constants and Critical Parameters of Flashover of Polluted Insulators. *IEEE Trans. Dielectr. Electr. Insul.* **2010**, *17*, 1764–1771. [CrossRef]
20. Sima, W.; Guo, F.; Yang, Q.; Yuan, T. Calculation of the Arc Velocity Along the Polluted Surface of Short Glass Plates Considering the Air Effect. *Energies* **2012**, *5*, 815–834. [CrossRef]
21. Nekahi, A.; Farokhi, S.; Farzaneh, M.; Stewart, B.G. Arc Energy and Temperature during Its Propagation Over Ice-Covered Surfaces. *IEEE Trans. Plasma Sci.* **2014**, *42*, 114–119. [CrossRef]
22. Nekahi, A.; Farzaneh, M. Excitation Temperature Determination of an Arc Formed over An Ice Surface Using Optical Emission Spectroscopy. *IEEE Trans. Dielectr. Electr. Insul.* **2011**, *18*, 1829–1834. [CrossRef]
23. Li, S.; Zhang, R.; Tan, K. Measurement of the Temperature of A the local arc Propagating along A Polluted Dielectric Surface. *Proc. CSEE* **1990**, *10*, 1–9. [CrossRef]
24. IEC. *Artificial Pollution Tests on High-Voltage Ceramic and Glass Insulators to Be Used on A.C. Systems*; IEC 60507; International Electrotechnical Commission: Geneva, Switzerland, 2013.
25. Griem, H.R. *Principles of Plasma Spectroscopy*; Cambridge University Press: Cambridge, UK, 1997; pp. 259–265.
26. Zollweg, R.J.; Liebermann, R.W. Electrical Conductivity of Nonideal Plasmas. *J. Appl. Phys.* **1987**, *62*, 3621. [CrossRef]
27. Li, X.; Liu, X.; Zeng, F.; Yang, H.; Zhang, Q. Study on Resistance and Energy Deposition of Spark Channel under the Oscillatory Current Pulse. *IEEE Trans. Plasma Sci.* **2014**, *42*, 2259–2265. [CrossRef]
28. Yang, H.; Ren, Y.; Wen, R.; Zhang, L.; Chen, Y. Study on the time-varying characteristics of AC local arc over the contaminated insulation surface. *AIP Adv.* **2021**, *11*, 025028. [CrossRef]
29. Rizk, F.M.; Rezazada, A.Q. Modeling of Altitude Effects on Ac Flashover of Polluted High Voltage Insulators. *IEEE Trans. Power Deliv.* **1997**, *12*, 810–822. [CrossRef]
30. Wilkins, R. Flashover voltage of high-voltage insulators with uniform surface-pollution films. *Proc. Inst. Electr. Eng.* **1969**, *116*, 457–465. [CrossRef]

energies

MDPI

Article

Research on Mechanical Defect Detection and Diagnosis Method for GIS Equipment Based on Vibration Signal †

Liang He [1], Jie Yang [2,*], Ziwei Zhang [1], Zongwu Li [3], Dengwei Ding [1], Minghu Yuan [1], Rong Li [1] and Mao Chen [1]

1 Tsinghua Sichuan Energy Internet Research Institute, Chengdu 610312, China; helxjtu@163.com (L.H.); ziwei.z@outlook.com (Z.Z.); sunnyall123@163.com (D.D.); tigervyuan@126.com (M.Y.); 18982123156@163.com (R.L.); chenmao@tsinghua-eiri.org (M.C.)
2 Chengdu Aeronautic Polytechnic, Chengdu 610100, China
3 State Grid Sichuan Maintenance Company, Chengdu 610041, China; lizongwu16@126.com
* Correspondence: yjandz@163.com; Tel.: +86-28-8845-9369
† This paper is an extended version of our paper published in International Conference on High Voltage Engineering and Application 2020, Beijing, China, 6–10 September 2020.

Abstract: Gas insulated switchgear equipment (GIS) is widely used in power system, and more attention has been paid to discharge defects than mechanical defects. However, since mechanical defects are a major cause of the failure in GIS, it is of great significance to carry out relevant research on mechanical defects. Detection and diagnosis methods of mechanical defects based on vibration signal are studied in this paper. Firstly, vibration mechanisms of GIS are analyzed. Due to structural differences between single phase insulated type GIS and three phase insulated type GIS, there are big differences in vibration mechanisms between the two types of GISs. Secondly, experimental research on mechanical defects is carried out based on a 110 kV GIS equipment and a self-developed vibration detection system; results show that mechanical defects can be diagnosed by analyzing signal amplitude, frequency spectrum and waveform distortion rate, and a large current is more beneficial for diagnosing mechanical defects. Lastly, field application has been carried out on 220 kV GIS equipment, and a poor contact defect is found, demonstrating that abnormal diagnosis can be realized by method proposed in this paper. Experimental research and field application demonstrate the feasibility and effectiveness of detection and diagnosis method for mechanical defects based on vibration signal and provide experience for subsequent engineering application.

Keywords: gas insulated switchgear; mechanical defect; vibration; detection and diagnosis

Citation: He, L.; Yang, J.; Zhang, Z.; Li, Z.; Ding, D.; Yuan, M.; Li, R.; Chen, M. Research on Mechanical Defect Detection and Diagnosis Method for GIS Equipment Based on Vibration Signal †. *Energies* **2021**, *14*, 5507. https://doi.org/10.3390/en14175507

Academic Editor: Pawel Rozga

Received: 21 June 2021
Accepted: 31 August 2021
Published: 3 September 2021

1. Introduction

Gas insulated switchgear (GIS) contains many components, including circuit breaker, isolating switch, CT, PT and busbar. In recent years, as the power grid develops rapidly, increased GIS equipment has been widely used due to many advantages, including compact structure, less occupied space, reliable operation and less maintenance work [1]. It is necessary to discover the hidden danger and potential defects of GIS equipment in a timely and effective manner.

Although the failure rate of GIS is relatively low, it is hard to ignore the fault of GIS with extensive use of GIS equipment. Statistics show that GIS defects mainly include partial discharge (PD) fault, mechanical fault and overheating fault; the mechanical fault of GIS occupy the largest proportion. Mechanical fault includes loose contact of the conductor, loose shield and loose bolts, which may cause abnormal vibration [2]. The continuous abnormal vibration may result in SF6 leakage and result in insulation decrease in GIS. Abnormal vibration may also cause PD fault and even breakdown event t. There are many research studies on detection and diagnosis of GIS, including UHF method [3], ultrasonic method [4], chemical method [5] and infrared temperature measurement method. While they are mainly used for studying PD faults and overheating faults, there is relatively little research work of mechanical faults [6,7]. Thus, it is important to study monitoring and diagnosis technology for mechanical faults.

With the advantage of high sensitivity, non-intrusive and anti-electromagnetic interference, a vibration test is a good detection and diagnosis method for the mechanical faults of power equipment, and it has been widely applied on circuit breaker [8], transformers and high voltage reactor [9]. In recent years, vibration test for GIS mechanical faults detection has been studied and applied. Evaluation method on contact faults by measuring vibration signals on the enclosures of GIS is reported in literature [10], demonstrating the feasibility of this method. Under the action of alternating electric field force in GIS, the vibration signal caused by mechanical fault is different from that of the normal situation. In order to detect and diagnose mechanical fault in GIS, it is important to study the vibration characteristics of GIS in depth. Vibration signals of a bus are studied, and the ratio between 1400 Hz component and 100 Hz component is used as characteristic quantity for loose defect in literature [11]. Time domain characteristics and power spectrum are used to analyze vibration signals. It was found that 200 Hz component of GIS equipment with abnormal noise is very big in literature [12]. The influence of air pressure, electrode shape and electric field intensity on vibration spectrum of GIS equipment is studied. However, the research mainly focuses on PD fault and not about mechanical fault in the literature [13]. Time domain and frequency domain features of vibration signals of the switch contact under different contact state are studied in literature [14], it is found that there are obvious 300 Hz and 600 Hz components when the contact condition is poor. The empirical mode decomposition is used for analyzing nonlinear and nonstationary time series in the literature [15]. It provides a reference for the GIS vibration signal. A new algorithm based on the neighbor algorithm and FCM algorithm was adopted to evaluate mechanical condition of GIS in the literature [16]; the key is to calculate cluster centers. A correlation function is adopted to extract characteristic quantity, and a multilayer classifier was adopted to evaluate condition in the literature [17]; however, intelligent detection technology is based on massive data, and a lot of research is needed to enrich the expert database. In summary, it can be observed that the above documents have made significant contributions to the detection and diagnosis technology on mechanical fault in GIS. However, due to the complex structure of GIS and the complicated excitation and transmission of vibration signals in GIS, it is difficult to obtain a general diagnostic process, including characteristic quantity selection and extraction and fault identification. Most of the research studies are carried out in laboratory, and there is less field application. It is necessary to study vibration detection and diagnosis technology for GIS in depth, and carry out field application.

In this paper, research on detection and diagnosis method of mechanical defects based on vibration signal is studied. Section 2 analyzes the characteristics of circulating current and vibration mechanism of GIS and analyzes the vibration character of single phase insulated type GIS and three phase insulated type GIS. Section 3 introduces experimental research on mechanical defects, including GIS platform, detection system, detailed experiments and vibration signal analysis. Section 4 presents field application on 220 kV GIS equipment. Lastly, the full paper is summarized in Section 5.

2. Vibration Mechanism of GIS Equipment

Alternating electrodynamic forces caused by AC current and magnetostriction effect in core coil equipment, such as a voltage transformer, are main reasons for the mechanical vibration of GIS. However, due to structural differences, the characteristics of circulating current in the three-phase insulated type GIS's enclosure and single-phase insulated type GIS's enclosure are completely different, resulting in different vibration mechanisms. Thus, the characteristics of circulating current in GIS's enclosure is analyzed firstly in this section, and the vibration mechanism is then analyzed with two methods: electrodynamic force and magnetostriction.

2.1. Characteristics of Circulating Current

Equivalent circuit diagram of GIS is shown in Figure 1 [18]. When AC current flows through an inner conductor, a circulating current is formed in GIS.

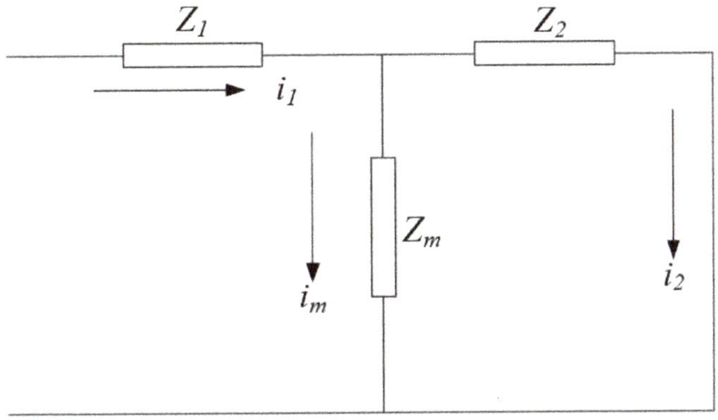

Figure 1. Circuit model of GIS.

In the equivalent circuit, Z_1 stands for leakage impedance of primary side (i.e., inner conductor), Z_2 stands for leakage impedance of secondary side (i.e., enclosure of GIS) and Z_m stands for excitation impedance. i_1 refers to as primary side current, i_2 refers to as secondary side current and i_m stands for excitation current. As the electromagnetic coupling between primary side and secondary side is relatively poor, this results in i_2 being smaller than i_1; however, i_2 and i_1 are in the same magnitude [18]. Literature [19] points out that the circulating current of 500kV GIL enclosure is about 75% of the conductor current.

Single-phase insulated type structure is often used in GIS equipment with high voltage levels (for example, 500kV GIS equipment) under normal operation conditions, and the circulating current is generated in the closed loop formed by the enclosure of GIS, frame and grounding grid. The loop includes two kinds: one is composed of the enclosure and grounding grid as illustrated in Figure 2; and one is composed of the enclosure of different phases and grounding grid as illustrated in Figure 3 [20]. In this case, i_2 is on the same number order as i_1, as analyzed above.

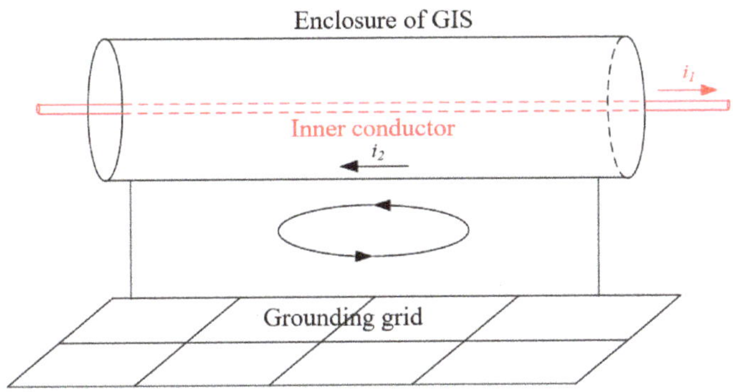

Figure 2. Loop composed of enclosure and grounding grid.

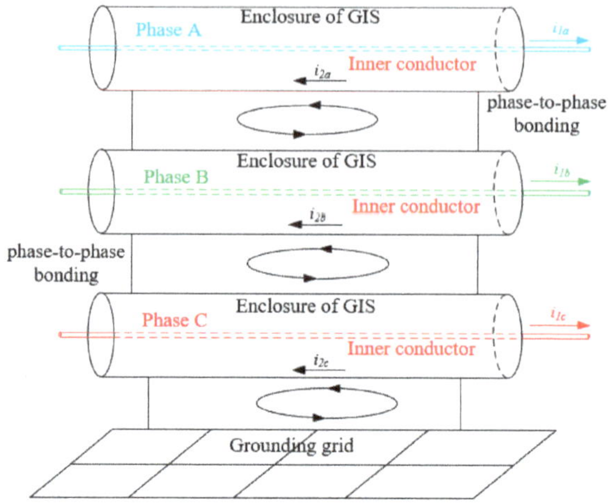

Figure 3. Loop composed of different phase and grounding grid.

Three-phase insulated type structure is often used in GIS equipment with low voltage level, for example, 110 kV GIS equipment. In an ideal situation, the current in phase A, B and C is balanced; this means that the sum of current vectors is nearly zero. Thus, there is almost no circulating current. Although unbalance induced current always exists when GIS is in operation, as shown in Figure 4, in this case i_2 is still far smaller than i_1 [20].

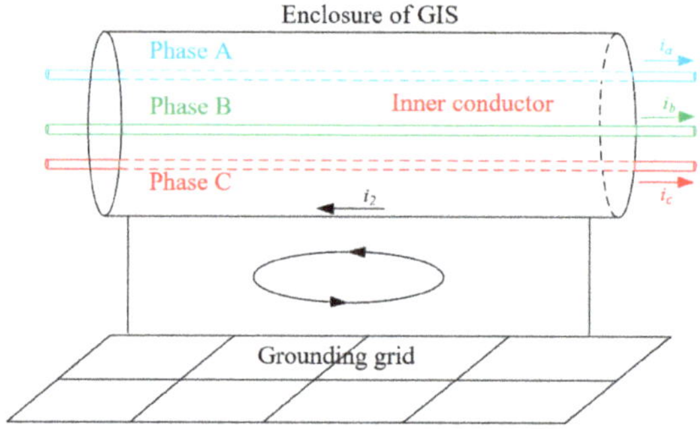

Figure 4. Circulating current of three phase insulated type GIS.

To summarize, the circulating current of single-phase insulated type GIS is in the same magnitude as the current of the inner conductor, while the circulating current of three-phase insulated type GIS is much less than the current of inner conductor. One can observe that circulating current of the two types of GIS is completely different, which results in different vibration mechanisms.

2.2. Vibration Mechanism of Single Phase Insulated Type GIS

As the enclosure of GIS possesses good shielding effects, the inner conductor is not affected by other the two-phase conductor. This means that vibration of inner conductor itself is very small. However, many electrical components in GIS are connected by contacts, such as circuit breaker and isolating switch. When current flows through contacts, the direction of current line is changed; in other words, the current line shrinks near the contacts' surface to generate electrodynamic force, as shown in Figure 5, thereby causing vibrations of GIS.

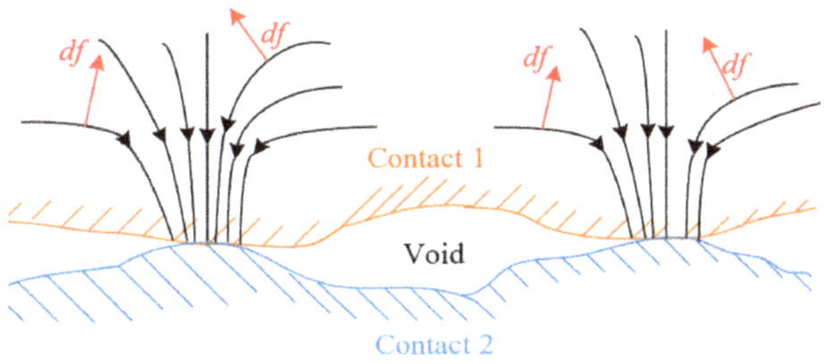

Figure 5. Force analysis of contacts.

Assuming that the current flowing through contacts is $i_0 = I_0\sin(\omega t)$, in which ω is current frequency and I_0 is current amplitude, the electrodynamic force caused by contacts can be expressed as Equation (1) in which a is the radius of contacts spot, D refers to the diameter of contacts surface, F denotes electrodynamic force, P_j denotes initial pressure acting on contact, H_b denotes cloth hardness of the material, ζ denotes the deformation coefficient of material between 0.3 and 1, and μ_0 is the permeability of vacuum [2].

$$F = \frac{\mu_0 I_0 \sin^2(\omega t)}{4\pi} \ln\left(\frac{D}{2a}\right) \tag{1}$$

$$a = \sqrt{\frac{P_j}{\pi\zeta H_b}} \tag{2}$$

On the other hand, enclosure is in the magnetic field generated by the inner conductor. As mentioned above, the circulating current in the enclosure (i.e., i_1) is in the same magnitude as the current of the inner conductor (i.e., i_0); that is, electromotive force exists between enclosure and inner conductor, as shown in Figure 6.

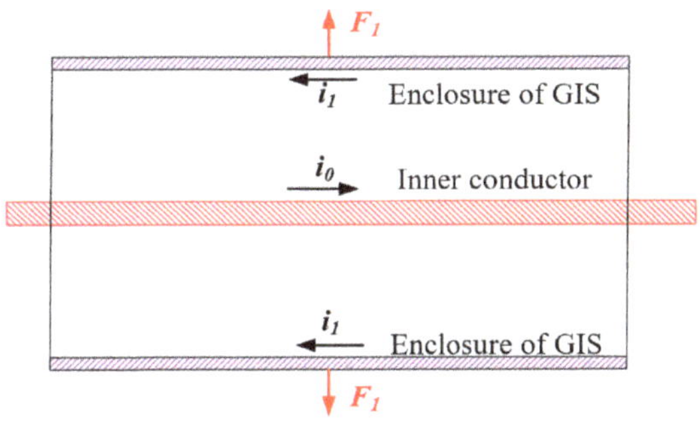

Figure 6. Force analysis of single phase insulated type GIS.

Assuming that radius of enclosure is R, which is far larger than the thickness of enclosure, magnetic field B_R in enclosure can be considered equal as expressed in Equation (3). The electrodynamic force is shown in Equation (4) [20].

$$B_R = \frac{\mu_0 i_0}{2\pi R} = \frac{\mu_0 I_0 \sin(\omega t)}{2\pi R} \tag{3}$$

$$F_1 = B_R i_1 L = \frac{\mu_0 k L I_0^2 \sin^2(\omega t)}{2\pi R} \tag{4}$$

It is observed that vibration caused by electrodynamic force consists of two parts: contacts and enclosure. The frequency is twice of ω; that is, when frequency of the current is 50 Hz, frequency of the vibration signal is 100 Hz. Meanwhile, experimental results show that harmonic components will appear in vibration signals when there is mechanical defect in GIS equipment, such as 200 Hz, 300 Hz, 400 Hz, 500 Hz and so on [2,9–17].

2.3. Vibration Mechanism of Three Phase Insulated Type GIS

An alternating magnetic field is generated by a three-phase inner conductor, and as mentioned above the circulating current is very small; that is, the electromotive force between the enclosure and the inner conductor is small. However, the electrodynamic force exists between three phase inner conductor. This is the significant difference of the vibration mechanisms between these two kinds of GIS.

Phase A is taken as an example to analyze electrodynamic force, as illustrated in Figure 7 [2,21]. The current of the three-phase can be expressed as Equations (5)–(7).

$$i_a = I_0 \sin(\omega t) \tag{5}$$

$$i_b = I_0 \sin(\omega t + 120^\circ) \tag{6}$$

$$i_c = I_0 \sin(\omega t + 120^\circ) \tag{7}$$

l_{ab}, l_{bc} and l_{ac} denote the distance between phase A–B, phase B–C and phase A–C, respectively. θ is angle between f_{ab} and f_{ac}, and it can be obtained according to cosine theorem, as shown below.

$$\cos\theta = \frac{l_{ab}^2 + l_{ac}^2 - l_{bc}^2}{2 l_{ab} l_{ac}} \tag{8}$$

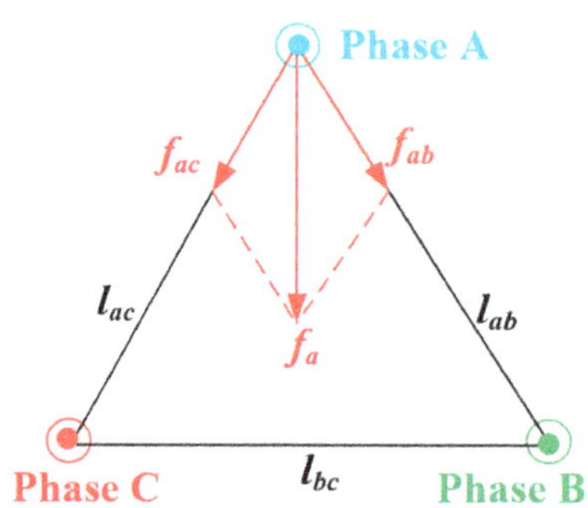

Figure 7. Force analysis of three phase insulated type GIS.

The electrodynamic force of phase A is shown below.

$$f_{ab} = \frac{\mu_0 L I_0^2 \sin(\omega t)\sin(\omega t - 120^\circ)}{2\pi l_{ab}} \tag{9}$$

$$f_{ac} = \frac{\mu_0 L I_0^2 \sin(\omega t)\sin(\omega t + 120^\circ)}{2\pi l_{ac}} \tag{10}$$

$$f_a^2 = f_{ab}^2 + f_{ac}^2 + 2 f_{ab} f_{ac} \cos\theta \tag{11}$$

Considering the actual situation, let $l_{\mathrm{ab}} = l_{\mathrm{ac}} = l_{\mathrm{bc}} = l$, and the force of L length inner conductor is shown in Equation (12). It is observed that the frequency of the electrodynamic force is also twice of ω. Experimental results show that harmonic components will appear in vibration signals when there is a mechanical defect.

$$f_a = \frac{\sqrt{3}\mu_0 L I_0^2 |\sin(\omega t)|}{4\pi l} \tag{12}$$

On the other hand, when current flows though inner conductor, the current line also shrinks near the contacts' surface to generate electrodynamic force, thereby causing vibration of GIS.

To summarize, vibration caused by electrodynamic force consists of two parts: contacts and inner conductor. This is the biggest difference of single-phase insulated type GIS.

2.4. Vibration Mechanism of Magnetostriction

Magnetostriction effect in core coil equipment can also cause vibration in GIS. Set $U_1 = U_0\sin(\omega t)$, and the magnetic flux intensity in iron core can be expressed as Equation (13) in which N refers to coil turns and S refers to core cross-sectional area [21].

$$B = \frac{U_0 \cos(\omega t)}{\omega NS} = B_0 \cos(\omega t) \tag{13}$$

$$B_0 = \frac{U_0}{\omega NS} \tag{14}$$

Considering that the magnetic flux density varies linearly with the magnetic field strength, the magnetic field strength can be expressed as Equation (15) in which B_s is saturation magnetic induction intensity, and H_c is coercivity.

$$H = \frac{B}{\mu} = \frac{BH_c}{B_s} = \frac{B_0 H_c \cos(\omega t)}{B_s} \tag{15}$$

When an alternating magnetic field exists, the micro deformation of ferromagnetic material is expressed as Equations (16) and (17) in which ε and ε_s refer to axial magnetostriction rate and saturation magnetostriction ratio, respectively, and L and ΔL refer to the original axial dimension and extension of silicon steel in the axial direction, respectively.

$$\varepsilon = \frac{\Delta L}{L} \tag{16}$$

$$\frac{\Delta L}{L}\frac{1}{dH} = |H|\frac{2\varepsilon_s}{H_c^2} \tag{17}$$

ΔL is shown in Equation (18), and the vibration acceleration caused by magnetostriction can be obtained by Equation (19).

$$\Delta L = L\int_0^H |H|\frac{2\varepsilon_s}{H_c^2}dH = \frac{\varepsilon_s L U_0^2 \cos^2(\omega t)}{(\omega N B_s S)^2} \tag{18}$$

$$a = \frac{d^2\Delta L}{dt^2} = -\frac{2\varepsilon_s L U_0^2 \cos(2\omega t)}{(N B_s S)^2} \tag{19}$$

It is observed that the frequency of vibration acceleration is twice of ω; that is, when the frequency of the current is 50 Hz, the frequency of the vibration signal is 100 Hz. However, due to the nonlinearity of core material, there are many higher harmonic components in vibration signal, including 200 Hz, 300 Hz, 400 Hz and so on.

To summarize, the vibration of single phase insulated type GIS includes three parts: enclosure vibration, contacts vibration and vibration caused by magnetostriction. The vibration of three-phase insulated type GIS includes three parts: inner conductor vibration, contacts vibration and vibration caused by magnetostriction. This is the biggest difference between two type GIS.

3. Experiments and Vibration Data Analysis

3.1. Experiment Platform and Detecting System

The test is conducted on a 110 kV GIS experiment platform filled with SF_6, including two bushing, two bus and one isolation switch; all components of the platform are true equipment used for simulating the actual situation, as shown in Figure 8. The current generator can generate a continuously current from 0 A to 3000 A. A vibration detection system is developed to collect and process data, including acceleration sensor, acquisition card and PC. PCB333B50 is selected as acceleration sensor, and its frequency response is from 0.5 Hz to 3000 Hz, sensitivity is 1000 mV/g and measurement range is from −5 g to +5 g. The model of acquisition card is NI9234, and it is a four-channel dynamic signal acquisition module with a precision of 24 bits, and the sampling rate is set as 25.6 kS/s. Vibration data are transmitted to PC for further analysis after being recorded by the acquisition card.

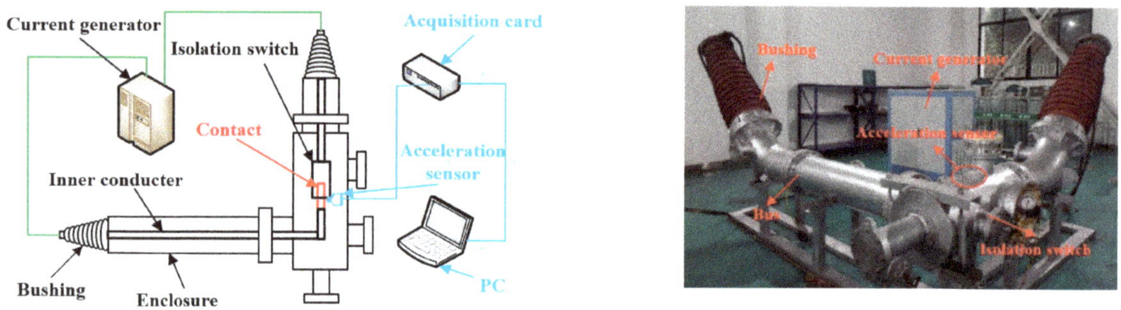

(**a**) (**b**)

Figure 8. GIS vibration experiment system. (**a**) Experimental setup. (**b**) Physical object of test platform.

In order to study the vibration characteristics under different contact states, three types of cases are set up by changing the state of the isolation switch, including normal, mild poor contact and serious poor contact, as shown in Figure 9. As the vibration wave is a mechanical wave, which decays gradually in the process of propagation, it is necessary to install the acceleration sensor closer to the defect source in order to obtain better detection effects [22]. Thus, all vibration tests are carried out on the top of the enclosure, directly above the defect.

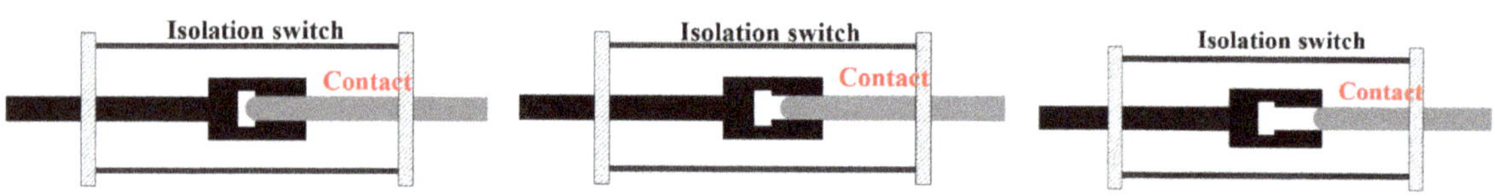

(**a**) (**b**) (**c**)

Figure 9. Mechanical defects setup. (**a**) Normal case; (**b**) mild poor contact; (**c**) serious poor contact.

3.2. *Vibration Data Analysis*

3.2.1. Time Domain Analysis

As the current affects vibration characteristics of GIS equipment seriously, four kinds of current are set for each contact state, including 500 A, 1000 A, 1500 A and 2000 A. It should be noted that the vibration test is susceptible to external interference, and there exists some discreteness in the test data. External vibration sources should be excluded during the test, and abnormal data should be eliminated. Finally, for each state, five groups of vibration data are saved in the steady state, and the acquisition time of each group is 1 s.

The first 0.2 s part of typical time domain waveform is extracted to study vibration characteristics under different conditions, as shown in Figure 10. It is obvious that amplitude of vibration signal increases with the current, and vibration signal shows strong periodicity. When the current is 500 A, the different between three kinds of contact state is small. When current increases to 1000 A, 1500 A and 2000 A, it can be found that the vibration signal under a serious poor contact state is the largest and, under normal state, is the smallest; it is feasible enough to distinguish the contact state of GIS equipment from the time domain waveform when the current is big.

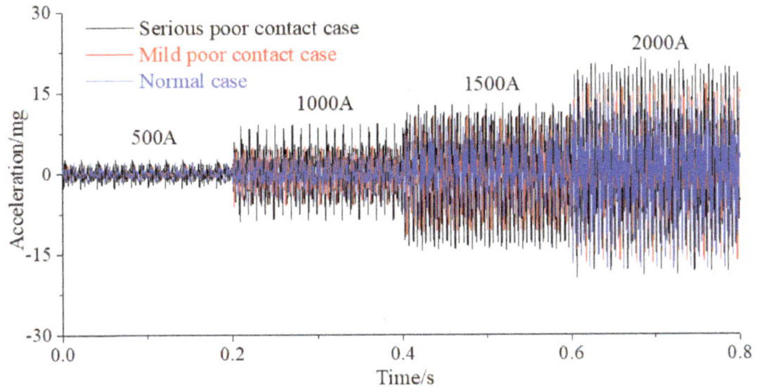

Figure 10. Time-domain waveform under different case.

Considering the dispersion of vibration signal, the mean and standard deviation of five groups of data are counted, the amplitude of vibration signal under different states is shown in Figure 11. It is illustrated that the amplitude increases gradually with the current, and three different contact states have the same increasing trend. When current is 500 A, the amplitude is small and the difference is relatively small. However, when the current is larger, it is easier to diagnose the contact state by comparing the amplitude of vibration signal. That is, a large current is more beneficial for diagnosing mechanical defects.

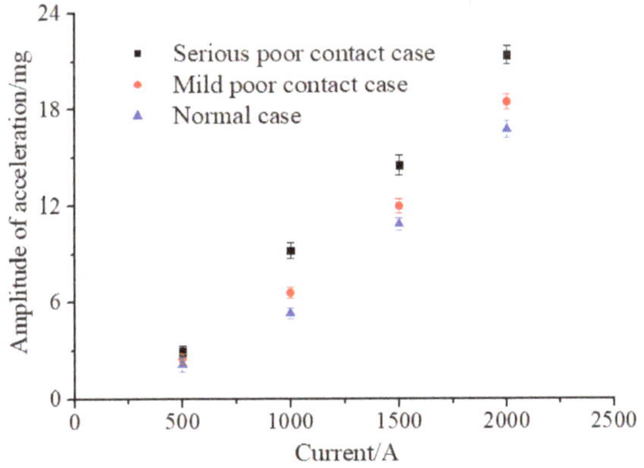

Figure 11. Comparison of acceleration under different cases.

3.2.2. Frequency Domain Analysis

Figure 12 shows the typical frequency domain waveforms of vibration signal under a normal case. It can be observed that, when the current is 500 A, the main energy of vibration signal is concentrated below 600 Hz. When the current is 1000 A, there are some higher frequency components such as 700 Hz and 800 Hz. When the current increases to 1500 A and 2000 A, the frequency distribution becomes wider towards 1200 Hz. At the same time, the amplitude of each frequency component increases with the current. Figure 13 shows the frequency domain waveforms under mild poor contact case. It is observed that frequency range is beyond 600 Hz when current is 500 A, and the frequency distribution becomes wider towards 1300 Hz when current increases to 1500 A and 2000 A, and there is an increasing trend in amplitude. The frequency domain waveforms under serious poor contact case are shown in Figure 14, and frequency distribution range and amplitude show the same trend as the former.

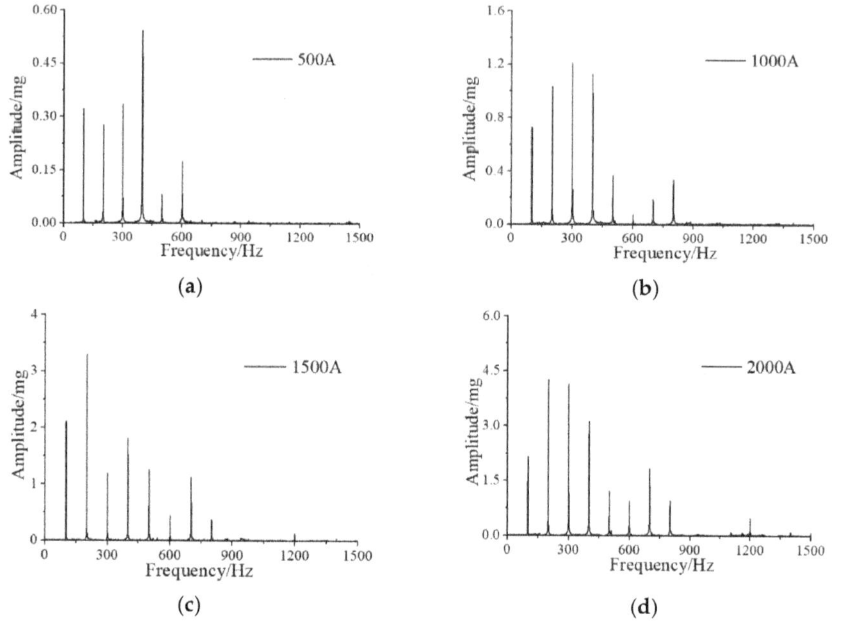

Figure 12. Frequency domain waveforms under the normal case: (**a**) 500 A; (**b**) 1000 A; (**c**) 1500 A; (**d**) 2000 A.

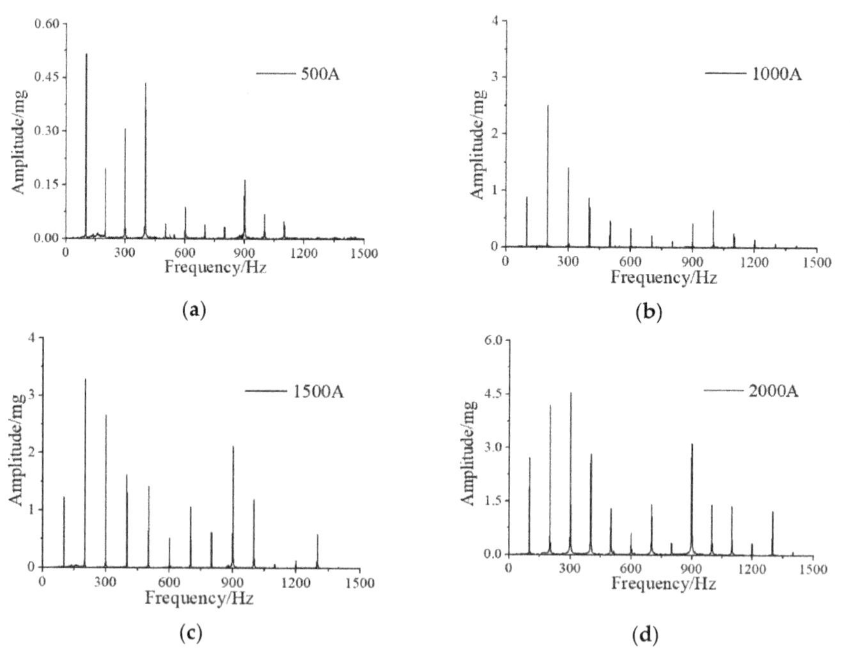

Figure 13. Frequency domain waveforms under the mild poor contact case: (**a**) 500 A; (**b**) 1000 A; (**c**) 1500 A; (**d**) 2000 A.

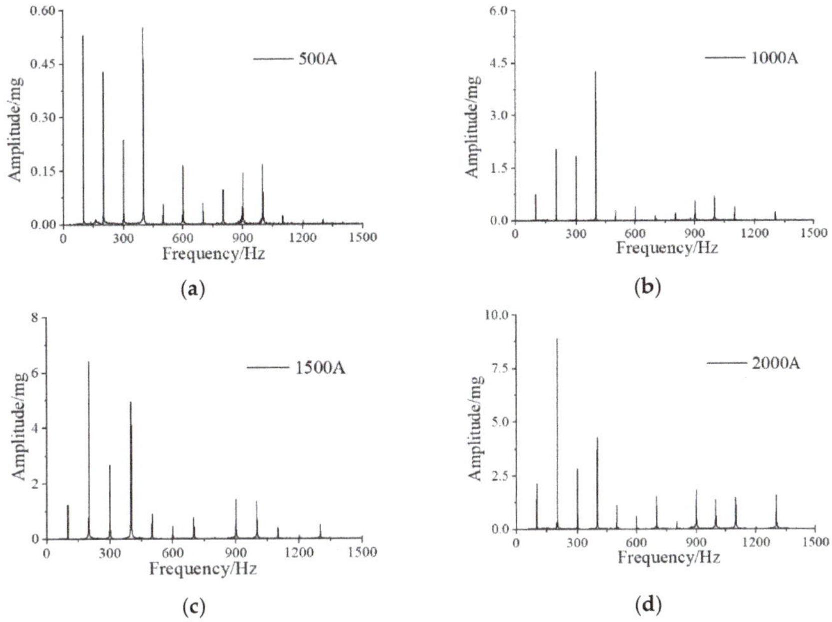

Figure 14. Frequency domain waveforms under the serious poor contact case: (**a**) 500 A; (**b**) 1000 A; (**c**) 1500 A; (**d**) 2000 A.

To summarize, when the contact condition deteriorates, the frequency spectrum range becomes wider, and increased energy is concentrated at higher frequency components. The amplitude of each frequency component becomes bigger. When the current increases gradually, there are more high frequency components, and the amplitude also becomes bigger.

The waveform distortion rate (THD) reflects the distortion degree of a signal waveform relative to the sinusoidal waveform, which can be used to describe the influence of higher harmonics component on vibration signal waveform [23]. In this paper, THD can be calculated by power spectrum of vibration signal, as shown in Equations (20)–(24). In the formulas, $v(t)$ ($0 \leq t \leq 1$ s) represents vibration signal, τ refers to displacement variable ($0 \leq \tau \leq 1$ s), $F(\tau)$ denotes the autocorrelation function of $v(t)$ and $P(f)$ refers to power spectrum.

$$F(\tau) = \lim_{T \to \infty} \frac{1}{T} \int_{-\infty}^{+\infty} v(t)v(t+\tau)dt \tag{20}$$

$$P(f) = \int_{-\infty}^{+\infty} F(\tau)e^{-i2\pi f\tau}dt \tag{21}$$

$$A_{100} = P(f)|f = 100 \tag{22}$$

$$A_k = P(f)|f = 100k \tag{23}$$

$$THD = 20\log\left(\frac{1}{A_{100}}\sqrt{\sum_{k=2}^{\frac{f1}{100}-1}(kA_k)^2}\right) \tag{24}$$

Considering the dispersion of vibration signal, the mean and standard deviation of five groups of THD are counted, as illustrated in Figure 15. THD increases with the increase in contact deterioration degree. When current becomes large, the growth trend of THD slows down gradually. That is, it is feasible to distinguish different contact states by comparing THD.

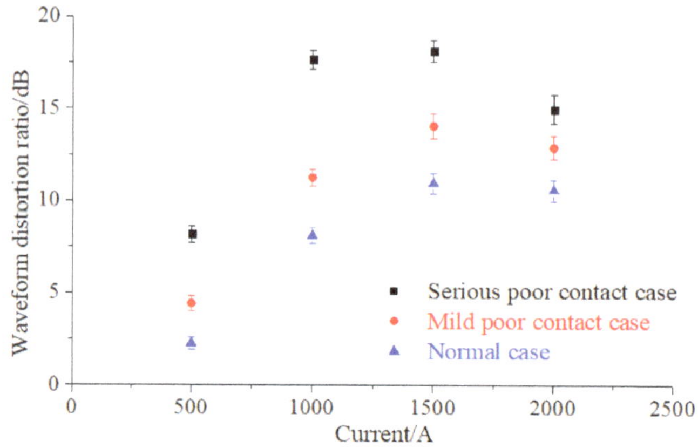

Figure 15. Comparison of waveform distortion ratio under different case.

To summarize, mechanical defects can be diagnosed by analyzing signal amplitude, frequency spectrum and waveform distortion rate. The greater the current, the better the effect.

4. Field Application

Vibration detection on GIS equipment was carried out in one 220 kV substation, as mechanical defects are mainly caused by poor contact, unbalanced alignment of the conductor and components' looseness and so on. Thus, vibration test points are often located near joint parts, such as the circuit breaker, isolation switch and contact. One abnormal signal is found in an isolation switch, and on-site detection is shown in Figure 16.

Figure 16. The location of the vibration measuring point.

Figure 17 shows the typical time domain waveform of vibration signal of a 220 kV isolation switch, and it can be observed that the vibration signal shows strong periodicity, and there is a big difference between phase A, B and C. In order to eliminate the dispersion of test results, five groups of vibration data are recorded, and the acquisition time of each group is 1 s. The amplitudes of phase A and phase B are similar, which are smaller than that of phase C, as shown in Figure 18. At the same time, there is obvious distortion in the waveform of phase A and phase C, and the distortion of phase of B is relatively small. It can be concluded that there may be abnormality in phase C.

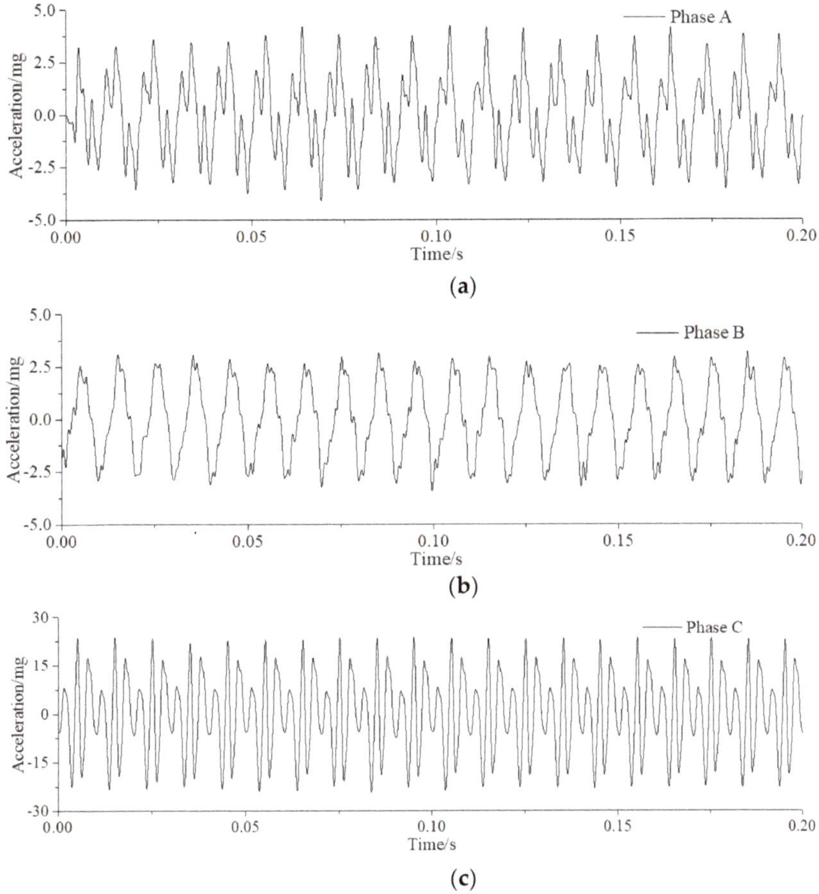

Figure 17. Time-domain waveform of isolation switch. (**a**) Phase A; (**b**) Phase B; (**c**) Phase C.

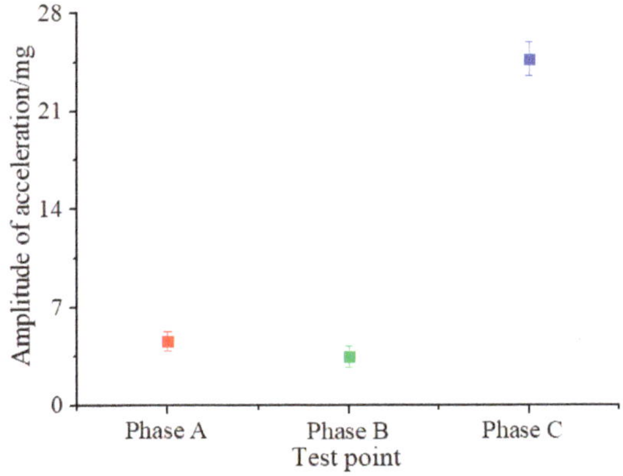

Figure 18. Comparison of acceleration.

Figure 19 shows the typical frequency domain waveforms of a vibration signal, and it can be observed that the main energy of phase A is concentrated at 100 Hz and 300 Hz, and there are some components at 500 Hz, 700 Hz and 800 Hz. The main energy of phase A is concentrated at 100 Hz, and the amplitudes of 200 Hz, 300 Hz, 500 Hz and 600 Hz are relatively small. However, the frequency spectrum of phase C is relatively wide, and the amplitude is relatively large. It contains many higher components, such as 300 Hz, 400 Hz, 500 Hz and 800 Hz. The 300 Hz component is the largest, and the 100 Hz component is larger than that of phase A and phase B. The THD of phase A, B and C is illustrated in Figure 19d, and it is observed that phase B is the smallest (close to −28 dB), phase A is the second (nearly −8 dB) and phase C is the largest (close to 18 dB); that is, the situation of phase C is the worst.

As phase A and phase C are symmetrically distributed, the structure of the three-phase is the same, and the load across all of them are similar; thus, the vibration characteristics of the three-phases are comparable. By comparing time domain and frequency domain waveform, it can be concluded that there is some abnormality in phase C. In the outage maintenance test, results show that the loop resistance of phase C is obviously higher than the other two phases, which is confirmed by vibration test results, demonstrating that abnormal diagnosis can be realized by the method proposed in this paper.

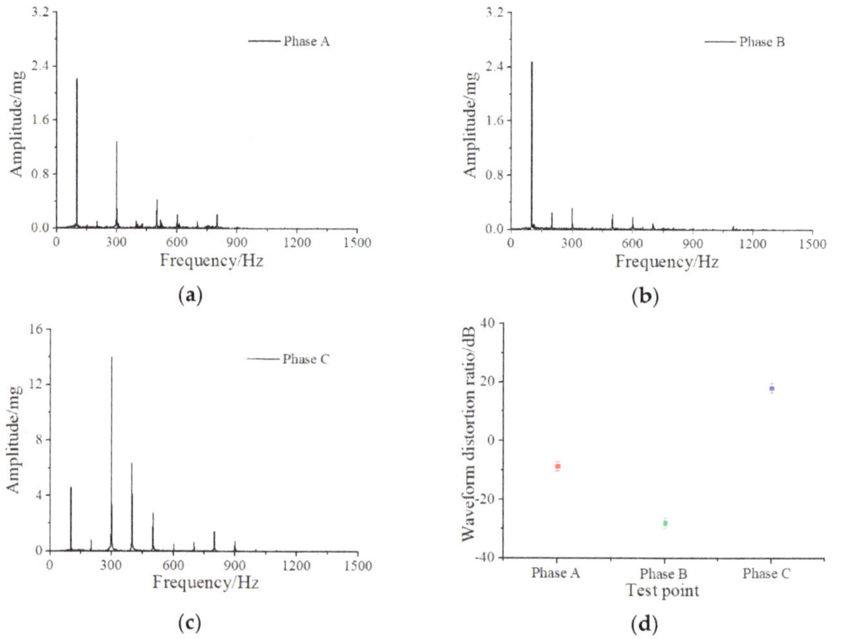

Figure 19. Frequency domain waveform of isolation switch. (**a**) Phase A; (**b**) Phase B; (**c**) Phase C; (**d**) comparison of all three phases.

5. Conclusions

Detection and diagnosis technologies of mechanical defects based on vibration signal are studied in this paper, including theoretical analysis, experimental research and field application. The results show the following.

1. GIS equipment structure lends to different vibration mechanism. The vibration of single-phase insulated type GIS includes three parts: enclosure, contacts and magnetostriction. The vibration of three-phase insulated type GIS includes three parts: inner conductor, contacts and magnetostriction.
2. Mechanical defects can be diagnosed by analyzing signal amplitude, frequency spectrum and waveform distortion rate. When the current is small, the change of signal

amplitude is not as obvious as the frequency spectrum and waveform distortion rate. A large current is more beneficial for diagnosing mechanical defects.

3. Field application demonstrates the feasibility and effectiveness of detection and diagnosis technology for mechanical defects based on vibration signal. Abnormal diagnosis can be realized by the methods proposed in this paper. In the future, we will carry out more field applications and accumulate data in order to lay a foundation for applying artificial intelligence to mechanical fault diagnosis.

Author Contributions: Investigation, L.H.; methodology, L.H., J.Y. and D.D.; software, M.Y. and Z.Z.; data curation, M.C. and Z.L.; writing—original draft preparation, J.Y. and L.H.; writing—review and editing, D.D. and R.L. All authors have read and agreed to the published version of the manuscript.

Funding: Project supported by Science and Technology Project of State Grid Sichuan Electric Power Company. Research and development of the intelligent sensor of vibration and voiceprint based on mechanical contact perception and evaluation and diagnosis system.

Conflicts of Interest: The authors declare no conflict of interest.

References

1. Wang, H.M.; Yang, J.C.; Wang, X. Feature Fingerprint Extraction and Abnormity Diagnosis Method of the Vibration on the GIS. In Proceedings of the 2020 IEEE International Conference on High Voltage Engineering (ICHVE 2020), Beijing, China, 6–10 September 2020; pp. 1–4. [CrossRef]
2. Liu, Y.; Yang, J.G.; Jia, Y.Y. Connection state diagnosis method of gis disconnector based on mechanical vibration. *High Volt. Eng.* **2019**, *45*, 1591–1599.
3. Li, T.; Rong, M.; Zheng, C.; Wang, X. Development simulation and experiment study on UHF Partial Discharge Sensor in GIS. *IEEE Trans. Dielectr. Electr. Insul.* **2012**, *19*, 1421–1430. [CrossRef]
4. Lundgaard, L.E.; Runde, M.; Skyberg, B. Acoustic diagnosis of gas insulated substations: A theoretical and experimental basis. *IEEE Trans. Power Deliv.* **2002**, *5*, 1751–1759. [CrossRef]
5. Tang, J.; Li, T. Device of SF6 dissociation apparatus under partial discharge and gaseous decomposition components analysis system. *High Volt. Eng.* **2008**, *34*, 52–57.
6. Schichler, U.; Koltunowicz, W.; Endo, F. Risk assessment on defects in GIS based on PD diagnostics. *IEEE Trans. Dielectr. Electr. Insul.* **2013**, *20*, 2165–2172. [CrossRef]
7. Ji, H.X.; Ma, G.M.; Li, C.R. Influence of voltage waveforms on partial discharge characteristics of protrusion defect in GIS. *IEEE Trans. Dielectr. Electr. Insul.* **2016**, *23*, 1058–1067. [CrossRef]
8. Landry, M.; Leonard, F.; Landry, C. An improved vibration analysis algorithm as a diagnostic tool for detecting mechanical anomalies on power circuit breakers. *IEEE Trans. Power Deliv.* **2008**, *23*, 1986–1994. [CrossRef]
9. Zhao, L.H.; Zhang, Z.D. Transformer Fault Diagnosis Method Based on Vibration Signal under Fluctuating Operating Condition. *High Volt. Eng.* **2020**, *336*, 214–222.
10. Youichi, O.; Yukio, K. A diagnostic technique to detect abnormal conditions of contacts measuring vibrations in metal enclosures of gas insulated switchgear. *IEEE Trans. Power Deliv.* **1989**, *4*, 2090–2094.
11. Huang, Q.; Wei, X. Research of GIS Bus Loose Contact Fault Diagnosis Techniques Based on Mechanical Vibration. *High Volt. Appar.* **2017**, *53*, 0099–0104.
12. Cheng, L. Research on vibration diagnosis method of ultra high voltage GIS/HGIS equipment. *Power Constr.* **2009**, *30*, 17–19.
13. Guo, B.H.; Zhang, H.H. Prediction of internal faults of GIS by frequency spectrum of typical vibration on the enclosure. In Proceedings of the 2nd International Conference on Properties and Applications of Dielectric Materials, Beijing, China, 12–16 September 1988; pp. 148–151.
14. Li, X.G.; Wu, X.T. Research on Contacting Fault of GIS Contact Based on Vibration Signal. *High Volt. Appar.* **2016**, *52*, 0165–0169.
15. Huang, N.E.; Zheng, S.; Long, S.R. The empirical mode decomposition and the Hilbert spectrum for nonlinear and non-stationary time series analysis. *Proc. Math. Phys. Eng. Sci.* **1998**, *454*, 903–995. [CrossRef]
16. Shen, X.; Lin, Z.; Peng, G. Research on mechanical fault diagnosis of Ultra high voltage GIS based on the combination of neighbor algorithm and FCM. In Proceedings of the Power and Energy Engineering Conference, Xi'an, China, 25–28 October 2016; pp. 2045–2049.
17. Yang, Y.; Ma, S.L.; Wu, J.W.; Jia, B.W.; Li, W.X.; Luo, X.W. Frequency feature learning from vibration information of GIS for mechanical defect detection. *Sensors* **2019**, *19*, 1949. [CrossRef]
18. Xu, G.Z.; Guan, Y.G. Simplified Analysis and Calculation on Circulating Current and Power Loss in Enclosure of GIS and GIL. *High Volt. Eng.* **2009**, *35*, 247–249.
19. Tang, L.L.; Cai, W. Circulation Characteristics and Influencing Factors of GIL Grounding System. *High Volt. Eng.* **2009**, *46*, 2098–2105.

20. Xue, J.L.; Xu, H.S.; Cheng, J.R. Research on Circulating Current Characteristics and the Vibration Mechanism on the GIS enclosure. In Proceedings of the 2019 2nd International Conference on Electrical Materials and Power Equipment (ICEMPE), Guangzhou, China, 7–10 April 2019; pp. 619–622.
21. Zhu, Y.Y.; Ji, S.C. Vibration Mechanism and Influence Factors in Power Transformers. *J. Xi'an Jiaotong Univ.* **2015**, *49*, 115–125.
22. Zhong, Y.; Hao, J.; Liao, R.; Wang, X.; Jiang, X.; Wang, F. Mechanical defect identifcation for gas-insulated switchgear equipment based on time-frequency vibration signal analysis. *High Volt.* **2020**, *6*, 531–542. [CrossRef]
23. Ding, D.W.; Zhang, X.H.; Lan, X.S. Effects of HVDC monopole operation on vibration of a 500 kV AC transformer. *J. Vib. Shock.* **2016**, *35*, 201–206.

energies

MDPI

Article

Study on Evaluation Method of Impulse Dispersion Performance of Grounding Device Using X-ray Imagining System †

Xiaoyan Han [1], Donghui Luo [1,2], Xin Zhang [3,*], Yongxing Cao [1], Yu Zhang [1] and Gege Ban [3]

1 State Grid Sichuan Electric Power Company, Chengdu 610041, China; hanxiaoyansc@163.com (X.H.); dhluo@foxmail.com (D.L.); 87085306@163.com (Y.C.); zy863129@163.com (Y.Z.)

2 Department of Electrical Engineering, Tsinghua University, Beijing 100084, China

3 State Key Laboratory of Power Transmission Equipment and System Security and New Technology, Department of Electrical Engineering, Chongqing University, Chongqing 400044, China; 13364069369@163.com

* Correspondence: 20191113048t@cqu.edu.cn; Tel.: +86-199-3607-8561

† This paper is an extended version of our paper published in International Conference on High Voltage Engineering and Application 2020, Beijing, China, 6–10 September 2020.

Abstract: When the lightning current enters the ground through the grounding system, the impulse dispersion performance can be observed by the phenomenon of soil spark discharge, which is fundamentally determined by the nearby soil. At present, engineers use an empirical formula to convert the soil spark discharge to the impulse coefficient of impulse grounding resistance. Therefore, there is no available quantitative analysis method to evaluate soil impulse dispersion performance. To solve this problem, this paper proposes an evaluation method for the impulse discharge efficiency of soil by using X-ray images, define V_I as the parameter, which is the ratio of the volume of the discharge area to the peak current. Then, the rationality and validity of the method are verified. Finally, the variation rules of impulse discharge efficiency are analyzed in different soils. Results show that the V_I can reflect the change rules of impulse dispersion performance more clearly under different soil conditions, and this parameter provides a new idea for enhancing the impulse dispersion performance of soil near the grounding electrode.

Keywords: high soil resistivity; soil discharge; X-ray imaging; super absorbent polymer; impulse discharge performance

Citation: Han, X.; Luo, D.; Zhang, X.; Cao, Y.; Zhang, Y.; Ban, G. Study on Evaluation Method of Impulse Dispersion Performance of Grounding Device Using X-ray Imagining System. *Energies* **2021**, *14*, 5124. https://doi.org/10.3390/en14165124

Academic Editors: Issouf Fofana, Bo Zhang and Alberto Geri

Received: 29 June 2021
Accepted: 16 August 2021
Published: 19 August 2021

1. Introduction

Soil is the key to affect the impulse dispersion performance of grounding devices [1]. Especially, spark discharge will occur when the impulse current is injected into the ground electrode, which significantly enhances the impulse dispersion performance of the grounding device [2,3]. Therefore, the accurate evaluation of soil impulse dispersion performance can provide a basis for the selection of soil resistance reduction measures [4].

At present, due to the lack of effective observation methods, it is impossible to carry out accurate quantitative analysis on the real topography of soil discharge area. Reference [5] conducted impulse breakdown experiments on the soil between spherical electrodes by building a cylinder test tank in the laboratory, and the breakdown discharge path of soil was obtained by X-ray film. In [6], the impulse discharge experiment of a suspended electrode on the soil surface was carried out in a hemispherical iron pot with a diameter of 10 cm, and the soil spark discharge channel was recorded by the conductive paper. Reference [7] obtained the actual soil discharge channel through field test excavation. In the above studies, due to the fuzzy measured results, the discharge channels obtained cannot be used as a quantitative index to evaluate the soil impulse dispersion performance.

The impulse dispersion performance of the grounding device is determined by the structure and size of the grounding electrode, the impulse dispersion performance of the soil, and the buried depth of the grounding electrode. However, when the impulse current flows into the ground, the intensity of the instantaneous electric field in the ground near the electrode will exceed the intensity of the soil's critical breakdown field, and the spark discharge area will appear at the end of the electrode. A large number of experimental studies found that the soil discharge area is affected by the distribution of the electric field in the ground and that soil discharge is unpredictable. The existence of a soil discharge area can greatly improve the impulse dispersion capacity of the soil near the grounding electrode [8–12]. At present, there is still a lack of a visualized index to evaluate the impulse dispersion performance of soil, which is mainly characterized by the critical breakdown field strength and residual resistivity of typical soil, which are not easy to be applied in practical engineering [13–15].

However, at present, there is no unified understanding of the mechanism of soil impulse discharge among scholars at home and abroad, and different researchers have proposed a variety of viewpoints on soil impulse discharge based on experimental observations [16]. In reference [17], the characteristics of soil ionization were firstly taken into account in the soil dispersion model, and it was proposed that there is an ionization region around the grounding device under the effect of impulse current, and its structure was considered to be an axisymmetric cylindrical structure about the electrode. Based on the same idea, a hemispherical discharge model was proposed in reference [18]. In reference [19], a cylindrical discharge model was proposed after considering the spatial inhomogeneity of current in the earth. A penetrating X-ray imaging technology was first applied in reference [20] to study the soil ionization under surge current and to calculate the soil residual resistivity of discharge area by a finite-element method, according to experiment parameters and a 3-D diagram of the discharge area. Reference [21] calculated the volume of successive impulse discharge channels in the ground under various currents and soil conditions. It also analyzed the changing law with time intervals based on the gray information analysis method and improved the four-zone model of soil under impulse current. Reference [1] proposed a four-zone soil discharge model. In the process of impulse dispersion, the soil's electric field intensity near the grounding pole gradually decreases outward, which can be divided into four regions: arc region, spark discharge region, semiconductor region, and constant conductivity region. The above soil impulse discharge models are all based on the assumption that soil is isotropic. However, soil is a polymorphic composite medium containing solid soil particles, liquid water, and air. Therefore, ignoring the influence of anisotropy of the soil on the impulse discharge process will be different from the actual situation. Reference [7] found that the actual soil discharge area was different from the above hypothesis through field experiment excavation, but was a channel structure with fewer branches. It can be seen that when the impulse current is injected into the soil, the intensity of the soil's electric field at the end of the grounding will exceed the intensity of the critical breakdown field of the soil, and a concentrated discharge area will be generated, forming a breakdown channel structure.

In summary, there is still a lack of suitable quantitative methods to evaluate the soil impulse dispersion performance in the near-ground area according to the data from various sources. To solve this problem, based on the image observation of soil spark discharge [8,9], a quantitative evaluation index was proposed to combine the discharge area and impulse current amplitude of soil samples. The rationality and effectiveness of the method are verified, and the variation law of the impulse dispersion performance of grounding devices is analyzed in soil samples with different properties. This method can quantitatively evaluate the impulse dispersion performance of the soil near the ground electrode, which provides a new idea for enhancing the impulse dispersion capability of the soil near the ground electrode.

2. Theoretical Derivation of Characteristic Quantity of Soil Impulse Dispersion Performance

Generally, the common grounding device of transmission line towers consists of horizontal or vertical ground electrodes [22–24]. Taking the vertical electrode structure as an example, the process of soil dispersion is analyzed theoretically. In Figure 1, the radius of the ground electrode is r_0, the length is l, and it is buried in the ground vertically. The soil resistivity is ρ, and the soil discharge area is a slender cylindrical channel with a bottom radius of r', a length of l', and a current of I, which flows into the medium through the rod. In this paper, a mirror electrode and discharge region with a length of $(l + l')$ are assumed under the ground by means of mirroring, and the semi-infinite space in the calculation of the current field is transformed into infinite space.

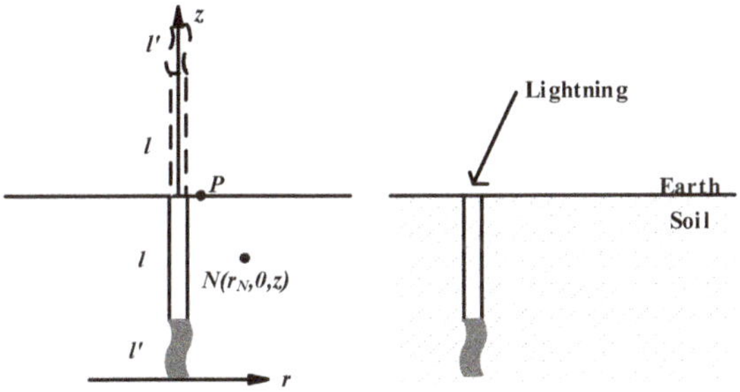

Figure 1. A round rod-shaped ground electrode embedded in the ground vertically.

When no discharge occurs in the soil, assuming that the current density is δ at any point in the soil, the potential of $N(r_N, 0, z)$ at any point is

$$u_N = \frac{\rho}{4\pi}\int_0^{2l} \frac{\delta dz}{\sqrt{(z_N - z)^2 + r^2}} = \frac{\rho\delta}{4\pi}\ln\frac{z_N + \sqrt{z_N^2 + r_N^2}}{z_N - 2l + \sqrt{(z_N - 2l)^2 + r_N^2}} \quad (1)$$

If the electrode potential u is equal to the potential u_p on the surface of the midpoint along the electrode length, that is $z_N = l$, $r_N = r_0$, then the electrode potential is:

$$u = u_p = \frac{\rho\delta}{4\pi}\ln\frac{l + \sqrt{l^2 + r_0^2}}{-l + \sqrt{(-l)^2 + r_0^2}} \approx \frac{\rho I}{2\pi l}\ln\frac{2l}{r_0} \quad (2)$$

The grounding resistance of the ground electrode in the semi-infinite medium is:

$$R = \frac{u}{I} = \frac{\rho}{2\pi l}\ln\frac{2l}{r_0} \quad (3)$$

When soil discharge occurs, the residual resistivity drops to about 0.003, 62% of the initial soil resistivity [8], which is close to the metal resistivity. The grounding resistance of the ground electrode is:

$$R' \approx \frac{\rho}{2\pi(l + l')}\ln\frac{2(l + l')}{r_0} \quad (4)$$

Grounding resistance is reduced by ΔR:

$$\Delta R = \frac{\rho}{2\pi l}\ln\frac{2l}{r_0} - \frac{\rho}{2\pi(l + l')}\ln\frac{2(l + l')}{r'} \quad (5)$$

According to (5), the reduction of ground resistance is related to the volume parameter of the area where soil ionization occurs. If the length of the ground electrode is 2.5 m, the section radius is 4 mm, the length of the breakdown area is 0.5 m, and the radius $r' = 2r_0$, then $\Delta R \approx 0.145R$, and the grounding resistance of ground electrodes is reduced by 14.5%. It can be seen that the soil impulse diffusivity is an important factor affecting the diffusivity of the ground electrode.

A larger discharge area leads to a more intense soil discharge process, which promotes the release of lightning energy in the soil [25–27]. Moreover, the volume of the discharge area increases with the rise of impulse current. V_I defined as the equivalent value of soil discharge volume per unit electric current, that is:

$$V_I = \frac{V}{I} \quad (6)$$

where, V is the volume of the soil impulse discharge area, cm^3; I is the amplitude of the ground current, kA.

A small value of V_I indicates that there is a space to further improve the impulse dispersion performance of the ground electrode by modifying soil. Therefore, the parameters can reflect the impulse dispersion performance of grounding device directly and clearly.

3. Test Method of Soil Impulse Dispersion near the Area of Ground Electrode

From the above analysis, it can be seen that it is necessary to put forward a method to obtain the specific structural parameters of the soil discharge area. The soil impulse discharge observation platform is shown in Figure 2. The impulse current is generated and applied to the ground electrode and recorded by oscilloscope. Then, an X-ray imaging system was used to image the discharge area in the soil; finally, the spatial structure of the soil discharge area is obtained by image processing.

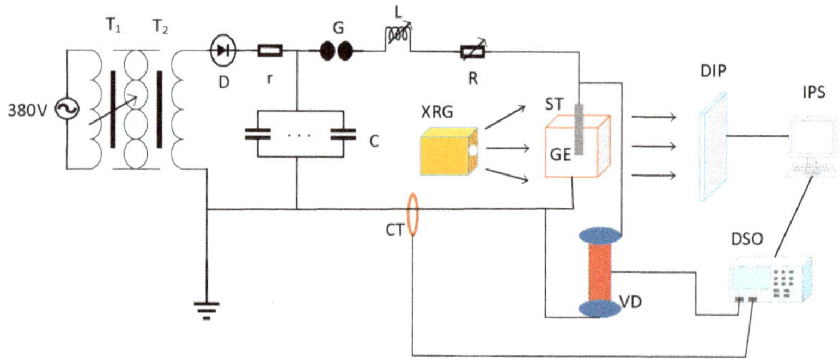

Figure 2. Sketch of the impulse current test setup: T_1: Voltage regulator; T_2: Voltage booster; D: Rectifier; r: Protective resistance; C: Charging capacitor bank; G: Sphere gap; L: Wave modulated inductance; R: Wave modulated resistance; XRG: X-ray generator; CT: Electric current transducer; ST: Sample sandbox; GE: Grounding electrode; VD: Voltage divider; DIP: Digital imaging plate; DSO: Oscilloscope; IPS: Image processing system.

In the experiment, glass fiber was prepared with a side length of 0.2 m and a thickness of 0.01 m. The cylindrical vertical ground electrode was selected with a diameter of 0.0025 m and a length of 0.15 m. Besides, the distance is 0.1 m between the electrode bottom and the sides of the test groove. The conductive copper foil is embedded in the inside of the test groove as the ground electrode, and the distances are equal between the ground electrode terminal and each side of the copper foil.

According to the image of the soil spark discharge area, the three-dimensional (3D) reconstruction algorithm of two-stage stretching is adopted to obtain the 3D structure of the soil discharge area [28,29]. According to the soil sample discharge image, the volume of soil spark discharge area was calculated using the method given above, and the value of V_I of soil sample discharge area could be calculated according to (6). The measured value of ground resistance R_{ch1} can be obtained by calculating the current and voltage waveform recorded by the oscilloscope.

4. Study on the Validity of the Paper's Method and the Main Influence Factors

It can be seen from the above analysis that it is feasible to use the equivalent value, which represents soil discharge volume per unit electric current (V_I), to describe the evaluation index of the near-area soil impulse dispersion ability. In order to further verify its validity and rationality, this paper carried out a comparative experiment on the near-area soil dispersal capacity.

4.1. The Validity Analysis of the Method for Evaluating the Near-Area Soil Impulse Dispersion Capacity

In actual engineering, the empirical formula is used by translating the soil spark discharge effect into the impulse coefficient of grounding device, as shown in Equation (7):

$$R_{ch} = \alpha R \tag{7}$$

where: R is the power frequency grounding resistance of the grounding device; α is the impulse coefficient, and the range is 0.5~3. In engineering, α is usually based on the geometric size of the grounding electrode, soil resistivity, and other parameters, which are set as 0.85 [1] in this paper.

In order to verify the effectiveness of soil discharge volume per unit electric current (V_I) as an evaluation index of the dispersion capacity of grounding device, this paper carried out a comparative test study on the near-area soil dispersion capacity. First of all, by adding super absorbent polymer (SAP) and regulating soil water content, soil samples with a resistivity of 50 Ω·m and different moisture content were prepared [30–32]. Then, an impulse current with an amplitude of 2.0 kA was applied to the soil samples. The image of the soil discharge channel could be obtained by X-ray projection imaging technology, and the structural parameters of the soil discharge area could be calculated by using the three-dimensional reconstruction method. Finally, the value of V_I of the soil sample's discharge area can be calculated by using Equation (6), combining the amplitude of measured impulse current. At the same time, the oscilloscope can be used to record the current and voltage waveform on the grounding conductor, so as to obtain the measured value of the impulse grounding resistance R_{ch}. Formula (7) can be used to calculate the value of empirical impulse grounding resistance R'_{ch}.

The images obtained in the experiment are shown in Figure 3, and the impulse current and voltage waveform on the grounding conductor are shown in Figure 4. V_I, R'_{ch}, R_{ch} are shown in Table 1.

The moisture in Table 1 was determined by the experimental measurements, and four typical results were chosen from the experiment due to space constraints. Since that the soil resistivity of the soil sample, the grounding resistance R, and the impulse coefficient values α of Formula (7) do not change, the calculated value of the impulse grounding resistance R_{ch2} does not change, while the measured grounding electrode's impulse grounding resistance R_{ch1} decreases slightly, this is because the impulse current applied in every soil sample was the same, but only the impulse voltage showed a slight difference, which were 39 kV, 38.42 kV, 37.2 kV, and 35 kV, respectively. This also results in very little difference in the values of R_{ch1} calculated in Table 1, but the V_I of the discharge area of the soil sample increases greatly. As can be seen, there was a large change in resistance when the soil moisture only changed by 1.36%, which means that when the moisture of the sample chosen was near this value, the change of impulse grounding resistance was more sensitive, and a small change of water content would cause a large change of impulse grounding resistance. The change of impulse grounding resistance was very slow before this value, so it showed that with the increase in water content, the impulse grounding resistance changed more and more obvious, and the value of V_I showed the same rule, which proved that the conclusion is correct and that V_I has a certain correlation with the impulse grounding resistance. The results also showed that V_I is better than the impulse grounding resistance when we evaluate the impulse dispersion performance of the grounding device because V_I is more obvious than the impulse grounding resistance calculated by the waveform recorded. Therefore, compared with the empirical formula calculation method commonly used in engineering, the V_I 's value, which is based on the image characteristics of soil impulse dispersion, can more clearly indicate the influence of the change of soil medium properties and describe the impulse dispersion capacity of the grounding equipment in the process of impulse dispersion.

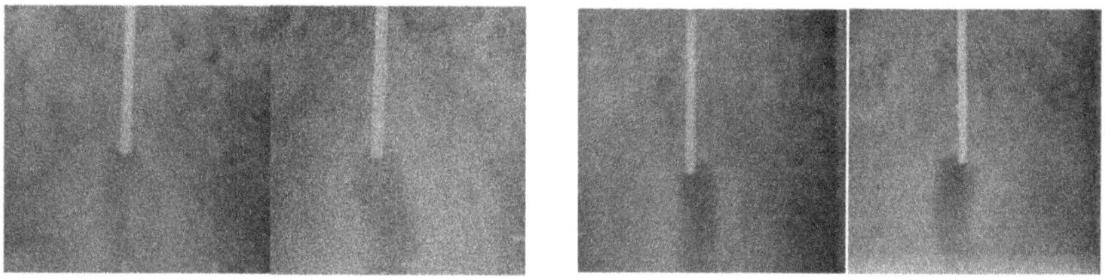

(**a**) (**b**) (**c**) (**d**)

Figure 3. Images of soil with the same resistivity and different water content: (**a**) The moisture is 23.56%; (**b**) moisture is 39.14%; (**c**) moisture is 40.50%; (**d**) moisture is 48.80%.

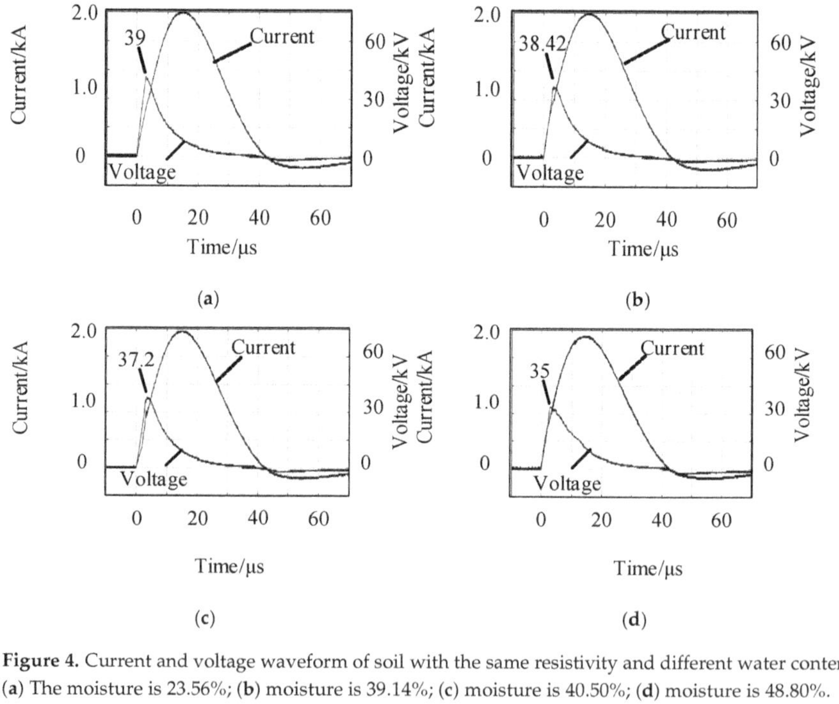

Figure 4. Current and voltage waveform of soil with the same resistivity and different water content: (**a**) The moisture is 23.56%; (**b**) moisture is 39.14%; (**c**) moisture is 40.50%; (**d**) moisture is 48.80%.

Table 1. V_I and R_{ch2} of soil with the same resistivity that is 50 Ω·m and different water content.

Soil Water Content/%	Calculated Impulse Grounding Resistance R_{ch2}/Ω	Measured Impulse Grounding Resistance R_{ch1}/Ω	Soil Discharge Area $V_I/(cm^3/kA)$
23.56	35.9	19.50	22.27
39.14	35.9	19.21	23.75
40.50	35.9	18.60	26.18
48.80	35.9	17.50	34.48

4.2. Correlation Analysis between V_I and R_{ch}

This paper also carried out the correlation analysis between V_I and impulse grounding resistance R_{ch}. According to the analysis above, there is a negative correlation between the two parameters. In order to facilitate the analysis, the correlation between the reciprocal of the impulse grounding resistance and V_I of the grounding device is analyzed.

In this paper, soil samples with a resistivity of 100, 300, 500, 700, and 1000 Ω·m are selected, which are impinged a current of 2.0 kA to carry out the comparative test of near area soil dispersal capacity. Some images obtained from the test are shown in Figure 5, the impulse current and the impulse voltage waveform on the grounding conductor are shown in Figure 6, and the variation rule of V_I and R_{ch}'s reciprocal is shown in Figure 7.

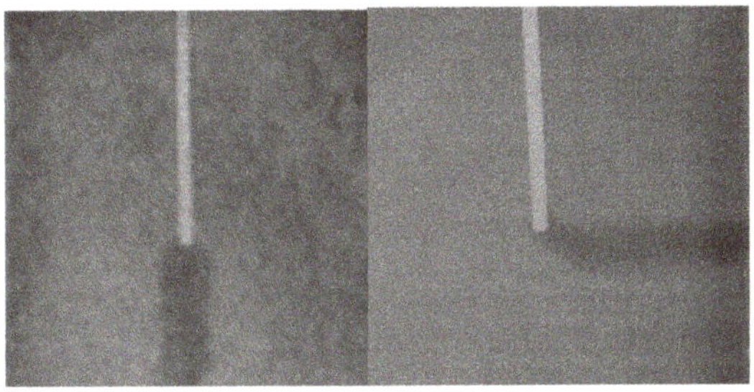

(a) (b)

Figure 5. Image of the soil discharge channel of different soil resistivity: (**a**) ρ = 300 Ω·m; (**b**) ρ = 700 Ω·m.

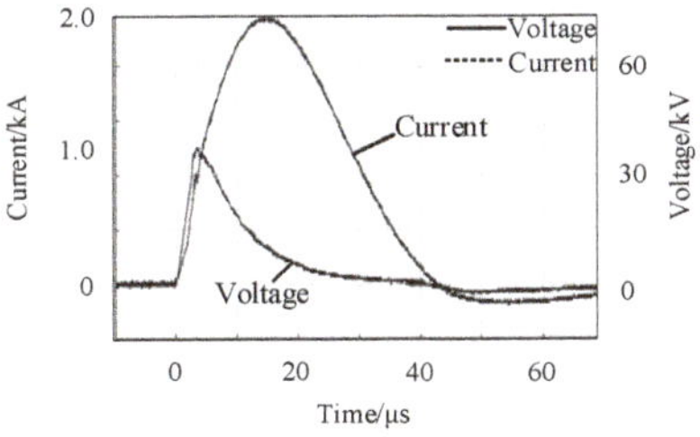

(a)

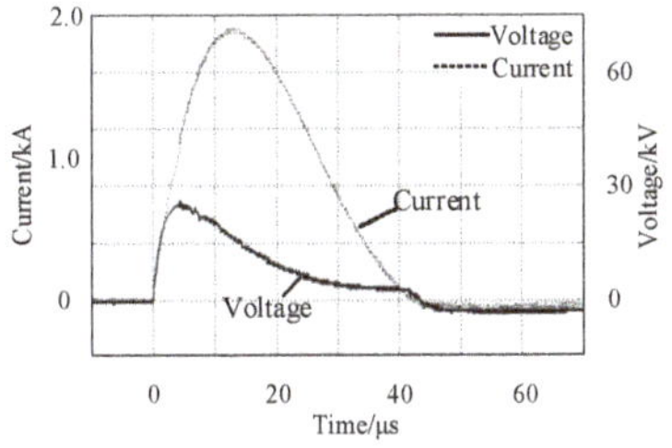

(b)

Figure 6. Current and voltage waveform of soil of different soil resistivity when injected into impulse current: (**a**) ρ = 300 Ω·m; (**b**) 700 Ω·m.

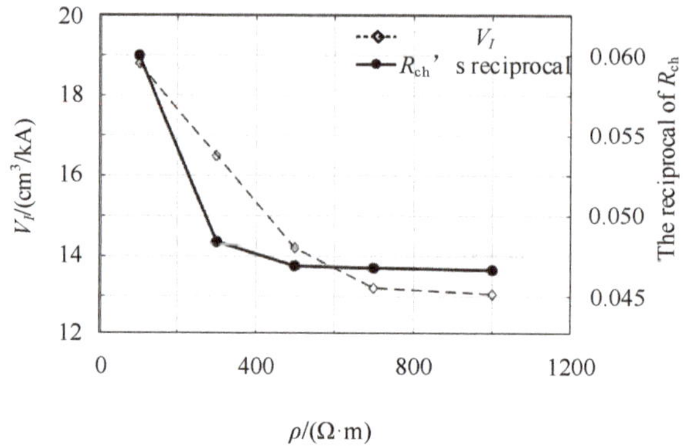

Figure 7. V_I and reciprocal of R_{ch} of different soil resistivity.

As can be seen from Figure 7, The change rule of the equivalent value of the soil discharge volume per unit electric current is the same as that of the reciprocal of the grounding resistance of the grounding electrode. The calculated correlation coefficient r is shown in Table 2.

Table 2. V_I and reciprocal of R_{ch} of different soil resistivity.

The Variables	Correlation	Impulse Grounding Conduntance
V_I and reciprocal of R_{ch}	Person correlation index r	0.985
	Significance (bilateral)	0.002
	Number of sample M	5

According to Table 2, under different soil resistivity, the correlation coefficient r of V_I and R_{ch}'s reciprocal is above 0.9, showing a significant correlation. Therefore, the evaluation method of the near-area soil impulse dispersion performance proposed in this paper can directly depict the impulse dispersion performance of soil media, it is proved to be feasible and effective enough to enrich the evaluation system of the impulse dispersion ability of a grounding device, complementing the evaluation method of the soil impulse dispersion performance.

4.3. Variation of V_I with the Content of SAP

In order to compare and analyze the change of the volume of the modified soil discharge area under the equivalent unit current under different content of SAP, soil moisture content was controlled to be 5% and soil density was controlled to be 1.25 g/cm^3. Modified soils with a content of SAP of 0%, 0.1%, 0.25%, and 0.5% were prepared, respectively. The amplitude of the impulse current was 2.0 kA when the ground electrode in the soil modified by SAP was injected. Figures 8–11 show the topography of the discharge area obtained during the impact diffuser test, the impulse current and impulse voltage waveforms flowing through the modified soil are shown in Figures 12–15.

(**a**) (**b**) (**c**) (**d**)

Figure 8. Image of the soil discharge channel before modification: (**a**) Before breakdown; (**b**) front view; (**c**) side view; (**d**) top view.

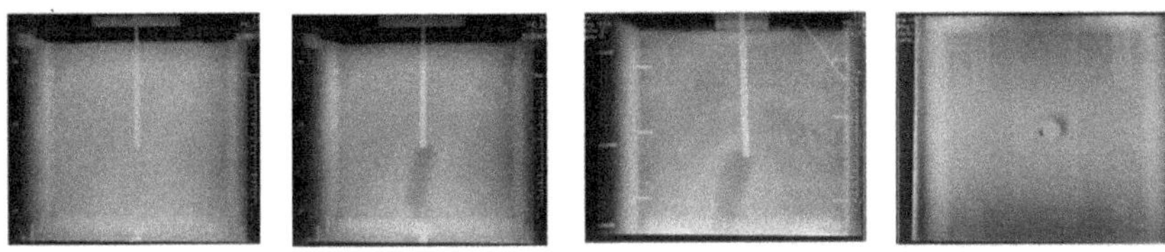

(**a**) (**b**) (**c**) (**d**)

Figure 9. Images of modified soil with the SAP content of 0.1%: (**a**) Before breakdown; (**b**) front view; (**c**) side view; (**d**) top view.

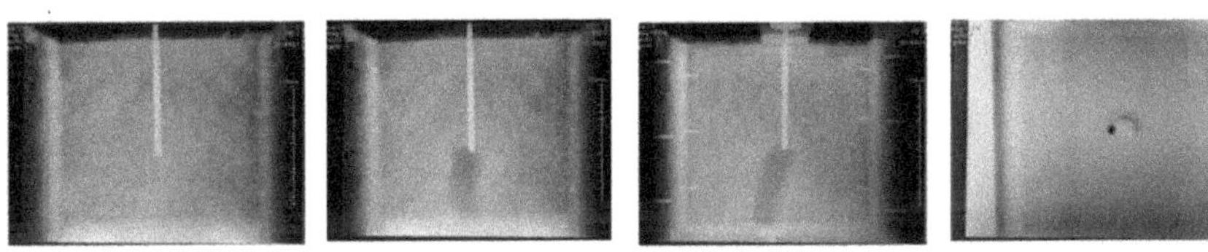

(**a**) (**b**) (**c**) (**d**)

Figure 10. Images of modified soil with the SAP content of 0.25%: (**a**) Before breakdown; (**b**) front view; (**c**) side view; (**d**) top view.

(**a**) (**b**) (**c**) (**d**)

Figure 11. Images of modified soil with the SAP content of 0.5%: (**a**) Before breakdown; (**b**) front view; (**c**) side view; (**d**) top view.

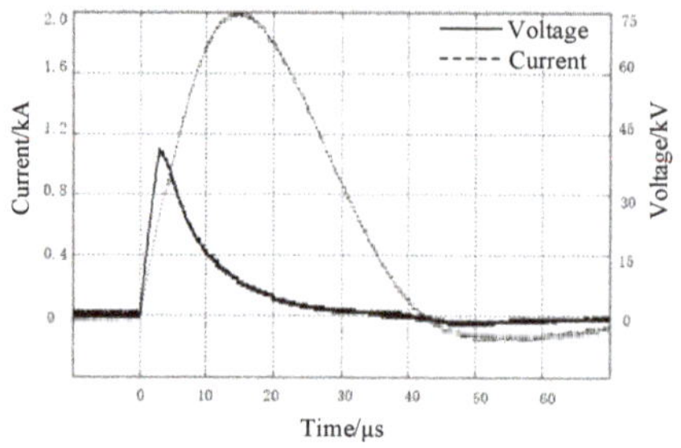

Figure 12. Current and voltage waveform of soil without SAP when injected into the impulse current.

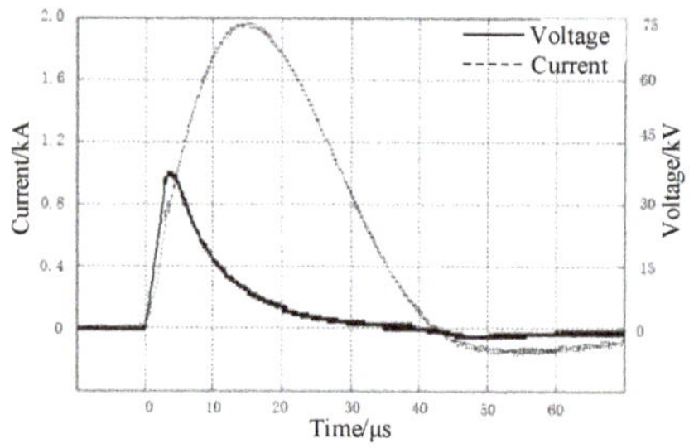

Figure 13. Current and voltage waveform of modified soil with the SAP content of 0.1% when injected into the impulse current.

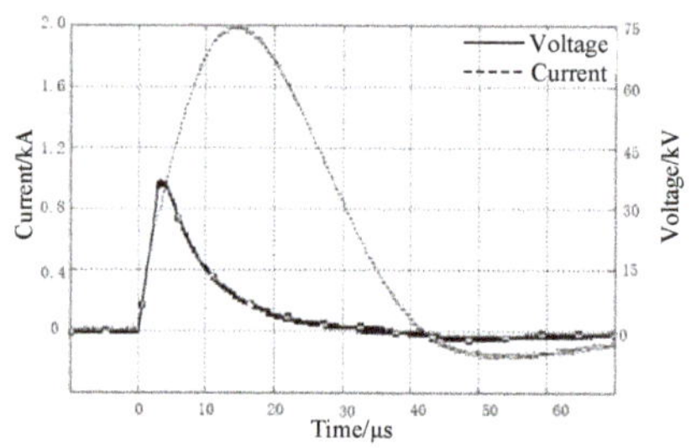

Figure 14. Current and voltage waveform of modified soil with the SAP content of 0.25% when injected into the impulse current.

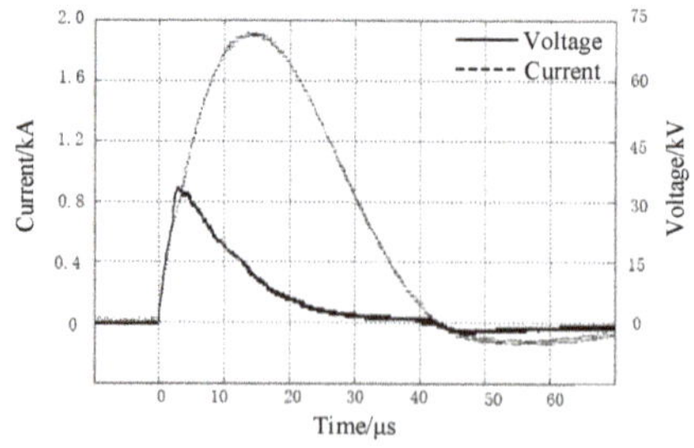

Figure 15. Current and voltage waveform of modified soil with the SAP content of 0.5% when injected into the impulse current.

Figure 16 shows that with the increase in the content of SAP in the soil, the impulse grounding resistance of the soil decreases slightly, and the value of V_I in the soil discharge area increases gradually. SAP changes the ratio of solid, liquid, and gas in the soil, and it reduces the resistivity of the soil dispersion area, so it makes the soil discharge process more intense, which is macroscopically represented by the increase in the volume of the discharge area. When the SAP's content was 0.5%, the value of V_I of the dispersion zone in the modified soil was 24.59 cm^3/kA. Compared with the modified soil without SAP, the value of impulse grounding resistance decreased by 10.26%, and the value of V_I increased by 3.54 times. Compared with the impulse grounding resistance's value of the grounding electrode, the change of the value of V_I of the soil is more obvious, and it is less affected by the test conditions. Therefore, in areas with high soil resistivity, when SAP is used to modify the soil near the grounding electrode, the long-term effectiveness of SAP's resistance reduction performance can be monitored based on the size of V_I's value.

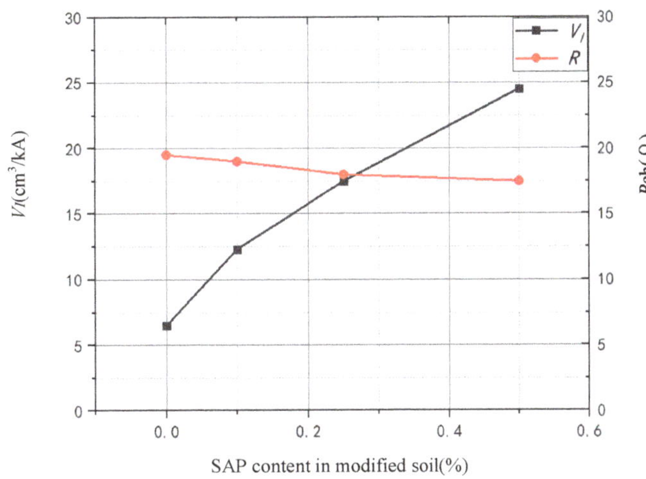

Figure 16. Curve of V_I with the content of SAP.

4.4. Variation of V_I with Water Content of Modified Soil

In order to compare and analyze the dispersion efficiency of the soil near the grounding conductor under different water contents, this paper selects the modified soil with different water contents to carry out the comparative test of the dispersion efficiency of the soil near the grounding electrode. Keeping the mass ratio of the modified material to the unmodified soil equal to 5:1000 and the soil density equal to 1.25 g/cm^3, the water content of the modified soil was adjusted to 10%, 15%, and 25%, respectively. The amplitude of the impulse current is 2.0 kA when the ground electrode of modified soil is injected. Figures 17–19 show the morphologic characteristics of the discharge area obtained during the impulse dispersion test process. Figures 20–22 show the impulse current and impulse voltage waveform when the current flows through the modified soil.

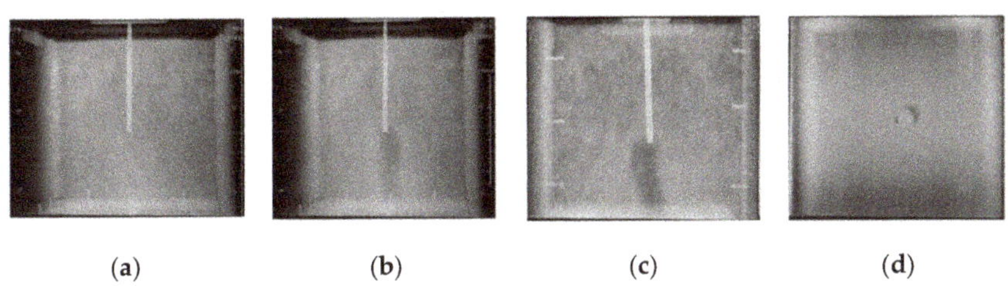

Figure 17. Images of modified soil with water content of 10%: (**a**) Before breakdown; (**b**) front view; (**c**) side view; (**d**) top view.

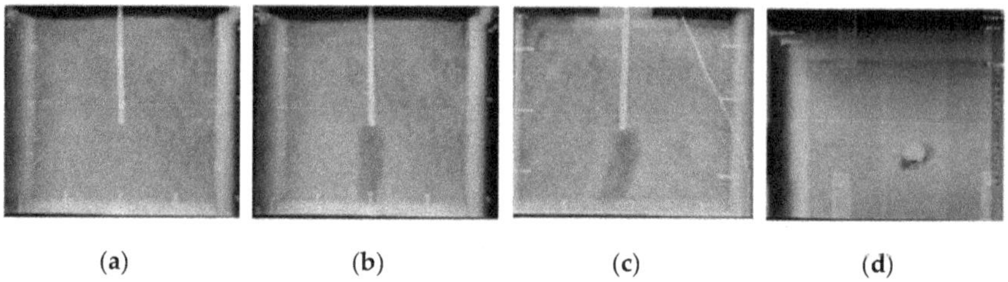

Figure 18. Images of modified soil with water content of 15%: (**a**) Before breakdown; (**b**) front view; (**c**) side view; (**d**) top view.

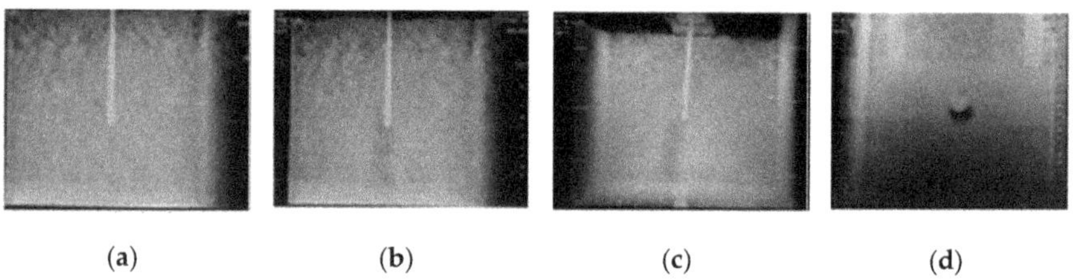

Figure 19. Images of modified soil with water content of 25%: (**a**) Before breakdown; (**b**) front view; (**c**) side view; (**d**) top view.

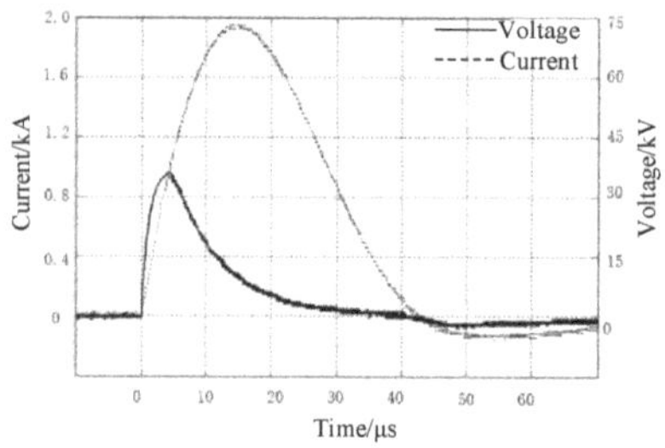

Figure 20. Current and voltage waveform of modified soil with the water content of 10% when injected into the impulse current

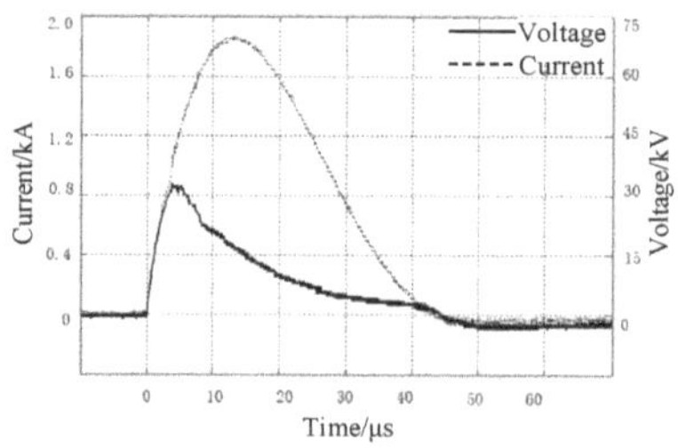

Figure 21. Current and voltage waveform of modified soil with the water content of 15% when injected into the impulse current.

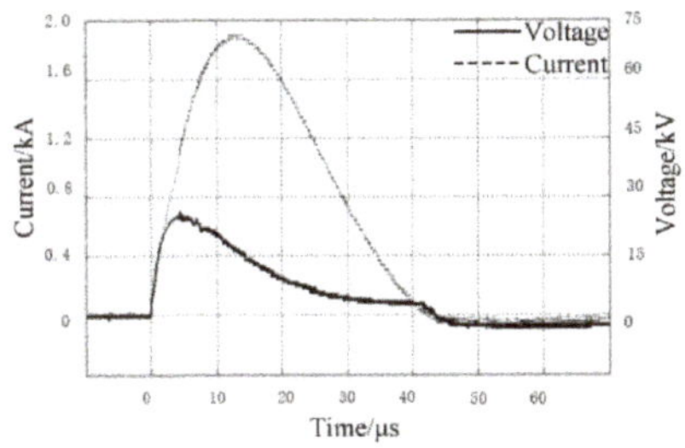

Figure 22. Current and voltage waveform of modified soil with the water content of 25% when injected into the impulse current.

As can be seen from Figure 23, with the change of water content in the modified soil, the value of V_I in the spark discharge area in the modified soil increases, and the value of impulse grounding resistance of the grounding electrode gradually decreases, indicating that the dispersion performance of the modified soil is also continuously enhanced. In addition, when the moisture content of the modified soil was 25%, V_I was 20.10 A/cm^3, which decreased by about 55.47% compared with that of 5%. This indicates that modified soil with higher moisture brings a discharge process with greater intensity and a better dispersion capacity of the impulse current. This is because SAP absorbs sufficient water in the modified soil, and a large number of water molecules fill the air gap between the soil particles, which reduces the resistivity of the modified soil near the grounding electrode. At the same time, with the increase in moisture in the soil, water film appears on the contact surface between the grounding conductor and the soil, which reduces the contact resistance, and the soil discharge process becomes more intense, which is macroscopically represented by the increase in the volume of the soil discharge area.

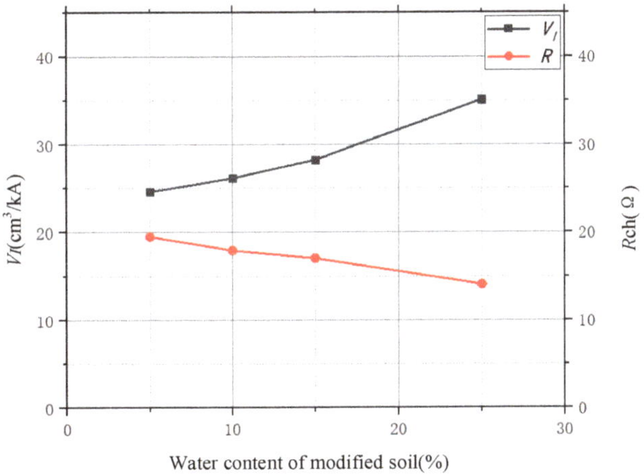

Figure 23. Curve of V_I with water content of modified soil.

5. Conclusions

In this paper, X-ray imaging technology is used to observe the discharge area in the soil, and the equivalent value of the discharge volume of the soil per unit electric current (V_I) is proposed as an index to evaluate the soil impulse dispersion performance in the near-area of the grounding device. The feasibility and effectiveness of the method were verified by the simulation test, and the variation law of impulse dispersal capacity of soil samples with different properties was analyzed. Specific conclusions are as follows:

- The proposed equivalent value of soil discharge volume per unit electric current, combined with the image characteristics of soil impulse discharge, can directly reflect the dispersion of the impulse current in the soil, and the results of the evaluation can reflect the change law more significantly than the results calculated by an empirical formula and measurement waveform. However, the limitation is that other types of soil and soil heterogeneity are not considered.
- The experimental results show that when the samples' soil resistivity is the same and the soil moisture content is different, the V_I of the soil medium is different, and increases with the increase of the moisture content, but the value of the empirical formula recommended by the regulations remains unchanged.
- In the area of high soil resistivity, when the V_I 's value is small, it indicates that there is space to further improve the impulse dispersion performance of the grounding electrode by modifying the soil.
- With the increase in the content of SAP, the V_I of the discharge zone of the soil increased gradually. When the content of the modified material was 0.5%, the V_I's value of the dispersion zone of the soil modified by SAP was 24.59 cm^3/kA. Compared with the modified soil without SAP, the impulse grounding resistance decreased by 10.26%, and the value of V_I increased by 3.54 times.
- With the increase in moisture content of the soil modified by SAP, the discharge process of soil becomes more intense, and the impulse dispersion performance is better. In the soil modified by SAP, V_I is 20.10 cm^3/kA when the water content is 25%. Compared with the modified soil with 5% water content, V_I increases by about 55.47%.

In this article, the sand was the only test sample to be modified, and then this paper explored the impulse dispersion performance of modified soil. Therefore, in the future, the modification of other types of soil is expected to be carried out to study their V_I characteristics, such as gravel soil, clay soil, frozen soil, etc. In addition, the experimental platform is going to be updated to increase the amplitude of the impulse current, to permit double impulse current, and to increase the size of soil samples, so as to study whether the change rule of V_I's value under the uneven soil model is still applicable. Finally, on the basis of all the research, the evaluation method of the impulse dispersion performance of various existing grounding resistance reduction materials and resistance reduction schemes will be studied, so as to select the most effective resistance reduction measurement to minimize the grounding resistance in high soil resistivity areas and provide a safer grounding environment.

Author Contributions: Investigation, X.H.; methodology, X.Z. and G.B.; project administration, D.L., Y.Z. and Y.C. All authors have read and agreed to the published version of the manuscript.

Funding: This work is supported by the Science and Technology Project of State Grid Corporation of China ("Study on the Transient Characteristics of Grounding System and the Test and Evaluation Method of Current Dispersion Performance Under the Successive Impulse Current", 5500-202026088A-0-0-00).

Institutional Review Board Statement: Not applicable.

Informed Consent Statement: Not applicable.

Data Availability Statement: The study did not report any data.

Acknowledgments: Thank Xiejun Du and Xiaochuan Li for their advice and help in the process of paper.

Conflicts of Interest: The authors declare no conflict of interest.

References

1. He, J.L.; Zeng, R. *Power System Grounding Technology*; Science Press: Beijing, China, 2007; pp. 130–137.
2. Yuan, T.; Li, T.X.; Sima, W.X.; Liu, S.W.; Luo, D.H.; Li, X.C. Analysis of soil discharge image characteristics under impulse currents. *Proc. CSEE* **2018**, *38*, 5266–5274.

3. Chen, M.M.; Cao, X.B.; Yang, L.; Wu, H.; Zhang, M.; Wu, G.N. Experimental study on soil impulse characteristics surrounding Coaxial cylindrical electrode. *Power Syst. Technol.* **2013**, *37*, 2291–2296.
4. Yuan, T.; Ban, G.G.; Luo, D.H.; Li, J.L.; Zhang, X.; Zuo, S.J. Evaluation Method of Impulse Discharge Performance of Ground Electrode Based on Soil discharge Image. In Proceedings of the 2020 IEEE International Conference on High-Voltage Engineering (ICHVE 2020), Beijing, China, 6–10 September 2020.
5. Gao, Y.Q. Study on Soil Impulse Breakdown Mechanism and Transient Characteristics of Grounding System. Ph.D. Thesis, Tsinghua University, Beijing, China, 2003.
6. Wang, J.P.; Liew, A.C.; Darveniza, M. Extension of dynamic model of impulse behavior of concentrated grounds at high currents. *IEEE Trans. Power Delivery* **2005**, *20*, 2160–2166. [CrossRef]
7. Rakov, V.A. *Triggered Lightning*; Springer: Berlin, Germany, 2009.
8. Liu, S.W. Study on X-ray Digital Image Characteristics of Soil Impulse Discharge and Transient Performance of Grounding Device. Ph.D. Thesis, Chongqing University, Chongqing, China, 2017.
9. Feng, Z.Q. Study on the Impulse Characteristics of the Grounding Devices and the Morphology Features of the Soil Sparkover. Ph.D. Thesis, Wuhan University, Wuhan, Hubei, China, 2016.
10. Zhu, B.; Sima, W.X.; Yuan, T. Structure parameter optimization of grounding device with needle-shaped conductors based electric field distribution in soil. *Power Syst. Technol.* **2015**, *39*, 2907–2914.
11. Li, J.L.; Jiang, J.D.; Li, L.L. Simulation and experiment study on resistance-reducing mechanism of grounding device with spicules. *Power Syst. Technol.* **2013**, *37*, 211–217.
12. He, J.L.; Zhang, B: Progress in Lightning Impulse Characteristics of Grounding Electrodes with Soil Ionization. *IEEE Trans. Ind. Appl.* **2015**, *51*, 4924–4933. [CrossRef]
13. Yang, L.; Wu, G.N.; Tian, X.F. Analysis of impulse characteristic grounding electrode in frequency and time domain based on EMTP. *Trans. China Electrotech. Soc.* **2011**, *26*, 194–198.
14. Jose, C.; Antonio, F.O.; Carlos, G. Nodal frequency analysis of grounding systems considering the soil ionization effect. *IEEE Trans. Power Deliv.* **2000**, *15*, 103–107.
15. Lu, H.L.; Feng, Z.Q.; Wen, X.S. Full-time electrical network model of the transient characteristics of the grounding devices considering soil sparkover. *Proc. CSEE* **2017**, *37*, 7058–7065+7098.
16. Neknoul, B.; Labie, P.; Zgainski, F.X.; Meunier, G. Caculating the impedance of the grounding system. *IEEE Trans. Nucl. Sci.* **1996**, *32*, 1509–1512.
17. Snowden, D.P.; Erler, J.W. Initiation of electrical breakdown of soil by water vaporization. *IEEE Trans. Nucl. Sci.* **1983**, *30*, 4568–4571. [CrossRef]
18. Bellaschi, P.L.; Armington, R.E. Impulse and 60-cycle characteristics of driven grounds—III. effect of lead in ground installation. *Trans. Am. Inst. Electr. Eng.* **1943**, *62*, 334–345. [CrossRef]
19. Liew, A.C.; Darveniza, M. Dynamic model of impulse characteristics of concentrated earths. *Electr. Eng. Proc. Inst.* **1974**, *121*, 123–135. [CrossRef]
20. Sima, W.X.; Liu, S.W.; Yuan, T.; Luo, D.H.; Wu, P.; Zhu, B. Experimental study of the discharge area of soil breakdown under surge current using X-ray imaging technology. *IEEE Trans. Ind. Appl.* **2015**, *51*, 5343–5351. [CrossRef]
21. Luo, D.H.; Cao, Y.X.; Zhang, Y.; Xie, S.J.; Zhang, C.M.; Cao, S.P. Study on Structural Parameters and Analysis Method of Soil Successive Impulse Discharge Channel. *Energies* **2021**, *14*, 877. [CrossRef]
22. Radakovic, Z.R.; Kostic, M.B. Behaviour of grounding loop with bentonite during a ground fault at an overhead line tower. *IEEE Trans. Gener. Transm. Distrib.* **2001**, *148*, 275–279. [CrossRef]
23. Bellaschi, P.L. Impulse and 60-Cycle Characteristics of Driven Grounds. *Trans. Am. Inst. Electr. Eng.* **1941**, *60*, 123–128. [CrossRef]
24. Zhu, S.Y.; Yuan, T.; Zhu, B. Finite element model of impulse dispersing characteristics of grounding equipment in layered soil. *Power Syst. Technol.* **2014**, *38*, 2304–2309.
25. Guo, Z.H.; Xing, T.F.; Wu, G.N. Grounding resistance change rule of vertical grounding electrode in frozen soil. *High Volt. Eng.* **2014**, *40*, 698–706.
26. Wang, J.J.; He, J.L.; Zou, J. Influence of seasonal frozen soil layer on current division factor of substation grounding system. *Power Syst. Technol.* **2006**, *30*, 41–45.
27. Chen, X.L.; Huang, Y.; Zhang, J.Y. Caculation formula of impulse grounding resistance for grounding device of power transmission tower. *Power Syst. Technol.* **1996**, *20*, 9–12.
28. Li, T.X. Study on Grounding Resistance Reduction Performance of Super Absorbent Polymer Material. Ph.D. Thesis, Chongqing University, Chongqing, China, 2018.
29. Xiao, Y.M. Preliminary study of standard system construction of new medical imaging devices. *China Med. Device Inf.* **2015**, *21*, 41–44+47.
30. Yamane, H.; Ideguchi, T.; Tokuda, M. Stability against temperature and environment for ground resistance reducing material using watered absorbent polymer. *Electron. Commun.* **1994**, *77*, 100–110.
31. Yamane, H.; Ideguchi, T.; Tokuda, M. A new ground resistance reducing material based on water-absorbent polymer. *Electron. Commun.* **2010**, *77*, 68–78. [CrossRef]
32. Khan, Y.; Malik, N.H.; Arainy, A.A. Efficient use of low resistivity material for grounding resistance reduction in high soil resistivity areas. In Proceedings of the Tencon IEEE Region 10 Conference, Fukuok, Japan, 21–24 November 2011.

energies

MDPI

Article

Transient Voltage UWB Online Monitoring System for Insulation Failure Analysis and Fault Location of GIL †

Ziwei Zhang [1], Dengwei Ding [1], Liang He [1], Weidong Liu [1], Cuifen Bai [2,]* and Junjun Liu [3]

1 Sichuan Energy Internet Research Institute, Tsinghua University, Chengdu 610213, China; ziwei.z@outlook.com (Z.Z.); sunnyalll23@163.com (D.D.); helxjtu@163.com (L.H.); lwd-dea@mail.tsinghua.edu.cn (W.L.)
2 Hebei Key Laboratory of Safety Monitoring and Mining Equipment, North China Institute of Science and Technology, Langfang 065201, China
3 Maintenance Center, State Grid Chengdu Power Supply Company, Chengdu 610041, China; liujj1696@sc.sgcc.com.cn
* Correspondence: cuifenbai@163.com
† This paper is an extended version of our paper published in 2020 IEEE International Conference on High Voltage Engineering and Application (ICHVE), Beijing, China, 6–10 September 2020.

Abstract: Transient voltage generated in ultra-high voltage (UHV) transmission system has a severe impact on the insulation state of gas-insulated transmission lines (GIL). In order to monitor the transient voltage process occurring in UHV GIL during operation, this paper constructs a transient voltage ultra-wideband (UWB) online monitoring system based on capacitive voltage division. This system has been applied in an 1100 kilovolt (kV) GIL utility tunnel project. It can be used to analyze the characteristics of the recorded transient voltage waveforms for distinguishing different types of insulation failure. In this paper, through the case studies in time domain and time–frequency domain, the case of SF_6 gap breakdown and that of post insulator flashover have differentiated characteristics in instantaneous frequency. Additionally, a case of secondary discharge is successfully estimated through the time–frequency distribution of the transient voltage. In order to find the malfunctioning chamber of GIL rapidly, a two-terminal TW-based fault location method based on this monitoring system is developed in this paper. The case study validates the locating accuracy which directly support for shortening the maintenance time of GIL.

Keywords: UHV GIL; transient voltage; on-line monitoring system; UWB voltage sensing

Citation: Zhang, Z.; Ding, D.; He, L.; Liu, W.; Bai, C.; Liu, J. Transient Voltage UWB Online Monitoring System for Insulation Failure Analysis and Fault Location of GIL. *Energies* **2021**, *14*, 4863. https://doi.org/10.3390/en14164863

Academic Editor: Pawel Rozga

Received: 5 July 2021
Accepted: 4 August 2021
Published: 9 August 2021

1. Introduction

In order to raise the efficiency of clean energy and reduce the coal consumption, a power transmission system with large capacity is necessary to deliver electricity from the renewable resource-rich areas to the areas with high electricity consumption [1]. Ultra-high voltage (UHV) gas-insulated transmission lines (GIL) are an excellent technology for a high-power underground transmission system with long delivering distances and low energy losses [2]. A general structure of GIL consists of two concentric aluminum tubes [3,4]. The inner tube is the conductor with high current-carrying capability, and the outer aluminum enclosure offers a mechanical and electromagnetic encapsulation [5–7]. The conductor is held in the center of the enclosure by using basin-type insulators and post insulators. A gas mixture of nitrogen (N_2) and sulfur hexafluoride (SF_6) is filled in the tubes as insulating medium [8,9]. Usually, a UHV power transmission system is constructed across every long distance, and overhead transmission lines cannot be built in some circumstances, such as crossing a wide river, a mountain or metropolitan areas [10]. Thus, the GIL technology is tending to be used as an underground solution for UHV transmission systems. Moreover, due to the advantage of space saving, the GIL technology is widely used to connect large-scale hydropower plants or nuclear power plants to the public transmission grid [11]. In recent years, an 1100 kilovolt (kV) GIL project is built under the Yangtze River to transmit clean energy from west to east China.

The operating reliability of GIL is directly related to the security and stability of the power system. In highly intensive electric field inside UHV GIL, latent insulation defects are prone to occur on the insulators, because insulators of UHV GIL require enhanced dielectric strength and more critical manufacturing techniques [12,13]. Additionally, any scratching or dislocation of installation during transportation and installation of GIL could lead to latent dangers, such as electrode surface defects, insulator cracking and left inside small metal particles [14–16]. These latent dangers inside GIL possibly lead to insulation breakdown in the process of voltage withstand test or normal operation [17,18]. For on-site test of 1100 kV GIL, the power frequency withstand voltage is 1100 kV (r.m.s. value), and the lightning impulse withstand voltage is 2400 kV (peak value). Once a breakdown occurs in GIL, transient voltage with steep change will be excited [17]. The transient voltage will travel back and forth between the fault point and the port of the bushing, forming the typical traveling wave (TW) process. Due to small resistance of GIL, the TW energy cannot vanish quickly. The attenuation of transient voltage will slow down, which will threaten to the insulation of other parts of GIL equipment. Therefore, it is important to interpret the characteristics of transient voltage traveling inside UHV GIL.

In previous research, the characteristics of transient voltage traveling inside UHV GIL are usually studied by simulation. However, the simulation results have limited support for practical engineering applications. Capturing accurate transient voltage in real operating GIL offers an alternative way to study the characteristics of transient voltage traveling inside long-distance UHV GIL [2]. Furthermore, in order to minimize the impact of outage, once a breakdown occurs in GIL, it is necessary to quickly and accurately locate the discharge location [19]. A gas chamber of UHV GIL is typically 20 m, so the accuracy of the fault location technique should keep within 20 m to recognize the malfunctioning chamber. Previous research on TW-based fault location technology is mostly aimed at conventional overhead lines and cables [20,21]. The accuracy is around 300 m (a typical tower span), which is difficult to meet the requirements of the fault location for UHV GIL [22].

This paper presents a novel structure of a transient voltage on-line monitoring system based on ultra-wideband (UWB) capacitive voltage sensor. This system is applied in an 1100 kV GIL project. Depending on this on-line monitoring system, a fault location method that can satisfy the required precision of UHV GIL is developed. This method is simple to automatically implement with respect to rapidly locating the malfunctioning chamber of typical UHV GIL. Two real-world cases of insulation breakdown happened in the GIL project are recorded by this monitoring system and delivered in this application paper. Relying on these two insulation failure cases that occurred in dielectric withstand voltage test and actual operation respectively, the time–frequency distributions of transient voltage excited in a real UHV GIL is investigated in discussion. Additionally, in the case of insulation flashover during operation, the fault location method is implemented in practice to validate the accuracy of fast fault locating function of this system.

2. Materials and Methods

2.1. Construction of Transient Voltage UWB Online Monitoring System

As a part of the Huainan-Nanjing-Shanghai 1000 kV alternating current (AC) power transmission system, the 1100 kV Sutong GIL utility tunnel laid under the Yangtze River measured 5.4 km long and 75 m deep [23,24]. This project is a worldwide breakthrough for UHV GIL technology with the longest distance, the highest voltage level and the largest transmission capacity. The transient voltage online monitoring system with ultra-wide bandwidth is designed innovatively and applied in this project. The structure of this monitoring system is shown in Figure 1. In this system, transient voltage measurement points are set at the ports of bushings which are located at both ends of each phase of the GIL tunnel. As illustrated in Figure 1, the system is divided into three levels: voltage sensor, monitoring terminal, and storage control unit.

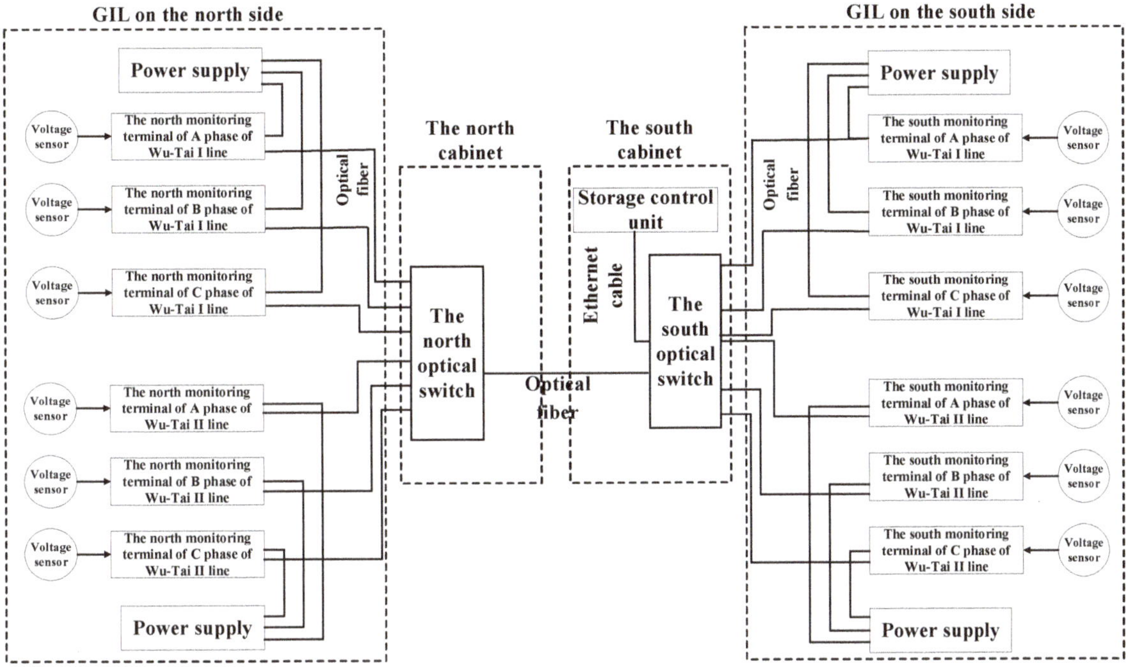

Figure 1. Structure diagram of transient voltage UWB online monitoring system for 1100 kV Sutong GIL project.

2.1.1. Ultra-Wideband Voltage Sensor

The UWB voltage sensor is used to measure the voltage in an expected frequency range. The principle is based on the theory of capacitive voltage division. Its basic components include hand hole cover plate, induction electrode and insulating film, as described in Figure 2a. The original cover plate on the GIL is modified to install this voltage sensor. As shown in Figure 2b, the induction electrode is installed inside the hand hole of the GIL, and a bespoke shield cover is added to control the electric field distribution around the induction electrode. The structural parameters of voltage sensors are estimated according to the detailed structure of the hand hole by finite element method (FEM). In this system, a shield plate with 110 mm inner diameter and the 83 mm height is also fabricated to achieve an ideal voltage division ratio. In this case, the voltage division ratio is about 1,360,000.

The induction electrode is installed in the shield cover as depicted in Figure 2a. The induction electrode is a cylinder with 100 mm diameter and 20 mm thickness. Teflon film with 100 μm thickness is chosen as the insulating medium between the electrode and the hand hole cover plate [25]. The electrode and the high-voltage conductor constitute the high-voltage arm capacitance of 0.001 pF level, and the electrode and the cover plate constitute the low-voltage arm capacitance of nF level. The effective bandwidth of the voltage sensor is verified at 2.1 Hz ~ 230 MHz calibrated by a standard resistance divider in the calibration experiment.

(a) (b)

Figure 2. Ultra-wideband voltage sensor. (**a**) Schematic structure of the voltage sensor; (**b**) entity of voltage sensor and its installation position.

2.1.2. Monitoring Terminal

A monitoring terminal includes an impedance conversion unit, a high-speed acquisition unit, a power module, an isolation transformer, a GPS module and an antenna. Figure 3a shows the schematic structure. Except for the antenna, other modules are installed in stainless steel shielding box, which provides 220 V AC power supply externally. Through an isolation transformer, the power module converts the AC power into DC power for acquisition unit and impedance conversion unit.

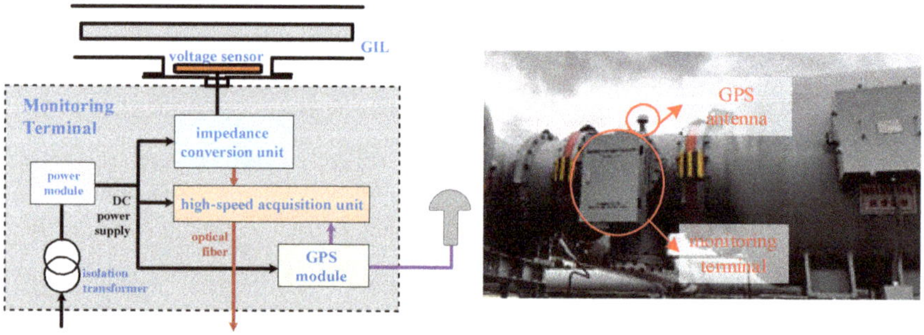

(a) (b)

Figure 3. Monitoring terminal. (**a**) Structure of the monitoring terminal; (**b**) Entity of monitoring terminal and its installation position.

To ensure a sufficient measuring accuracy of power frequency voltage, an impedance conversion circuit is added at the output of the low-voltage arm capacitance. The input resistance of the impedance conversion circuit is set at giga-Ω (GΩ) range, so that the low cut-off frequency of the measurement system is expanded sufficiently. The output resistance is set around 0.1 Ω to enhance the driving ability to the acquisition unit. The sampling rate of the data acquisition unit is 250 MS/s, the analog bandwidth is 100 MHz. The acquisition unit supports First Input First Output (FIFO) mode of continuous acquisition and storage. Additionally, the acquisition unit supports gradient triggering mode. When a sudden change in monitoring voltage exceeds the presupposed gradient value, the acquisition unit will be triggered for long-term recording. The monitoring terminal is fixed on the flange outside the voltage sensor, as shown in Figure 3b.

A GPS module with 10 ns resolution is applied to meet the requirement of this monitoring system. To ensure the soundness of the power supply for the monitoring terminal, the isolation transformer is used to suppress the interference of ground potential rise caused by the operation of isolation switch and circuit breaker. The insulation withstand voltage of iron core (iron core grounding) between the primary and secondary winding can reach 20 kV. The common mode rejection attenuation characteristic curve of the isolation transformer is shown in Figure 4. It can be found that the isolation transformer can greatly restrain the interference signal above 20 kH.

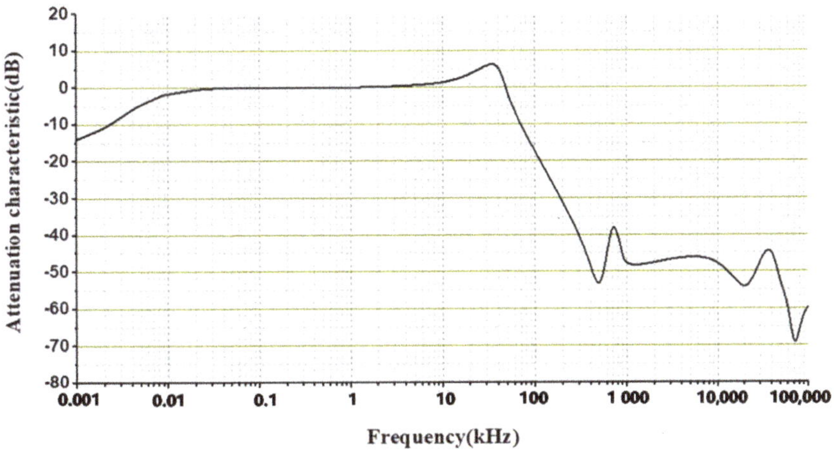

Figure 4. Common-mode suppression attenuation characteristic curve of the isolation transformer.

2.1.3. Storage Control Unit

The storage control unit is placed in a cabinet in a relay protection room. It includes an optical switch, control host (server) and other equipment. Twelve monitoring terminals are connected with optical switch through single-mode optical fiber. The basic function of the storage control unit it to store and communicate the transient voltage data. Additionally, an automatic analysis for location of insulation breakdown fault is accomplished relying on the storage control unit. The location accuracy is expected to attain at least 20 m. This analysis function can realize fast location of breakdown point of the operation GIL. The applied fault location method is based on principle of two-terminal TW theory with synchronous UWB measurement of transient voltage.

2.2. Two-Terminal TW-Based Fault Location for GIL

TW-based fault location methods have been widely applied for adaptive auto-reclosing, line protection and disturbance diagnosis of transmission lines. It is also applicable for GIL in theory since the transient voltage can be accurately captured by the UWB online monitoring system. The principle of TW-based fault location for GIL is depicted in Figure 5. When an insulation breakdown occurs in GIL, a pair of transient voltage TWs as the same amplitude as the breakdown voltage and opposite polarity will propagate along the GIL in opposite directions. The distance between two UWB voltage sensors which are adjacent to the bushings at the ends of GIL is marked as L. The distance between the breakdown position and the sensor A in the south is denoted as d. The moments that the pair of TWs arrive at the installing positions of sensor A and B are symbolized as t_a and t_b, respectively.

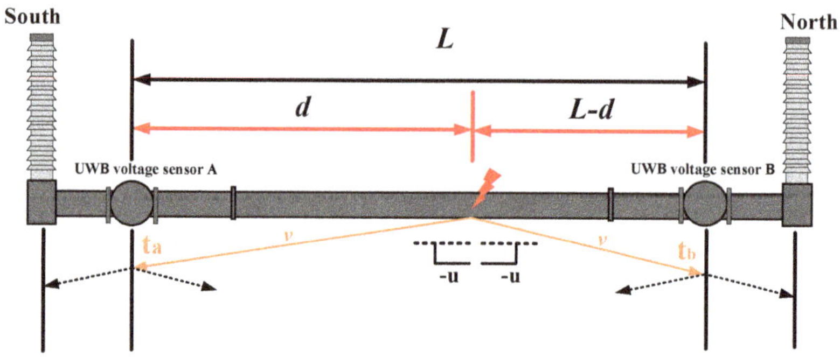

Figure 5. Schematic diagram of two-terminal traveling wave location principle of GIL.

The distance L can be calculated by the following equation:

$$d = \frac{L + (t_a - t_b)v}{2}, \quad (1)$$

where v represents the wave propagation velocity along the GIL. The propagation velocity can be obtained by on-site measurement or calculation based on structural parameters of GIL. The formula is:

$$v = \frac{\omega}{\beta} = \frac{\omega}{\mathrm{Im}(\sqrt{(R + j\omega L)(G + j\omega C)})}, \quad (2)$$

where ω is frequency in radians/second of the transient wave. β is called phase constant, which is the imaginary part of the propagation constant. R, L, G and C are the line parameters, which represent resistance, inductance, conductance and capacitance per unit length of GIL, respectively.

Regarding the UHV GIL tunnel used in the Sutong project, the manufacturers provide the parameters that C equals 45 pF/m, L equals 0.26 μH/m, R equals 3.33 μΩ/m and G can be neglected. Considering high frequency components of the generated transient voltage, the GIL is a case of low loss line ($\frac{R}{\omega L} \ll 1$ and $\frac{G}{\omega C} \ll 1$). Then, Equation (2) can be simplified as:

$$v \approx \frac{1}{\sqrt{LC}}. \quad (3)$$

Consequently, the calculated velocity is around 292.4 m/μs by Equation (3) in the application. Based on three cases of actual breakdown positions during the dielectric withstand voltage tests of this project, the propagation velocity is justified as 292.9 m/μs.

3. Results

3.1. Record of TW during Dielectric Withstand Voltage Test

During the dielectric withstand voltage test, a case of SF_6 gap breakdown was captured by the transient voltage UWB online monitoring system. When the test voltage reached 900 kV (effective value), a phase of GIL broke down. Figure 6 shows the record of the transient voltage waveforms in the north and south measuring points.

Figure 6a shows that the voltage rapidly drops to zero after a short oscillation, and then lasts for nearly 20 ms. After the arc at the fault point was extinguished, the power supply of the test system had not been cut off. A voltage around 300 kV at the test frequency resumed. As shown in Figure 6b, the transient voltage waveforms on both sides last nearly 400 μs. For the TW measured in the south, half a period of the initial square wave is 28.01 μs, and the half-wave steep time of this first wave is around 120 ns. After one cycle, the regular transient voltage is distorted, due to external flashover of reactor in the test circuit overlapped with the original voltage traveling wave. For the TW measured in the north, half a period of the initial square wave is 10.38 μs. After nearly 52 μs, a new traveling wave process appears, half the period is around 1.19 μs. After 250 μs, it gradually attenuates to zero. Additionally, this waveform head decreases from 1182 kV to −1702 kV, taking about 180 ns, and the half-wave steep time is about 90 ns.

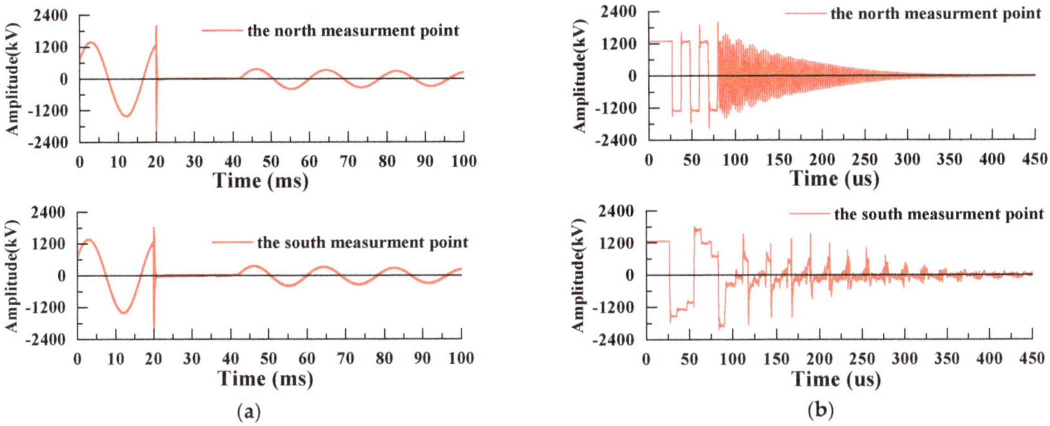

Figure 6. Transient voltage waveform generated by the SF_6 gap breakdown in 1100 kV GIL during dielectric withstand voltage test. (**a**) Full recorded waveform within 100 ms; (**b**) an expanded view of the transient waveform.

3.2. Record of TW during Operation

A case of internal insulation breakdown was captured in a phase of operating GIL. Figure 7 displays the recorded voltage waveforms of this case. Damped square waves were generated due to refraction and reflection of the travelling wave between the breakdown point and the connection point. For the TW measured in the north, half a period of the initial square wave is around 26.8 μs, and the half-wave steep time is about 760 ns. For the TW measured in the south, half a period of the initial square wave is around 12.1 μs, and the half-wave steep time is about 640 ns. Comparing the voltage waveform in Figure 6, the transient voltage caused by GIL insulation flashover during operation jumps much slower.

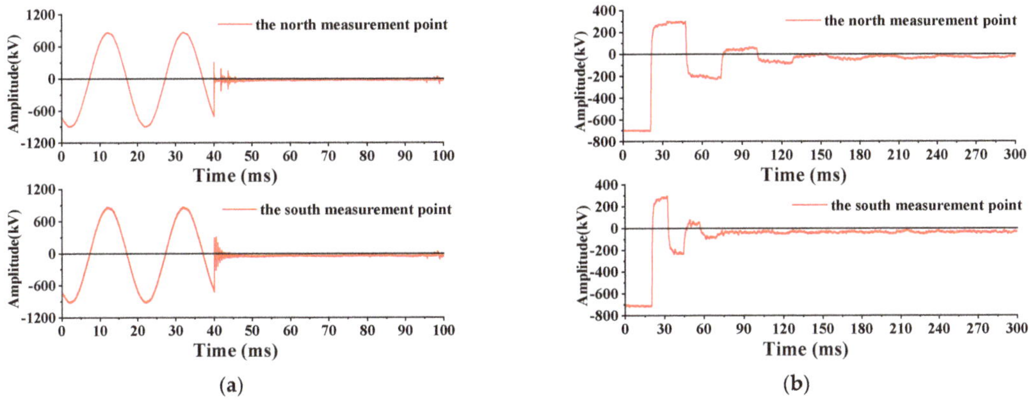

Figure 7. Transient voltage waveform generated in a phase of operating 1100 kV GIL. (**a**) Full recorded waveform within 100 ms; (**b**) an expanded view of the transient waveform.

4. Discussion

4.1. Time–Frequency Analysis

To further investigate the characteristics of the transient voltage excited in different conditions, the transient voltage in time–frequency domain is discussed in this section. The recorded waveforms are transferred by using continuous wavelet transform (CWT) method. Figure 8 shows the time–frequency distributions of SF_6 gap breakdown. The instantaneous high-frequency components excited by the voltage mutation are framed.

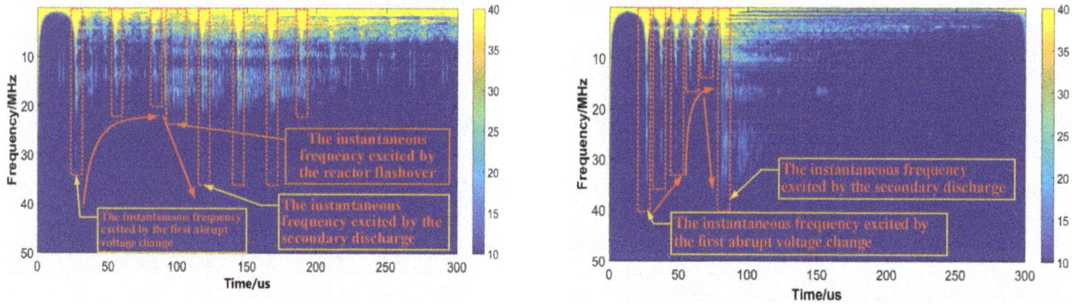

(**a**) (**b**)

Figure 8. Time–frequency distributions of the transient voltage excited by SF_6 gap breakdown during dielectric withstand voltage test. (**a**) Spectrum measured by sensor A (south); (**b**) spectrum measured by sensor B (north).

As illustrated in Figure 8a, the instantaneous frequency of the transient voltage measured in the south is near to 35 MHz at the first abrupt change. The instantaneous frequency component of the next voltage abrupt change appears to be an obvious attenuation. Then, the instantaneous frequency component, which is less than 25 MHz, is observed due to a new transient voltage caused by the reactor flashover. Combining the observation of Figures 6b and 8a, a high frequency component close to 35 MHz appears again after 75 μs, and exists in the following two square half-wave periods, and the square half-wave period is shorter at this time. It is speculated that a secondary breakdown happened between the first breakdown point and the end of the south bushing of GIL. In Figure 8b, the instantaneous frequency of the transient voltage measured in the north is close to 35 MHz at the first voltage abrupt change, and then the instantaneous frequency component gradually attenuates. After 52 μs, a high frequency component with more than 35 MHz appears again. It lasts for nearly 25 μs and then decays to less than 10 MHz. Similar to on the south side, it indicates that there may be a secondary breakdown between the first breakdown point and the north bushing. During the disassembly inspection, it was found that SF_6 gap breakdown occurred between the high voltage conductor and the shell flange at 3460 m, 4110 m and 5459 m away from the end of the south bushing. Therefore, it verified the results of the time–frequency analysis above.

Figure 9 presents the time–frequency distributions excited by flashover in a post insulator of an operating GIL. It can be observed that the time–frequency distributions measured at both sides are similar. The instantaneous high frequency component excited by the first voltage sudden change is close to 3.5 MHz. After three square half-wave cycles, the high-frequency components are all attenuated to less than 1 MHz. From the observation of the time–frequency spectrum, only one breakdown happened in this case. Compared with the case of SF_6 gap breakdown, the instantaneous frequency component in this case is much lower. Consequently, it is considered that this failure case is caused by a different type of insulation breakdown. As a result, a post insulator of this GIL broke down, leading to this failure case.

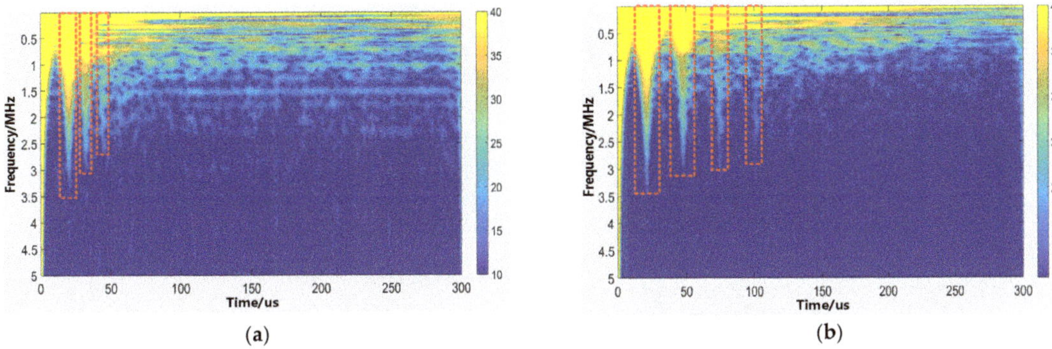

Figure 9. Time–frequency distributions of the transient voltage excited by flashover in a post insulator of an operating GIL. (a) Spectrum measured by sensor A (south); (b) spectrum measured by sensor B (north).

4.2. Fault Location of GIL Insulation Flashover during Operation

Regarding to the case of GIL insulation flashover during operation, the two-terminal TW-based fault location method is applied. The distance between these two UWB voltage sensors installed in this phase of GIL is 5657.7 m. The time difference that the reciprocal TWs arrive at positions of sensor A and B is estimated automatically by identifying the abrupt moments of TW heads. Figure 10 depicts the expanded view of the recorded TW heads, and the time difference $(t_a - t_b)$ is about −7250 ns. The propagation velocity v is approximately 292.9 m/µs. According to Equation (1), the distance d between the location where insulation flashovers and the installation point of sensor A can be derived as follows:

$$d = \frac{L + (t_a - t_b)v}{2} = 1767.4 \text{ m}.$$

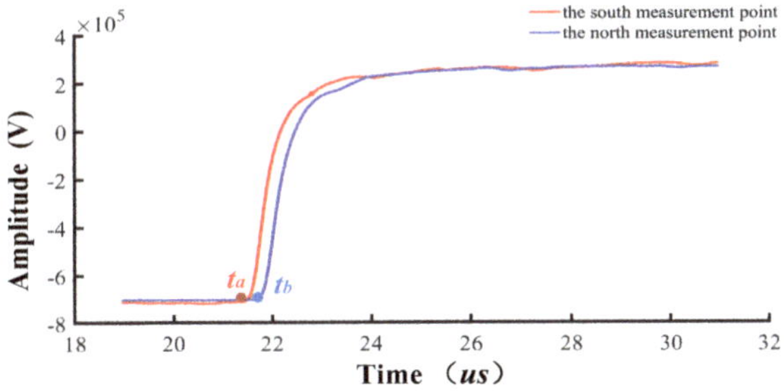

Figure 10. Expanded view of the recorded TW heads.

An inspection was taken after opening the suspected gas chamber of the GIL. A trace of flashover was found near a post insulator of the GIL, as shown in Figure 11. It was 1763.6 m far away from the installation position of sensor A. Therefore, the location error of this case is 3.8 m. The result satisfies the accuracy of fault location for recognizing the malfunctioning chamber of the 1100 kV GIL tunnels.

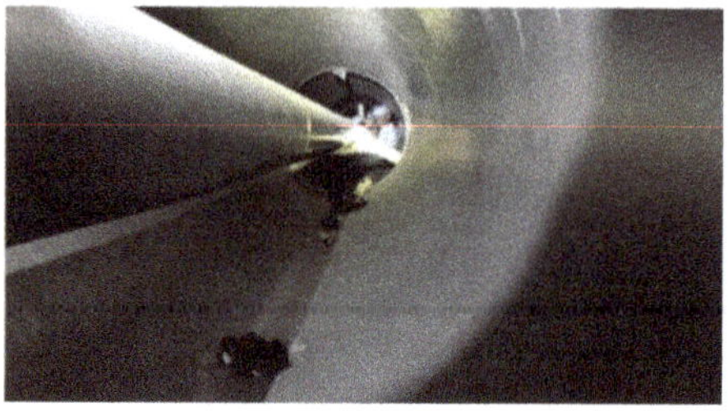

Figure 11. Inspection view of the fault location caused by flashover of a post insulator in GIL.

5. Conclusions

In this application paper, a novel transient voltage on-line monitoring system, which is applied in an kV GIL project, is introduced. The effective measurement bandwidth ranges from 2.1 Hz to 230 MHz. The sampling rate is 250 MS/s, and the analog bandwidth is 100 MHz. In order to fast locate the malfunctioning GIL chamber and shorten the maintenance time, a two-terminal TW-based fault location method relying on this monitoring system is developed and implemented. Based on two real-world cases of insulation breakdown captured by the online monitoring system, the characteristics of the transient voltage waveforms in both time domain and time–frequency domain are presented and discussed. The following conclusions are drawn:

(1) The instantaneous frequency of the transient voltage caused by a SF_6 gap breakdown case during withstand test exceeds 35 MHz, whereas the instantaneous frequency excited by post insulator flashover during operation is around 3.5 MHz, which is much lower. Therefore, the characteristics of waveforms could indicate different types of insulation breakdown.
(2) Time–frequency analysis, can be used to estimate whether the GIL has secondary discharge by the time–frequency distribution of transient voltage.
(3) The accuracy of the applied fault location method is validated in the case of insulation flashover during operation. The location error of this case is 3.8 m and successfully recognizes the malfunctioning chamber.

Author Contributions: Conceptualization, Z.Z. and W.L.; methodology, D.D.; software, Z.Z.; validation, L.H.; formal analysis, Z.Z.; investigation, J.L.; resources, C.B.; writing—original draft preparation, D.D.; writing—review and editing, Z.Z.; visualization, Z.Z.; supervision, W.L.; project administration, L.H.; funding acquisition, C.B. All authors have read and agreed to the published version of the manuscript.

Funding: This research was funded in part by National Natural Science Foundation of China (51977120), National Natural Science Foundation of China (52067021) and Hebei Key Laboratory of Safety Monitoring and Mining Equipment Funding (SM202005).

Conflicts of Interest: The authors declare no conflict of interest.

References

1. Huang, D.; Shu, Y.; Ruan, J.; Hu, Y. Ultra High Voltage Transmission in China: Developments, Current Status and Future Prospects. *Proc. IEEE* **2009**, *97*, 555–583. [CrossRef]
2. Jiao, C.; Ding, D.; Liu, W.; He, L.; Zhang, Z.; Yuan, M. Construction of transient voltage UWB on-line monitoring system in UHV GIL based on capacitive voltage division. In Proceedings of the 2020 IEEE International Conference on High Voltage Engineering and Application (ICHVE), Beijing, China, 6–10 September 2020; pp. 1–4.
3. Koch, H. *Gas Insulated Transmission Lines (GIL)*, 1st ed.; John Wiley & Sons: Chichester, UK, 2011; p. 4.
4. Koch, H. Basic information on gas insulated transmission lines (GIL). In Proceedings of the 2008 IEEE Power and Energy Society General Meeting-Conversion and Delivery of Electrical Energy in the 21st Century, Pittsburgh, PA, USA, 20–24 July 2008; pp. 1–4.
5. Volcker, O.; Koch, H. Insulation co-ordination for gas-insulated transmission lines (GIL). *IEEE Trans. Power Deliv.* **2001**, *16*, 122–130. [CrossRef]

6. BS EN IEC 62271-204. *High-Voltage Switchgear and Controlgear: Part 204. Rigid Gas-Insulated Transmission Lines for Rated Voltage above 52 Kv*; BSI: London, UK, 2019.
7. IEEE PC37.122.4/DG. *December 2012-IEEE Draft Guide for Application and User Guide for Gas-Insulated Transmission Lines (GIL), Rated 72.5 kV and above*; IEEE: Piscataway, NJ, USA, 2013; pp. 1–48.
8. Kuroyanagi, Y.; Toya, A.; Hayashi, T.; Araki, T. Construction of 8000A class 275 kV gas insulated transmission line. *IEEE Trans. Power Deliv.* **1990**, *5*, 14–20. [CrossRef]
9. Hillers, T. Gas insulated transmission lines (GIL): Ready for the real world. In Proceedings of the 2000 IEEE Power Engineering Society Winter Meeting, Singapore, 23–27 January 2000; pp. 575–579.
10. Chen, X.; Hu, Y.; Xin, Y.; Pan, Y.; Liu, F.; Yin, T. Prospect of high voltage long distance compressed-air insulated transmission lines. *High Volt. Eng.* **2009**, *35*, 3137–3142. (In Chinese)
11. Benato, R.; Mario, C.D.; Koch, H. High-capability applications of long gas-insulated lines in structures. *IEEE Trans. Power Deliv.* **2007**, *22*, 619–626. [CrossRef]
12. Zhou, Y.; Qin, Y.; Du, Q.; Wu, Z.; Ma, J.; Li, X.; Wen, T.; Zhang, Q. Flashover characteristics of basin-type insulator with metallic particles on its surface under standard lightning impulse. In Proceedings of the 2017 IEEE Electrical Insulation Conference (EIC), Baltimore, MD, USA, 11–14 June 2017; pp. 27–54.
13. Guo, Z.; Wu, Z.; Wang, H.; Tian, H.; Liu, L.; Peng, Z.; Li, H.; Wang, Q. Experimental and numerical study on formation of interface separation and interfacial dielectric strength of GIL insulator. *IEEE Trans. Dielectr. Electr. Insul.* **2019**, *26*, 1738–1746. [CrossRef]
14. Ueta, G.; Wada, J.; Okabe, S.; Miyashita, M.; Nishida, C.; Kamei, M. Insulation characteristics of epoxy insulator with internal crack-shaped micro-defects-fundamental study on breakdown mechanism. *IEEE Trans. Dielectr. Electr. Insul.* **2013**, *20*, 1444–1451. [CrossRef]
15. Ueta, G.; Wada, J.; Okabe, S.; Miyashita, M.; Nishida, C.; Kamei, M. Insulation performance of three types of micro-defects in inner epoxy insulators. *IEEE Trans. Dielectr. Electr. Insul.* **2012**, *19*, 947–954. [CrossRef]
16. Okubo, H.; Hoshino, T.; Takahashi, T. Insulation design and on-site testing method for a long distance gas insulated transmission line (GIL). *IEEE Electr. Insul. Mag.* **1998**, *14*, 13–22. [CrossRef]
17. Xiong, H.; Jiang, H.; Zhou, B.; Zhou, G. Research on an accident of 500 kV gas insulated transmission line. *Appl. Mech. Mater.* **2014**, *492*, 178–181. [CrossRef]
18. Khan, Q.; Refaat, S.S.; Abu-Rub, H.; Toliyat, H.A. Partial discharge detection and diagnosis in gas insulated switchgear: State of the art. *IEEE Electr. Insul. Mag.* **2019**, *35*, 16–33. [CrossRef]
19. Poehler, S.; Rudenko, P. Directly buried gas-insulated transmission lines (GIL). In Proceedings of the PES T&D 2012, Orlando, FL, USA, 7–10 May 2012; pp. 1–5.
20. Li, Y.; Gao, H.; Du, Q.; Qi, X.; Pang, Q.; Zhu, G. A review of single-phase-to-ground fault location methods in distribution networks. In Proceedings of the 2011 4th International Conference on Electric Utility Deregulation and Restructuring and Power Technologies (DRPT), Weihai, China, 6–9 July 2011; pp. 938–943.
21. Baù, M.; Benato, R.; Dambone Sessa, S.; Poli, M.; Quaciari, C. Phase-to-ground fault location methods in unearthed sub-transmission networks: A review. In Proceedings of the 2016 AEIT International Annual Conference (AEIT), Capri, Italy, 5–7 October 2016; pp. 1–6.
22. Lopes, F.V.; Lima, P.; Ribeiro, J.P.G.; Honorato, T.R.; Silva, K.M.; Leite, E.J.S.; Neves, W.L.A.; Rocha, G. Practical Methodology for Two-Terminal Traveling Wave-Based Fault Location Eliminating the Need for Line Parameters and Time Synchronization. *IEEE Tran. Power Deliv.* **2019**, *34*, 2123–2134. [CrossRef]
23. Du, N.; Xu, W.; Xiang, Z.; Han, Y.; Zhou, P.; Yang, D.; Xiang, C. Research on Key Technical Problem of System Commissioning of Sutong GIL Utility Tunnel Project. In Proceedings of the 2020 5th Asia Conference on Power and Electrical Engineering (ACPEE), Chengdu, China, 4–7 June 2020; pp. 1928–1932.
24. ABB Power Grids Commissions World's First Transmission Line under the Yangtze River. Available online: https://new.abb.com/news/detail/61647/abb-power-grids-commissions-worlds-first-transmission-line-under-the-yangtze-river (accessed on 11 February 2021).
25. Yue, G.; Liu, W.; Chen, W.; Guan, Y.; Li, Z. Development of Full Frequency Bandwidth Measurement of VFTO in UHV GIS. *IEEE Trans. Power Deliv.* **2013**, *28*, 2550–2557.

energies

MDPI

Article

Research on the Influence of Typical Soil Parameters on Critical Breakdown Field Strength and Residual Resistivity Based on Discharge Topography †

Donghui Luo [1,2], Jialun Li [3,*], Yongxing Cao [1], Bo Tan [4], Wei Li [5] and Hanyu Wang [6]

1 State Grid Sichuan Electric Power Research Institute, Chengdu 610041, China; l_dh@cqu.edu.cn (D.L.); cyx64010@gmail.com (Y.C.)
2 Department of Electrical Engineering, Tsinghua University, Beijing 100084, China
3 State Key Laboratory of Power Transmission Equipment & System Security and New Technology, Chongqing University, Chongqing 400044, China
4 China Electric Power Research Institute Co., Ltd., Wuhan 430000, China; tanbo6969@gmail.com
5 State Grid Shanxi Electric Power Research Institute, Xi'an 710000, China; lw05@tsinghua.org.cn
6 State Grid Sichuan Electric Power Company, Chengdu 610041, China; why99912@gmail.com
* Correspondence: lijialun@cqu.edu.cn

† This paper is an extended version of our paper published in 2020 IEEE International Conference on High Voltage Engineering and Application (ICHVE), Beijing, China, 6–10 September 2020.

Abstract: Partial discharge of soil occurs when a lightning current enters the ground, and the strength of partial discharge is closely related to the magnitude of its critical breakdown field strength. Therefore, how to accurately obtain the variation law of the typical soil critical breakdown field strength and residual resistivity is the key to realizing the safe operation of the grounding devices and cables in the ground. This paper first selects a variety of typical soils to study the influence of various factors on the morphology of the discharge channel, and then studies the calculation methods of the soil critical breakdown field strength and residual resistivity under the introduction of different discharge channel morphologies and structures, and further discusses the reason why typical soil media factors have a small impact on the critical breakdown field. The experimental results show that under the same conditions, the critical breakdown field strengths of different soils from small to large are sand soil, loam soil and Yellow cinnamon soil. The largest ratio of residual resistivity to initial resistivity of the three soils is sand soil.

Keywords: discharge channel morphology characteristics; critical breakdown field strength; residual resistivity; X-ray imaging technology

Citation: Luo, D.; Li, J.; Cao, Y.; Tan, B.; Li, W.; Wang, H. Research on the Influence of Typical Soil Parameters on Critical Breakdown Field Strength and Residual Resistivity Based on Discharge Topography. *Energies* **2021**, *14*, 4810. https://doi.org/10.3390/en14164810

Academic Editors: Issouf Fofana and Bo Zhang

Received: 28 June 2021
Accepted: 3 August 2021
Published: 6 August 2021

1. Introduction

At present, people have carried out a lot of research on dielectric breakdown, lightning protection and grounding [1–12]. In the process of studying the mechanism of soil impact breakdown, determining the critical breakdown field strength Ec and residual resistivity ρ_{res} of the soil is a very important goal of the research work. This is because in any transient calculation model of grounding devices that considers soil discharge, the electrical parameters of soil discharge characteristics (critical breakdown field strength Ec and residual resistivity ρ_{res}) are important and critical parameters. Summarizing the current status of Ec research, the numerical values calculated by different scholars vary widely, ranging from 29 kV/m to 1850 kV/m [1,3,6,13–17]. The large difference in the calculation results is mainly due to the differences in the soil discharge models used when calculating the critical soil breakdown field strength. The reason is mainly due to the lack of a unified understanding of the real discharge area structure of the soil, so the soil discharge model used in the calculation is also different, resulting in such a big difference in the calculation results. There are relatively few studies on the residual resistivity. In some studies, it is found that the resistivity of the soil discharge area has dropped significantly [4,11,16].

However, there are abundant soil types in the world, and various types of power facilities are distributed in different regions. Due to the different soil types, the impact discharge conditions of the soil around these facilities are also different. Therefore, it is not possible to simply use the impulse discharge characteristics of a certain type of soil to describe all impulse discharge characteristics. At present, most experiments on the research of soil impulse discharge use fine sand as simulated soil. However, there are differences between fine sand and actual soil, and the experimental results need to be improved. Some institutions have studied the comparison of impact characteristics of fine sand and red clay, which has certain reference significance, but the selected soil type is not sufficient [18]. At present, a large number of documents have studied the impact characteristics of grounding devices, and equivalently replaced them with some circuit models to obtain the value of impact grounding resistance or other parameters [19–25]. However, the number of articles on impulse discharge research on different soil types is relatively small, so it is necessary to study the impulse discharge characteristics of different typical soils.

The impact of soil discharge is related to many factors, the most important of which is the water content, salt content and density of the soil. When the values of the three parameters are different, the critical breakdown field strength and residual resistivity of the soil are different. Obtaining the minimum critical breakdown field strength of the soil has certain practical significance. Obtaining the influence of the three factors on the critical breakdown field strength and residual resistivity can better study the change law of them. At present, there are relatively few studies on the influence of these factors on the critical breakdown field strength and residual resistivity, so it is necessary to carry out relevant research.

In this paper, through the impulse current generator and the X-ray transmission imaging platform, relying on the three-dimensional reconstruction and inversion calculation methods, the influence of different typical soil media factors on the critical breakdown field strength and residual resistivity is studied, and the influence rules are analyzed. The soil breakdown mechanism is analyzed based on the experimental results.

2. Experimental Principle and Equipment

2.1. Impulse Current Generator

The schematic diagram of the soil impulse discharge observation platform based on X-ray digital imaging technology built in this paper is shown in Figure 1. The main equipment of the observation platform includes an impulse current generator (ICG) and an X-ray imaging system (XIS), which are connected by a time delay control unit (TDCU). The shortest exposure time of the X-ray machine is 1ms, and the impulse current wavelength is tens of microseconds, so TDCU is required to delay the triggering of the impulse current generator. When XIS is triggered, the TDCU unit gives ICG a delay signal to trigger the ICG to generate an impulse current. The delay accuracy of TDCU can reach the microsecond level, so the synchronous triggering of ICG and XIS can be guaranteed. The components of the ICG in Figure 1 are: voltage regulator T2, with a voltage regulation ratio of 380/45 kV; silicon stack rectifier D, with a maximum current of 1 A; capacitor bank C, with a capacitance value of 30 μF; the values of wave modulating resistance R and inductance L are 6 Ω and 32.97 μH, respectively; the current sensor CT has a maximum current capacity of 20 kA; VD is a voltage divider with a voltage division ratio of K = 691; a digital oscilloscope DSO with a sampling rate of 1.0 G samples/s and a bandwidth of 100 MHz; S is the ground electrode; ST is the sandbox. XRG is an X-ray generator; the model is MILLENNIUM (SEDECAL, Madrid, Spain); the shortest exposure time is 1 ms; the maximum output voltage is 150 kV; the output current is 500 mA; and the maximum output power is 80 kW. The digital flat-panel detector (FPD) uses Canon CXDI-50G (Canon Co., Ltd., Tokyo, Japan); the number of gray scales is 14-bit gray scale; the pixel size is

160 μm; and the pixel size is 5.9 million. The image processing system PC is used to adjust the output parameters of the X-ray generator and display the imaging results.

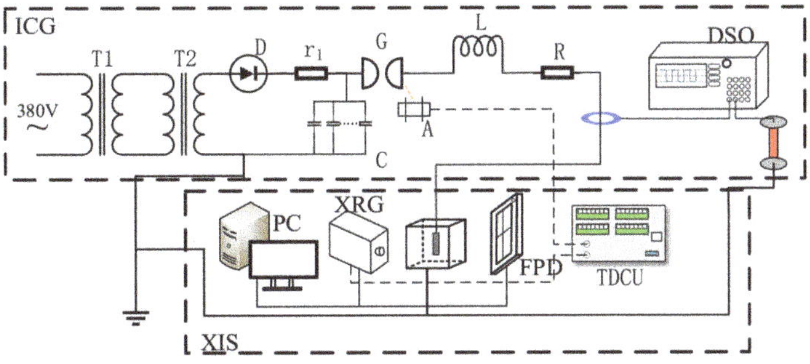

Figure 1. Impulse discharge and X-ray imaging test wiring diagram.

2.2. X-ray Imaging System

The X-ray imaging system (X-ray image system, XIS) is mainly composed of X-ray source, imaging board, a control system and an image processing system. The function of the X-ray source is to provide X-rays with suitable and stable radiation quality and transmit X-rays to soil samples.

Based on the principle of X-ray digital imaging, when X-rays are uniformly incident, the X-ray absorption capacity of the electrode, the simulated soil sample and the discharge area are different, resulting in different doses of rays passing through different areas. The flat-panel detector receives different doses of X-rays, thereby presenting images with different gray values, as shown in Figure 2.

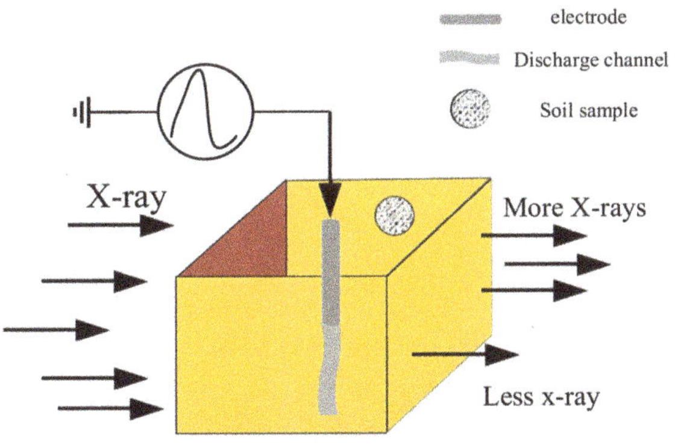

Figure 2. Schematic diagram of X-ray imaging of the soil discharge area.

Electrodes, discharge areas and simulated soil samples can be distinguished by different gray-scale images, which means that images of their discharge areas can be obtained by X-ray transmission imaging. The penetration thickness of X-rays depends on the nature of the penetrated medium and the power of the X-ray machine, according to Beer's law [26]

$$\frac{I}{I_0} = e^{-\mu_m \rho y} \tag{1}$$

where I is the amount of radiation after penetrating the substance, I_0 is the initial radiation amount of the X-ray machine, which depends on the power parameters of the X-ray machine itself, and μ_m is the attenuation coefficient of the soil. y is the thickness of the soil that can be penetrated, and the power of the X-ray machine is positively related to the thickness of the soil that can be penetrated.

The X-ray imaging board is also called a digital flat-panel detector. Its main function is to detect, collect and convert the digital signal of the X-ray irradiated on the detector. When the X-ray generated by the X-ray source is absorbed and attenuated by the soil sample, the remaining rays penetrate the soil and irradiate the digital flat-panel detector. The imaging board will convert the detected X-ray dose into a digital signal and transmit it to the image processing system, generating the corresponding gray-scale image after numerical operation. Based on the gray value information of different regions in the gray image, the condition of the discharge channel in the soil is obtained.

2.3. Preparation of Soil Samples

The type of experiment conducted in this manuscript is a small-scale simulation experiment carried out in the laboratory, so the soil samples need to be prepared using standardized methods during the experiment.

This article involves the preparation of three kinds of soils. The specific preparation steps of the three kinds of soils are as follows:

1. Separately screen and filter three kinds of soil with sufficient amounts to control the upper limit of the particle size of the three soils.
2. Add a sufficient amount of distilled water to the three types of soils, stir thoroughly until the salt in the soil is fully dissolved and when the natural precipitation is complete, pour out the water and measure the salt content with a refractometer. After repeating the same steps 6–9 times, when the salt content of the filtered soil exudate is less than 0.05%, it can be considered that the soil does not contain soluble salt.
3. Put the three kinds of soil into the drying box to dry. Each drying time is 5 h. After drying, take it out to measure its quality and make a record, until the difference between the two quality measurements is less than 0.5%, it is considered that the soil has been dried.
4. Configure the soil with different parameters according to the needs of the experiment.

3. Introduction to the Calculation Method of Discharge Channel Morphology and Characteristic Parameters

According to the principle of X-ray imaging, the electrode, as a solid metal, has the largest attenuation coefficient to X-rays, so it can absorb the most X-rays, and the place corresponding to the electrode has the highest gray value, showing a white image. The soil attenuation ability is inferior to the electrode and stronger than the discharge area, and its gray value will be somewhere in between. The discharge area has the weakest X-ray absorbing ability. The more X-rays pass through, the lower the gray value of the corresponding place and the black image is. The three views of the discharge channel are shown in Figure 3.

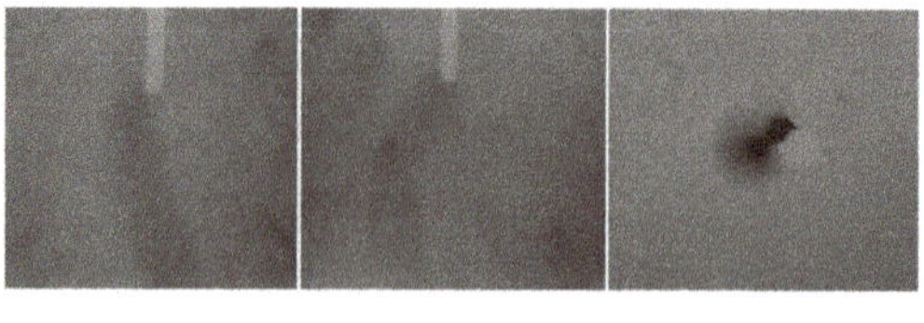

Figure 3. Three views of discharge channel.

According to Figure 3, on the basis of the known three views of the soil discharge channel, a three-dimensional reconstruction of the soil discharge channel is carried out, and the reconstructed model is exported to finite element simulation software. There is a more detailed description of this type of model by Luo et al. [27]. Figure 4 shows the model after three-dimensional reconstruction.

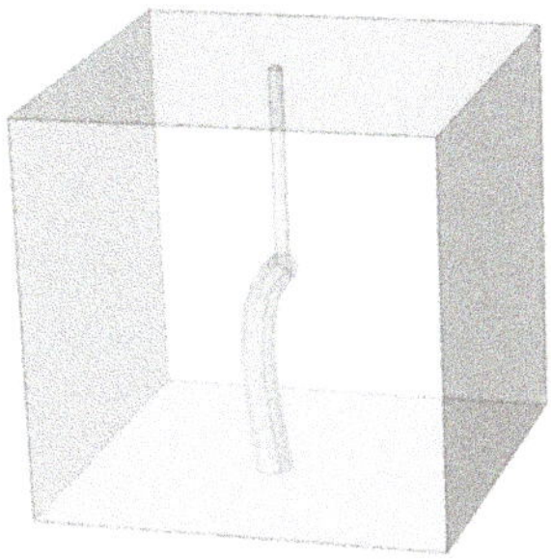

Figure 4. Three-dimensional reconstruction model.

The software used for simulation calculation is COMSOL Multiphysics 3.5a (COMSOL Inc., Stockholm, Sweden).

In this paper, the injected impulse current amplitude is 2.0 kA under the simulated test conditions, and the impulse voltage test measurement value is 23.5 kV. Using the method proposed in this section to combine the three-dimensional morphology of the discharge area with the finite element method, set the soil residual resistivity is 7% ρ_0 in the discharge area. The voltage value of the electrode injection point can be calculated by using the finite element method. After a certain step length iteration, the voltage calculated value and the actual measured value can be within the allowable 1% error range, then the value at this time is the convergence value.

When the soil residual resistivity is iterated to 000362% ρ_0, the impulse voltage simulation calculation value is 23.4 kV, and the error is 0.5%, which meets the allowable error requirements. At this time, take the average value of the field strength in the discharge area as the critical breakdown field strength value of 216.21 kV/m as the soil critical breakdown field strength value under the set of test conditions.

4. The Influence of Medium Factors on the Critical Breakdown Field Strength

Studies have shown that the soil water content, salt content and soil density have an impact on the value of soil critical breakdown field strength. In addition, different soils have different properties and their influence laws are also different. Therefore, in this section, we study the variation of the critical breakdown field strength of different typical soils under the conditions of different water content, salt content and Soil density.

4.1. The Influence Law of Water Content on Ec

In order to study the influence of the water content of sandy soil, loam soil and Yellow cinnamon soil on the electrical parameters of soil discharge characteristics in the laboratory, this paper, according to the actual situation, maintains the soil salt content at 0.5% and the soil density at 1.250 g/cm^3, and the water content is 3%, 5%, 7%, 9%, 11%, 13% and 15% of simulated soil samples, which were subjected to an impulse current with an amplitude of 2.0 kA. Soil moisture content refers to the ratio of the quality of water in the soil to the quality of dry soil [28,29]. The electrode used in the simulation experiment is a vertical electrode with a diameter of 3 mm and a length of 10 cm, and the following experimental electrodes maintain this parameter. Through an inversion calculation, we can obtain the law of three kinds of soil critical breakdown field strength with water content, as shown in Figure 5.

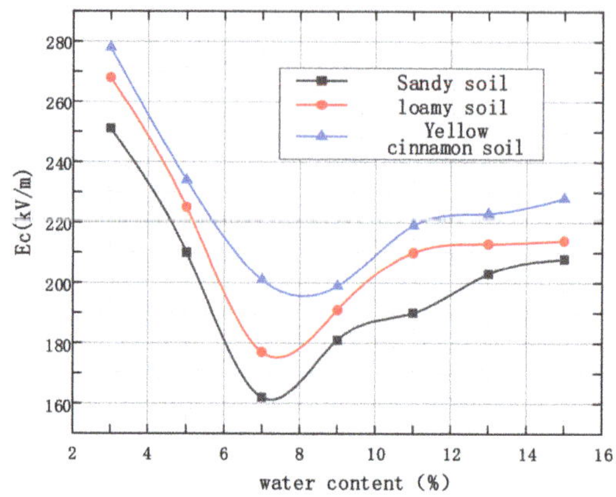

Figure 5. The influence of water content on the critical field strength of soil breakdown.

It can be seen from Figure 5 that under the condition of 3–15% water content, the three types of soil critical breakdown field strengths all show a trend of first decreasing and then increasing and finally becoming saturated with the increase in soil water content. The critical breakdown field strength of sandy soil varies between 161 kV/m~251 kV/m, the critical breakdown field strength of loam between 175 kV/m~268 kV/m and the critical breakdown field strength of Yellow cinnamon soil varies between 195 kV/m~278 kV/m. Additionally, under the same soil medium conditions, the relationship between the critical breakdown field strengths of the three soils is: Ec of sandy soil < Ec of loam soil < Ec of Yellow cinnamon soil.

The law of the critical breakdown field strength of sandy soil, loam soil and Yellow cinnamon soil with water content is as follows:

1. When the soil water content is between 3% and 7%, due to the increase in the water content, the salt in the soil is dissolved into conductive ions, which enhances the conductivity of the soil, and the critical breakdown field strength decreases with the increase in the water content.
2. When the water content is between 7% and 11%, as the soil water content further increases, the salt has been fully dissolved. At this time, as the water content increases, the soil ion concentration decreases, the soil conductivity becomes weak and the critical breakdown field strength increases.
3. When the water content exceeds 11%, Ec gradually becomes saturated, and the critical breakdown field strength tends to be stable with the change in water content.

4.2. *The Influence Law of Salt Content on Ec*

In order to study the influence of salt content on Ec, on the basis of maintaining soil water content of 5% and soil density of 1.250 g/cm^3, the salt contents were 0.1%, 0.3%, 0.5%, 0.7% and 1.0%, respectively. The impulse current with an amplitude of 2.0 kA was applied to the three soil samples. Through inversion calculations, we can acquire three kinds of soil critical breakdown field strengths with salt content, as shown in Figure 6.

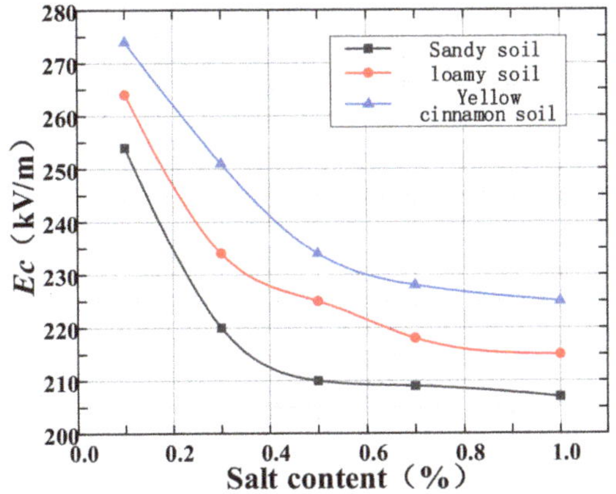

Figure 6. The influence of salt content on the critical breakdown field strength of soil.

In the case of 0.1~1% salt content, it can be seen from Figure 6 that the critical breakdown field strengths of sand, loam and Yellow cinnamon soil are between 207 kV/m~254 kV/m, 215 kV/m~264 kV/m and 225 kV/m~274 kV/m. As the salt content increases, the *Ec* of the three soils all show a change situation that first decreases and then tends to saturation. Additionally, under the same soil medium conditions, the relationship between the critical breakdown field strengths of the three soils is: *Ec* of sandy soil < *Ec* of loam soil < *Ec* of Yellow cinnamon soil.

The critical breakdown field strength of sandy soil, loam soil and Yellow cinnamon soil changes in the same way with the salt content, and the details are as follows:

1. When the salt content is between 0.1% and 0.5%, due to the unsaturation of salt dissolution, as the salt content increases, the dissolved ions in the solution increase, the soil conductivity increases and the soil critical breakdown field strength decreases.
2. When the salt content continues to increase, because the dissolution has reached saturation, as the salt content increases, the conductivity of the solution does not change. Therefore, in this interval, as the salt content increases, the critical breakdown field strength no longer has significant changes.

4.3. The Influence Law of Soil Density on Ec

In order to study the influence of soil density on the electrical parameters of soil discharge characteristics, according to actual conditions, on the basis of maintaining a soil water content of 5% and a salt content of 0.5%, the soil density samples are 1.042 g/cm^3, 1.145 g/cm^3, 1.250 g/cm^3, 1.354 g/cm^3 and 1.458 g/cm^3, respectively. Since the volume of the experimental sandbox selected in the experiment is fixed, the density of the soil can be changed by changing the quality of the soil filled in the sandbox of a fixed volume. After the above-mentioned soil density is normalized, the soil density can be expressed as: 1.0, 1.1, 1.2, 1.3 and 1.4. Applying an impulse current with an amplitude of 2.0 kA, through the inversion calculation, the variation law of the soil critical breakdown field strength with the salt content can be obtained, as shown in Figure 7.

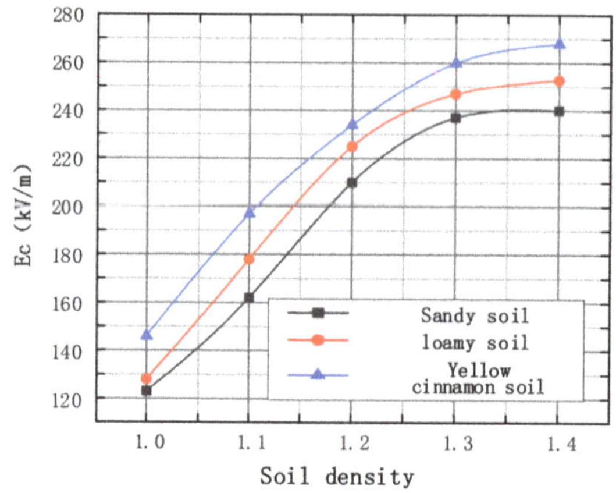

Figure 7. The influence of soil density on the critical breakdown field strength.

It can be seen from Figure 7 that when the soil density is 1~1.4, the critical breakdown field strengths of sandy soil, loam soil and Yellow cinnamon soil are 122 kV/m~240 kV/m, 128 kV/m~252 kV/m and 146 kV/m~268 kV/m, respectively. The soil density has a significant effect on the critical breakdown field strength. As the soil density increases, the *Ec* of the three soils gradually increases. Additionally, under the same soil medium conditions, the relationship between the critical breakdown field strengths of the three soils is: *Ec* of Yellow cinnamon soil < *Ec* of loam soil < *Ec* of sandy soil.

The variation law of the critical breakdown field strength of sandy soil, loam soil and Yellow cinnamon soil with soil density is as follows:

1. When the soil density is between 1 and 1.2, as the soil density increases, the soil becomes denser, and the gap between the particles in the soil is smaller, so that it is less likely to break down. Therefore, in this area, the soil is critically broken down The field strength increases as the tightness increases.
2. With the further increase in soil density, although the critical breakdown field strength continues to increase, its influence weakens, and the increasing trend of the critical breakdown field strength slows down.

It can be seen from Figures 5–7 that under the same soil parameter conditions, the magnitude relationship of the critical breakdown field strength value is: *Ec* of Yellow cinnamon soil < *Ec* of loam soil < *Ec* of sand soil.

4.4. Analysis of the Influence of Three Factors on the Critical Breakdown Field Strength

Soil is a complex mixture composed of three-phase media of gas, liquid and solid. The particle size of different soils is different. This factor will affect the content of water and air in the gaps between soil particles. The internal structure of the three soils is different. There are many clay particles in the loam soil and Yellow cinnamon soil, which can bond the surrounding soil particles to form larger particles [18]. The breakdown voltage between water, air and solid is: air < water < solid. Therefore, in the process of soil breakdown, water and solids can be regarded as insulators relative to air, and air can be regarded as conductors. However, in an unbreakable soil medium, the conductivity of water is greater than that of air. Therefore, whether it is broken down or not, soil particles are regarded as dielectrics, and air and water are regarded as conductors according to breakdown and non-breakdown, respectively.

When the impulse current flows into the soil, due to the conductivity of the water, a certain amount of leakage current flows through the water, so that the internal temperature of the soil rises rapidly. The temperature rise causes the air molecules in the soil to thermally dissociate, and electron jumps are formed under the action of the external electric field, which causes the soil to break down.

When the salt content and soil density are constant, at the beginning, the water content and the leakage current are small, so less heat is generated. At this time, a larger voltage is required to make the soil break down. In the process of increasing water, the leakage current in the soil increases, and the temperature rises rapidly, which promotes the thermal dissociation of the air and forms electron jumps to break down the soil. Therefore, the critical breakdown field strength decreases with the increase in water content. When the water content continues to increase, as the water content in the soil voids gradually increases, under the same voltage, the leakage current increases, and the current is more likely to be discharged in the form of a diffuse current rather than spark discharge. Therefore, with the increase in the water content at this time, the critical breakdown field strength of the soil gradually increases.

When the water content and soil density are constant, as the salt content increases, the ion concentration and the leakage current increase, so that the heat generated in the soil is higher, which promotes the thermal dissociation of air gap molecules, thereby promoting the formation of the electronic bounce. However, after the salt content increases to a certain level, the ion concentration no longer increases, and the critical breakdown field strength no longer drops significantly.

When the water content and salt content remain unchanged, the increase in soil density will reduce the water and air content per unit volume, thereby greatly reducing the content of conductive substances per unit volume, so the soil critical breakdown field strength increases.

It can be seen from Figures 5–7 that, compared with water content and soil density, salt content has the lowest impact on the critical breakdown field strength, because its impact is only the single factor of soil ion concentration. Compared with the water content, the maximum and minimum difference of the critical breakdown field strength caused by soil density is about 125 kV/m in this experiment, and under the water content factor, the maximum and minimum difference of the critical breakdown field strength is only 95 kV/m, so the soil density has the greatest impact on the critical breakdown field strength.

5. The Influence of Medium Factors on Residual Resistivity (ρ_{res})

For the residual resistivity of the soil discharge area, the general concern is its relationship with the initial resistivity of the soil, and the initial resistivity of the soil is affected by the soil water content, salt content, soil density and soil types. Therefore, this chapter starts from the perspective of soil initial resistivity changes, and comprehensively considers the influence of water content, salt content, and soil density on the relationship between ρ_{res} and initial resistivity ρ_0 under different soil conditions.

5.1. The Influence of Water Content on ρ_{res}

In order to study the influence of water content of sandy soil, loam soil and Yellow cinnamon soil on soil ρ_{res}, this paper, according to the actual situation, maintains a soil salt content of 0.5% and soil density of 1.250 g/cm^3, and the water content is 3%, 5%, 7%, 9%, 11%, 13% and 15%, respectively. Simulated soil samples were subjected to an impulse current with an amplitude of 2.0 kA. The electrode used in the simulation experiment is a vertical electrode with a diameter of 3 mm and a length of 10cm, and the following experimental electrodes maintain this parameter. Through inversion calculations, we can obtain three kinds of soil residual resistivity changes with water content, as shown in Figure 8.

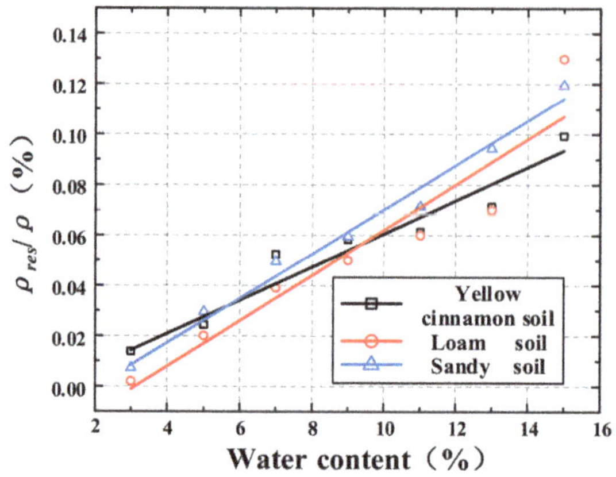

Figure 8. The influence of water content on ρ_{res} under different soil types.

As shown in Figure 8, the ρ_{res}/ρ of the three soils all increase with the increase in soil water content, and the relationship between the maximum values of the ρ_{res}/ρ of the three soils is Yellow cinnamon soil ρ_{res} < loam soil ρ_{res} < sandy soil ρ_{res}.

5.2. The Influence of Salt Content on ρ_{res}

In order to study the influence of salt content on ρ_{res}, on the basis of maintaining a soil water content of 7% and soil density of 1.250 g/cm^3, the salt content was 0.1%, 0.3%, 0.5%, 0.7% and 1.0%, respectively. Impulse currents with amplitude of 2.0 kA are applied to the three soil samples. Through inversion calculations, the change rule of the residual resistivity of the three soils with the salt content can be obtained, as shown in Figure 9.

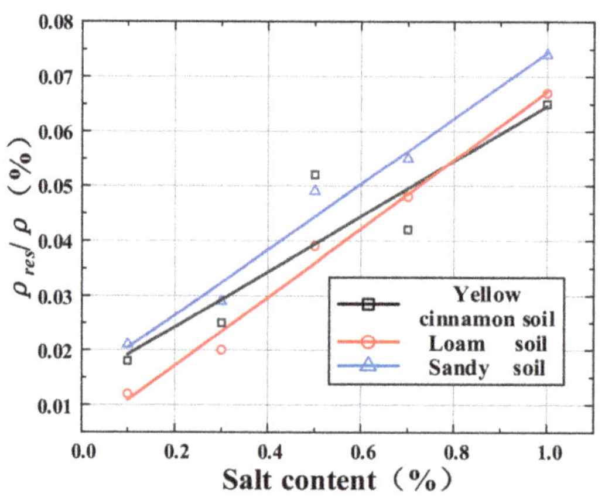

Figure 9. The influence of salt content on ρ_{res} under different soil types.

As shown in Figure 9, the ρ_{res}/ρ of the three soils all increase with the increase in soil salt content, and the relationship between the maximum values of ρ_{res}/ρ of the three soils is Yellow cinnamon soil ρ_{res} < loam soil ρ_{res} < sandy soil ρ_{res}.

5.3. The Influence of Soil Density on ρ_{res}

In order to study the influence of soil density on soil ρ_{res}, according to actual conditions, on the basis of maintaining a soil water content of 7% and salt content of 0.5%, the simulated soil density is 1.042 g/cm^3, 1.145 g/cm^3, 1.250 g/cm^3, 1.354 g/cm^3 and 1.458 g/cm^3, and the normalized soil density can be expressed as: 1.0, 1.1, 1.2, 1.3 and 1.4. Since the volume of the experimental sandbox selected in the experiment is fixed, the density of the soil can be changed by changing the quality of the soil filled in the sandbox of a fixed volume. Applying an impulse current with an amplitude of 2.0 kA, through inversion calculations, the residual resistivity of the soil can be obtained with the change in the soil density, as shown in Figure 10.

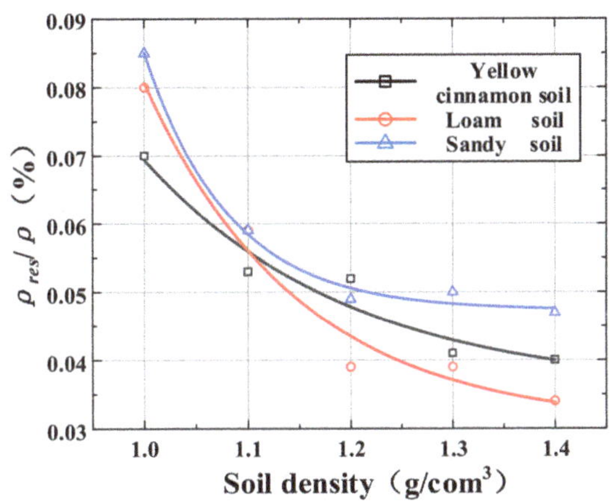

Figure 10. The influence of soil density on ρ_{res} under different soil types.

As shown in Figure 10, the ρ_{res}/ρ of the three soils all increase with the increase in soil density, and the relationship between the maximum values of ρ_{res}/ρ of the three soils is Yellow cinnamon soil ρ_{res} < loam soil ρ_{res} < sandy soil ρ_{res}.

5.4. Analysis of the Influence of Three Factors on Residual Resistivity

The graphical results of the analysis show that within the experimental conditions of this paper, with the increase in water content, the ρ_{res}/ρ of the three soils gradually increases. The range of ρ_{res}/ρ for sandy soil is between 0.007 and 0.120, for loamy soil it is between 0.001 and 0.129, and for Yellow cinnamon soil it is between 0.015 and 0.099. Analyzing the reason, as the water content increases, on the one hand, water increases the conductivity of the soil, and on the other hand, the increase in water content enables the salt in the soil to be fully dissolved. Therefore, the resistivity of the soil gradually decreases, and ρ_{res}/ρ gradually increases.

Similarly, when the salt content of sandy soil, loam soil and Yellow cinnamon soil changes, the range of ρ_{res}/ρ is 0.021~0.0735, 0.012~0.066 and 0.018~0.064. The increase in salt content increases the amount of dissolved salt in the soil, which makes the initial resistivity of the soil decrease, and ρ_{res}/ρ gradually increases.

With the increase in soil density, the range of ρ_{res}/ρ of sandy soil, loam soil and Yellow cinnamon soil was 0.047~0.085, 0.033~0.08, 0.04~0.0695. The increase in soil density means that the contact between soil particles is closer, the gap between soil particles will decrease, and the soil resistivity will increase. Therefore, it will be more difficult to generate discharge channels in the soil, so ρ_{res}/ρ will gradually decrease.

6. Analysis of Soil Impact Breakdown Mechanism

Based on the existing research results, it can be analogous to the air breakdown mechanism to explain the impact breakdown mechanism of the soil as follows:

6.1. "Electron Avalanche-Streamer Discharge-Pilot Discharge" Process

1. Electron avalanche stage: When using rod-plate electrodes, the electric field distribution between the electrodes is uneven. The maximum field strength of this uneven electric field appears near the vertical electrode with a small radius of curvature. Under the same conditions without the radius of curvature, the smaller the electrode curvature radius, the greater the maximum field strength, and the more uneven the electric field distribution. The free electrons existing in the soil near the vertical electrode are continuously accelerated along the direction of the electric field under the action of the electric field and they accumulate kinetic energy. When the voltage applied across the electrode reaches a certain value, the electric field near the vertical electrode provides free electron kinetic energy equal to or greater than the free energy of air molecules in the soil particle gap, and even the free energy of soil particle molecules, resulting in air molecules or the soil particle molecules split into electrons (or negative ions) and positive ions. The new electrons generated at this time, like the original free electrons, obtain kinetic energy from the electric field, and continue to move toward the positive electrode of the electric field to cause more collisions and dissociation, forming an initial electron avalanche.
2. Streamer discharge: With the continuous development of electronic avalanche, the number of electrons, negative ions and positive ions in the avalanche increases sharply with the distance of the development of the electronic avalanche. Due to the smaller mass of electrons, the acceleration obtained from the electric field is much greater than that of positive ions. Therefore, when electrons move quickly to the positive electrode, the positive ions slowly move to the negative electrode, and the positive ions can be considered to be static relative to the electrons. When the electron avalanche continues to develop toward the positive electrode, the positive ions remaining in the negative electrode act as positive space charges to distort and strengthen the electric field around the negative electrode, thereby emitting a large number of photons to the surroundings. These photons in turn cause the nearby molecules to undergo a process of light dissociation, generating secondary electrons. These secondary electrons form a new secondary electron avalanche under the action of the positive space charge distortion and the enhanced electric field. The electrons in the secondary electron avalanche move to the positive space charge region of the initial electron avalanche, and merge with it to form a mixed channel full of positive and negative charged particles, forming a streamer discharge channel.
3. Pilot discharge: Due to the long distance between the positive and negative electrodes, when the streamer discharge channel is not enough to penetrate the entire gap, the electrons move along the streamer channel due to the collision between the electrons and the molecules to make the temperature of the channel constantly increasing, up to thousands of degrees or higher, so that molecules produce thermal dissociation. This discharge process with the thermal dissociation process is called pilot discharge. Since the pilot discharge has a stronger process than the streamer discharge, it can promote the streamer to continue to grow until it penetrates the entire gap between the electrodes.

6.2. "Water Evaporation" Process

Compared with the breakdown process of air, in addition to the above-mentioned breakdown process of "electron avalanche-streamer discharge-leader discharge" in the breakdown of soil, there is also a process in which liquid water in the soil turns into gaseous water.

1. Under the effect of the collision of free electrons and new electrons generated by the collision, the distance between molecules becomes larger, and the water changes from liquid to gas.
2. Under the effect of the collision of electrons and molecules, the temperature of the discharge channel increases sharply, and a large amount of heat energy promotes

the evaporation of the surrounding water from a liquid to a gaseous state. The gas moisture produced by the process of "water evaporation" acts on the soil discharge channel, and together with the heated air, the discharge channel expands to the surroundings. It is precisely because of the distortion and strengthening of the electric field caused by the irregularity of the edge of the soil particles, and the existence of the "water" evaporation process, that the critical breakdown field strength of the soil is much smaller than the breakdown field strength of the equivalent air gap.

From the analysis of the "water evaporation" process, it can be inferred that the discharge channel is formed in the process of developing from the top to the bottom while also developing to the left and right. The soil impulse discharge mechanism analyzed in this paper is based on the image characteristics of the soil impulse discharge, and there is still a lack of direct verification results, such as the temperature change characteristics of the discharge channel. This problem still needs further research in the future.

7. Conclusions

The partial discharge of soil that occurs when the lightning current enters the ground through the grounding electrode poses a threat to the safety of personnel and equipment, but the difficulty of the formation of the soil discharge channel is closely related to the magnitude of the critical breakdown field strength, and the critical breakdown field strength of the soil residual resistivity are two key parameters of soil discharge. In this paper, a variety of typical soils were carried out to study the influence of soil water content, salt content, soil density and other factors on the morphological characteristics of the discharge channel, and then analyzed the critical breakdown field strength and the soil critical breakdown field strength with the introduction of different discharge channel morphologies and structures. The accurate measurement method of residual resistivity was carried out and we further discussed the minimum critical breakdown field strength of typical soil and its value conditions. The specific conclusions are as follows:

1. The three types of soil critical breakdown field strengths first decrease, then increase, and then become stable with the increase in water content; with the increase in salt content, they first decrease and then become stable; with the increase in soil density, the increase shows the rule of first increasing and then tending to be stable;
2. The residual resistivity of the three soils increases with the increase in soil water content and salt content and decreases with the increase in soil density.
3. Under the conditions of the same water content, salt content and soil density, the critical breakdown field strength value relationship is: Ec of sandy soil < Ec of loam soil < Ec of Yellow cinnamon soil.

Author Contributions: Investigation, D.L., B.T. and W.L.; methodology, J.L.; project administration, H.W. and Y.C. All authors have read and agreed to the published version of the manuscript.

Funding: This work is supported by the Science and Technology Project of State Grid Corporation of China ("Study on the Transient Characteristics of Grounding System and the Test and Evaluation Method of Current Dispersion Performance Under the Successive Impulse Current", 5500-202026088A-0-0-00).

Institutional Review Board Statement: Not applicable.

Informed Consent Statement: Not applicable.

Data Availability Statement: The study did not report any data.

Conflicts of Interest: The authors declare no conflict of interest.

References

1. Liew, A.C.; Darveniza, M. Dynamic model of impulse characteristics of concentrated earths. *Electr. Eng. Proc. Inst.* **1974**, *121*, 123–135. [CrossRef]
2. Mazzetti, C.; Veca, G.M. Impulse Behavior of Ground Electrodes. *IEEE Trans. Power Appar. Syst.* **1983**, *9*, 3148–3156. [CrossRef]
3. Mousa, A.M. The soil ionization gradient associated with discharge of high currents into concentrated electrodes. *IEEE Trans. Power Deliv.* **1994**, *9*, 1669–1677. [CrossRef]
4. Bellaschi, P.L.; Armington, R.E.; Snowden, A.E. Impulse and 60-cycle characteristics of driven grounds—II. *Electr. Eng.* **1942**, *61*, 349–363. [CrossRef]
5. Bellaschi, P.L.; Armington, R.E. Impulse and 60-Cycle Characteristics of Driven Grounds–III Effect of Lead in Ground Installation. *Am. Inst. Electr. Eng. Trans.* **1943**, *62*, 334–345. [CrossRef]
6. Towne, H.M. Impulse characteristics of driven grounds. *Gen. Electr. Rev.* **1929**, *31*, 605–609.
7. Geri, A. Behaviour of grounding systems excited by high impulse currents: The model and its validation. *IEEE Trans. Power Deliv.* **1999**, *14*, 1008–1017. [CrossRef]
8. Nekhoul, B.; Labie, P.; Zgainski, F.X.; Meunier, G.; Morillon, F.; Bourg, S. Calculating the impedance of a grounding system. *IEEE Trans. Magn.* **1996**, *32*, 1509–1512. [CrossRef]
9. Otani, K.; Baba, Y.; Nagaoka, N.; Ametani, A.; Itamoto, N. FDTD surge analysis of grounding electrodes considering soil ionization. *Electr. Power Syst. Res.* **2014**, *113*, 171–179. [CrossRef]
10. Viola, F.; Romano, P.; Miceli, R. Finite-Difference Time-Domain Simulation of Towers Cascade under Lightning Surge Conditions. *IEEE Trans. Ind. Appl.* **2015**, *51*, 4917–4923. [CrossRef]
11. Li, J.; Yuan, T.; Yang, Q.; Sima, W.; Sun, C.; Zahn, M. Numerical and Experimental Investigation of Grounding Electrode Impulse-Current Dispersal Regularity Considering the Transient Ionization Phenomenon. *IEEE Trans. Power Deliv.* **2011**, *26*, 2647–2658. [CrossRef]
12. Habjanic, A.; Trlep, M. The simulation of the soil ionization phenomenon around the grounding system by the finite element method. Magnetics. *IEEE Trans. Magn.* **2006**, *42*, 867–870. [CrossRef]
13. Chisholm, W.A.; Janischewskyj, W. Lightning surge response of ground electrodes. *IEEE Trans. Power Deliv.* **1989**, *4*, 1329–1337. [CrossRef]
14. Petropoulos, G.M. The high-voltage characteristics of earth resistances. *Electr. Eng. Proc. Inst.* **1948**, *95*, 59–70.
15. Korsuntcev, A.V. Application of the theory of similitude to the calculation of concentrated earth electrodes. *Elektrichestvo* **1958**, *5*, 31–35.
16. Oettle, E.E. A new general estimation curve for predicting the impulse impedance of concentrated earth electrodes. *IEEE Trans. Power Deliv.* **1988**, *3*, 2020–2029. [CrossRef]
17. Nor, N.M.; Haddad, A.; Griffiths, H. Determination of threshold electric field E_C of soil under high impulse currents. *IEEE Trans. Power Deliv.* **2005**, *20*, 2108–2113. [CrossRef]
18. Du, J. *Research on Typical Soil Erosion Characteristics of Yunnan and Analysis of Spark Discharge Phenomena*; Southwest Jiaotong University: Chengdu, China, 2017.
19. Guo, Y.; Wang, G.; Yang, M. Circuit method simulation analysis of impact characteristics of horizontal-vertical grounding body model. *Mod. Build. Electr.* **2010**, *1*, 56–59.
20. Li, J.; Jiang, J.; Li, L. Full-time domain electrical network model of grounding device transient characteristics considering soil spark discharge. *Chin. J. Electr. Eng.* **2017**, *37*, 7058–7065.
21. Li, H.; He, Z.; Chen, S.; Zhang, M.; Shu, Y.; Hu, C. Finite element calculation and analysis of impact dissipation characteristics of grounding device considering spark effect. *Electr. Porcelain Arrester* **2019**, *6*, 104–110.
22. Deng, C.; Zhou, W.; Wei, S. Analysis of the influence of inductance effect and spark effect on the lightning strike characteristics of grounding body and surrounding soil. *High Volt. Technol.* **2015**.
23. Feng, Z.; Tong, X.; Cai, H.; Li, J.; Liu, G.; Jia, L.; Lu, H. Simulation study on impact characteristics of tower grounding device considering nonlinear ionization of soil. *Hydropower Energy Sci.* **2016**, *34*, 173–177.
24. Zhu, C.; Xing, P.; Lu, H.; Lan, L.; Feng, Z. Calculation and analysis of impact characteristics of tower grounding body based on electromagnetic field. *China Electr. Power* **2014**, *47*, 83–87.
25. Xu, W.; Liu, X.; Huang, W. Calculation model of impact grounding resistance of tower grounding body based on ATP-EMTP. *Electr. Power Constr.* **2010**, *31*, 22–25.
26. Liu, Y. *Principles of Medical Imaging Imaging*; Jilin University Press: Changchun, China, 2012.
27. Luo, D.; Sima, W.; Yuan, T.; Sun, P.; Chen, W.; Wang, J. Influence-Factor Analysis and Parameter Calculation of Soil Discharge and Recovery Characteristics under Successive Impulse Currents. *IEEE Trans. Power Deliv.* **2019**, *34*, 514–523. [CrossRef]
28. Zhu, Z. *Soil Science*; Agricultural Press: Warsaw, Poland, 1983.
29. He, J.; Zeng, R. *Power System Grounding Technology*; Science Press: New York, NY, USA, 2007.

energies

MDPI

Article

Numerical Simulation on Charge Transport and DC Breakdown in Polyethylene-Based Micro-h-BN/Nano-SiO_2 with Filler Orientation Dependent Trap Energy †

Xuri Xu, Yu Gao *, Jing Li, Zheng Song, Huicun Zhao and Tao Han

School of Electrical and Information Engineering, Tianjin University, Tianjin 300072, China; xxr@tju.edu.cn (X.X.); jingli06@tju.edu.cn (J.L.); songzheng@tju.edu.cn (Z.S.); hcunzhao@tju.edu.cn (H.Z.); hant@tju.edu.cn (T.H.)
* Correspondence: hmgaoyu@tju.edu.cn

† This paper is an extended version of our paper published in 2020 IEEE International Conference on High Voltage Engineering and Application (ICHVE 2020), Beijing, China, 6–10 September 2020; pp. 1–4.

Abstract: In order to improve the thermal conductivity and the insulation properties of polyethylene (PE) used as cable insulation under DC stress, hexagonal boron nitride (h-BN) and inorganic particles have been considered as micro-filler and nano-filler, respectively. As a 2D material, the orientation of h-BN possibly affects the insulation properties of the polymer. It is important to understand the influence of the filler orientation on the insulation performance of the polymer. In this work, a numerical simulation has been performed to investigate the effect of orientation of micro-h-BN on charge transport and DC breakdown of PE-based micro/nano-composites and a comparison between the simulation result and previous literature data has been conducted. The h-BN was designated to be parallel, perpendicular to the normal sample surface vector (the direction of electric field in this work) or randomly distributed in the matrix, and the charge transport behavior and DC breakdown strength in the samples were discussed by using the bipolar charge transport (BCT) model. The results indicated that when the h-BN was perpendicular to the normal vector, the density of trapped charge was the largest and the DC breakdown strength was the highest among the three cases studied. It is suggested that the charge trapping/de-trapping processes and the electric field in the sample vary with the orientation of h-BN through tailoring the trap characteristics of the material.

Keywords: polyethylene; micro/nanocomposite; hexagonal boron nitride; charge transport; DC breakdown; orientation; trap characteristics

Citation: Xu, X.; Gao, Y.; Li, J.; Song, Z.; Zhao, H.; Han, T. Numerical Simulation on Charge Transport and DC Breakdown in Polyethylene-Based Micro-h-BN/Nano-SiO_2 with Filler Orientation Dependent Trap Energy. *Energies* **2021**, *14*, 4645. https://doi.org/10.3390/en14154645

Academic Editor: Ayman El-Hag

Received: 27 June 2021
Accepted: 29 July 2021
Published: 30 July 2021

1. Introduction

With the development of high voltage direct current (HVDC) transmission projects, the demand of energy delivery with high power density through underground cable has become an urgent issue [1]. Polyethylene (PE), as the base material of HVDC cable insulation, is exposed to high temperatures and high electric fields during operation. Due to the coaxial structure of a single core power cable, a temperature gradient appears across the insulating material, which leads to serious space charge accumulation and electric field distortion, thus threatening the safe operation of the cable [2,3]. From the viewpoint of safety, it is very important to improve the thermal conductivity of DC cable insulation so as to establish a compromise the thermal and the insulation performances.

Over the years, a number of investigations have been performed to improve the thermal conductivity of polymer insulating materials. One simple method is to add inorganic particles with high intrinsic thermal conductivity into a polymer, whereby a composite with high thermal conductivity could be formed [4,5]. Hexagonal boron nitride (h-BN) with an intrinsic thermal conductivity of 350–400 W/m·K has been widely considered as the best filler to tailor the thermal properties of polymers [6], and several studies regarding the enhancement of thermal conductivity of polymers by using h-BN as the filler have been reported. Zeng et al. revealed that the thermal conductivity of the composites was increased to 1.11 W/m·K, when h-BN with the average size of 1 μm was added into bismaleimide-triazine (BT) resin at 50 wt %, which represents a six-fold increase compared with pure BT resin [7]. Wang et al. reported that thermal conductivity with a value of 2.91 W/m·K could be obtained when 80 wt % micro-sized h-BN was added into epoxy (EP) [8]. Sato et al. employed polyimide (PI) and h-BN with an average particle size of 8 μm as the base material and filler to prepare a composite with high thermal conductivity. A thermal conductivity of 7 W/m·K was achieved with a filler loading of 60 vol%, which was 30 times higher than that of the neat PI [9]. It should be noted that the orientation of the added 2D material could affect the thermal conductivity of the composite. Karolina et al. [10] prepared highly aligned graphene-LDPE nanocomposites and measured the thermal conductivity from through plane and in-plane directions. It was found that the thermal conductivity along the extrusion direction for samples filled with 7.5 wt % of graphene nanoplatelets (GnP) reached 2.2 W/m·K whereas thermal conductivity measured through the plane direction decreased slightly. Such a phenomenon was also observed in polyimide (PI) films filled with h-BN by Tanimoto [11]. The obtained results indicated that the PI/h-BN films with in-plane directions filler had larger thermal conductivity than that with out-of-plane directions. In short, it is generally accepted that the addition of h-BN with appropriate content and orientation could obviously improve the thermal conductivity of polymer insulating materials.

However, it is also noticed that the insulation properties are usually reduced when micro-sized particles with high intrinsic thermal conductivity are added to a polymer [12]. Attempts have been made to add nano-sized inorganic particles into the material to form micro/nano-composites with balanced thermal and insulation properties. It was pointed out by Donnay [13] that when the content of micro and nano h-BN in the composite was 20 wt %, the breakdown strength could be increased by about 10%, and the thermal conductivity of the composite was three times of that of pure EP. Wang et al. [14] reported that EP composites co-filled with micro-h-BN and nano-SiO_2 could exhibit a thermal conductivity of 5.16 W/(m.K) and a breakdown strength of 102.9 kV/mm which was higher than that of neat EP. In addition, ultralow DC-conductivity that prevents the risk of thermal runaway and electrical failure is also an important insulation property, which is affected by the addition of nano-particles [15]. A large number of investigations have been performed with respect to this property. It was reported by Pourrahimi that the addition of poly(3-hexylthiophene) (P3HT) at 0.0005 wt % could reduce the DC electrical conductivity by a factor of up to 3 compared with LDPE [16]. Karolina et al. [17] studied the through-plane conductivity of a composite of LDPE filled with GnP exposed to low relatively low electric fields (<20 kV/mm). The results indicated that the DC conductivity in the composite was reduced slightly compared with the pure LDPE. Although previous works have been carried out to investigate the effect of the addition of micro/nano-particles on the thermal conductivity and the insulation properties of composites, when an 2D material with high intrinsic thermal conductivity like h-BN is employed, the effect of the filler orientation on the insulation performance of the micro/nanocomposite is rarely discussed.

At present, the orientation methods of thermal conductive fillers can be divided into two categories. One is the orientation driven by an external field including electric field orientation [18] and magnetic field orientation [19]. The method of electric field orientation is often applied to thermoset materials. The effect of the magnetic field orientation method depends on the encapsulation degree of h-BN in the polyethylene matrix. The other category is the orientation during processing, including pressure forming [20], tape casting [21], electrospinning methods [22]. The tape casting and electrospinning methods are suitable for thermoset polymers whereas polyethylene is a thermoplastic material. If we have the h-BN oriented in the melting state through a pressure forming method, the crystal growth in the cooling process may change the orientation. In summary, at this stage, it is still difficult to achieve an ideal orientation of h-BN in a polyethylene micro/nano-composite. Accordingly, it is worthwhile to perform numerical simulations with respect to this issue.

In our previous work, the preliminary result that the charge transport behavior and the DC breakdown strength in the composite is dependent on the orientation of micro-h-BN has been reported [23]. In this paper, as a follow-up and extension of our previous work, detailed and in-depth demonstrations about the charge dynamics and the electric field distortion have been performed to elucidate the mechanism of the filler orientation-dependent charge accumulation and DC breakdown behaviors through the bipolar charge transport (BCT) model using a novel method for threshold electric field extraction.

2. BCT Model

With the purpose of describing the charge transport behavior in polymer samples in detail during DC stressing, the BCT model [24] has been employed. The schematic diagram of the BCT model for a polymeric material under a DC voltage is shown in Figure 1. This figure is reproduced from [24]. The cathode and the anode are located at x = 0 and x = d, respectively. Electron and hole in the metal electrodes have to overcome the potential barriers between the electrode and the material to inject into the bulk. The injected charge can trap, de-trap, migrate and recombine in the polymer [25]. The mathematical equations of the BCT model are given in the sections that follow.

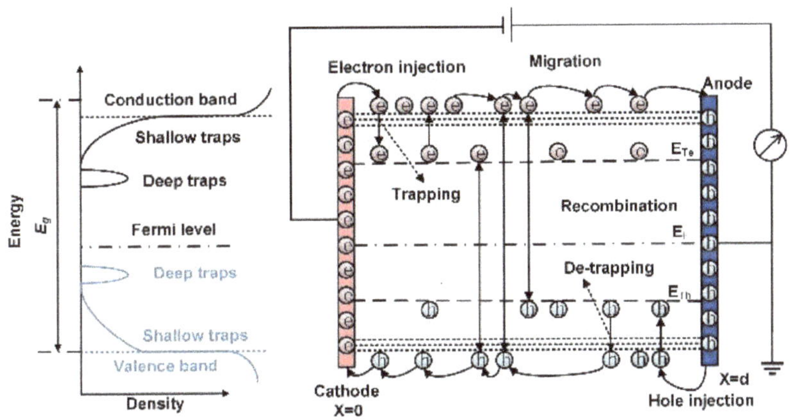

Figure 1. Scheme of BCT model of the polymeric materials, E_{Te} and E_{Th} are the trap energy of the electrons and holes, respectively.

2.1. Charge Injection

Electrons and holes are assumed to be injected into the sample from the cathode and anode, respectively, by Schottky thermionic emission [26]. The current densities induced by the charge injection are determined by the effective injection barrier between the sample and the electrode, electric field at the interface and temperature [27–29], as shown in Equations (1) and (2):

$$j_{in(e)}(t) = AT^2 \exp\left(\frac{\sqrt{eF(0,t)/4\pi\varepsilon_0\varepsilon_r} - E_{in(e)}}{k_B T}\right) \quad (1)$$

$$j_{in(h)}(t) = AT^2 \exp\left(\frac{\sqrt{eF(d,t)/4\pi\varepsilon_0\varepsilon_r} - E_{in(h)}}{k_B T}\right) \quad (2)$$

where, $j_{in(e)}$ (t) and $j_{in(h)}$ (t) are the current densities caused by electrons and holes injected from the cathode and anode in A/m^2, $E_{in(e)}$ and $E_{in(h)}$ are the injection barriers for electrons and holes in eV, e is the elementary charge in C, A and k_B are the Richardson constant and the Boltzmann constant, ε is the vacuum permittivity in F/m, ε_r is the relative permittivity of the sample, T is the temperature in K, $F(0,t)$ and $F(d,t)$ are the electric fields at the interfaces of x = 0 and x = d, respectively.

2.2. Self-Consistent Equations

The charge in polymeric materials is governed by a set of self-consistent equations [30], including charge continuity equation, charge transport equation and Poisson's equation, as depicted in Equations (3)–(6):

$$\frac{\partial q_{free}(x,t)}{\partial t} + \frac{\partial j_c(x,t)}{\partial x} = S_{free}(x,t) \tag{3}$$

$$\frac{\partial q_{trap}(x,t)}{\partial t} = S_{trap}(x,t) \tag{4}$$

$$j_c(x,t) = q_{free}(x,t)\mu_0 F(x,t) \tag{5}$$

$$\frac{\partial^2 \phi(x,t)}{\partial x^2} = -\frac{q_{free}(x,t) + q_{trap}(x,t)}{\varepsilon_0 \varepsilon_r} \tag{6}$$

where, q_{free} and q_{trap} are the free and trapped charge densities in the sample in C/m^3, j_c is the conducting current density in the material in A/m^2, μ is the carrier mobility in m^2/Vs, $F(x,t)$ is the electric field in the sample. $S_{free}(x,t)$ and $S_{trap}(x,t)$ are charge sources and are shown in detail below. The variation of charge in the material follows the charge continuity equations with the consideration of charge trapping, de-trapping and recombination processes [31]. Traps in this model are assumed to be with single trap energy when charges are trapping and de-trapping, and recombination occurs when electrons and holes encounter each other.

While solving the Poisson's equation (6), a boundary condition should be given. During the real DC breakdown experiment, a negative ramping voltage with a rising rate k_{ramp} is applied to the sample until the breakdown occurs. Accordingly, the potential is set to be zero on the anode and the same to the applied voltage on the cathode. The potential on the cathode is proportional to the time of voltage application. The boundary condition is expressed as follows:

$$\phi(0,t) = V_{appl}(t) = k_{ramp} t_{ramp} \tag{7}$$

where V_{appl} is the applied voltage in V, k_{ramp} is the ramping rate of the voltage in kV/s, and t_{ramp} is the time of voltage application in s. The electric field can be calculated by $F = -\Delta\phi$.

2.3. Charge Dynamics

Four species of carriers, namely, free and trapped electrons and holes are taken into account in the charge dynamics. The free carriers can hop in traps which determine the effective carrier mobility. The traps are usually caused by physical or conformational defects and chemical defects [17]. The trapping and de-trapping of charges are described in this work by considering the trap of single level. When the electron and the hole encounter each other in the material, recombination will occur. The following equations represent the first order charge dynamics in polymeric material, including charge trapping/de-trapping and recombination processes [28]:

$$S_{e\mu} = -P_{tr(e)} q_{free(e)} \left(1 - \frac{q_{trap(e)}}{q_e N_{T(e)}}\right) + P_{de(e)} q_{trap(e)} - R_{e\mu,h\mu} q_{free(e)} q_{free(h)} - R_{e\mu,ht} q_{free(e)} q_{trap(h)} \tag{8}$$

$$S_{et} = P_{tr(e)} q_{free(e)} \left(1 - \frac{q_{trap(e)}}{q_e N_{T(e)}}\right) - P_{de(e)} q_{trap(e)} - R_{et,h\mu} q_{free(h)} q_{trap(e)} \tag{9}$$

$$S_{h\mu} = -P_{tr(h)} q_{free(h)} \left(1 - \frac{q_{trap(h)}}{q_e N_{T(h)}}\right) + P_{de(h)} q_{trap(h)} - R_{e\mu,h\mu} q_{free(e)} q_{free(h)} - R_{et,h\mu} q_{free(h)} q_{trap(e)} \tag{10}$$

$$S_{ht} = P_{tr(h)} q_{free(h)} \left(1 - \frac{q_{trap(h)}}{q_e N_{T(h)}}\right) - P_{de(h)} q_{trap(h)} - R_{e\mu,ht} q_{free(e)} q_{trap(h)} \tag{11}$$

where $S_{e\mu}$, S_{et}, $S_{h\mu}$, and S_{ht} are the charge sources of mobile and trapped electrons and holes, respectively. $R_{e\mu,h\mu}$ is the recombination coefficient between free electrons and free holes in $m^3/C{\cdot}s$. $R_{e\mu,ht}$ and $R_{et,h\mu}$ are the recombination coefficient between free electrons and trapped holes and the recombination coefficient between trapped electrons and free holes in $m^3/C{\cdot}s$. Based on the Langevin recombination model [32], the recombination coefficient between free electrons and free holes can be expressed as $R_{e\mu,h\mu} = (\mu_e + \mu_h)/\varepsilon_0\varepsilon_r$. μ_e and μ_h are the carrier mobility for electrons and holes. According to the Shockley–Read–Hall model [33], trap-assisted recombination coefficient between free electrons and trapped holes and that between trapped-electrons and free-holes can be expressed as $R_{e\mu,ht} = \mu_e/\varepsilon_0\varepsilon_r$ and $R_{et,h\mu} = \mu_h/\varepsilon_0\varepsilon_r$. $P_{de(e)}$ and $P_{de(h)}$ are the de-trapping probability of trapped electrons and holes in 1/s. $P_{tr(e)}$ and $P_{tr(h)}$ are the trapping probability of free electrons and holes in 1/s. They are given by:

$$P_{de} = v_{ATE}\exp(-\frac{E_t}{k_B T}) \tag{12}$$

$$P_{tr} = \frac{q_e N_t \mu_0}{\varepsilon_0\varepsilon_r} \tag{13}$$

where, N_t is the density of the trap in $1/m^3$, V_{ATE} is the attempt-escape frequency in 1/s, and E_t represents the trap energy in eV.

3. Parameters for the BCT Model

In this work, PE-based micro/nano-composite has been employed as the sample for our numerical simulations. The fillers—20 nm-SiO_2 and 5 μm-h-BN particles—were selected with a weight ratio of 1 and 30 wt %, respectively, to ensure a high thermal conductivity according to our previous work [34]. It is considered that the orientation of h-BN in the real sample should be random, as shown in Figure 2. Three categories of the orientation of the h-BN could be summarized, which are perpendicular, parallel to the normal vector of the sample surface or randomly dispersed in the polymer. The normal vector of the sample surface in this paper is defined as the direction of the electric field as depicted in Figure 2.

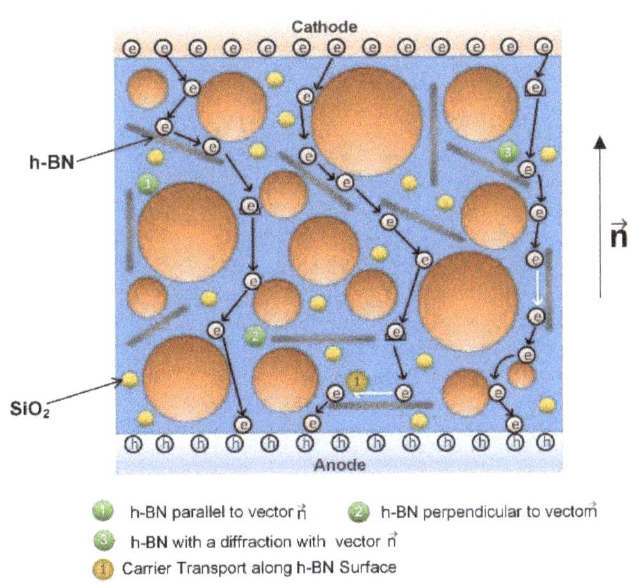

Figure 2. Scheme of various filler orientations and possible charge transport mechanism within the polyethylene micro/nano-composite.

Figure 2 shows various filler orientations and possible charge transport mechanisms in the material, taking the electron as an example. As discussed in [34], the de-trapped charge could be trapped again quickly, by which the overall time for the carrier within the trap is increased hence the trap depth would be deepened. Such an effect is strong while the orientation of the h-BN is perpendicular to the normal vector of the test sample, but it is weak when the h-BN is parallel to the vector. It is thereby proposed that the polymer with perpendicular orientation of the filler has the deepest trap energy, the material with filler distributing randomly follows, and the trap energy of a sample having a parallel filler orientation is the shallowest. According to our previous publication [34], the trap center of PE/h-BN/SiO_2 with random filler orientation is in the range of 0.84–0.95 eV. Therefore, the trap center used in this simulation is assumed to range from 0.7 eV to 1 eV. In our previous work [28], to keep the emphasis on the DC breakdown regulated by charge transport depending on trap center characteristics, the carrier mobility was set to be a constant for all samples. However, in fact, the carrier mobility is affected by the trap energy, namely, $\mu_{mol} = \mu_0 P_{de}/(P_{tr} + P_{de})$. μ refers to the shallow trap-controlled carrier mobility. For electrons and holes, μ_e and μ_{0h} are 1×10^{-14} and 3×10^{-15} m^2/Vs, respectively [35]. Table 1 shows the values of the parameters used in the BCT model [35–39]. In addition, the thickness of the samples used in our simulation is 200 μm, while the temperature is 298 K. It is assumed that insulation breakdown occurs when the maximum local electric field F_{max} inside the material exceeds a threshold F_c. In our previous work [23], the same threshold value was employed to discuss the DC breakdown strength of the composite with different filler orientations. At present, the threshold value can be extracted on the basis of the Poole-Frenkel effect. It is generally accepted that the energy that the charge requires to de-trap is reduced by the applied external electric field [40], which is referred to as the Poole-Frenkel effect. The corresponding reduction is depicted by Equation (14):

$$\Delta E_{PF} = \beta_{PF}\sqrt{F} = \sqrt{{q_e}^3 F/\pi\varepsilon_0\varepsilon_r} \quad (14)$$

Table 1. The values of the parameters used in BCT simulation [35–39].

Parameters	Value
Relative permittivity, ε_r	2.3
Carrier mobility for electron, μ_{emol}, ($\times 10^{-15}$ m^2/Vs)	8–9
Carrier mobility for hole, μ_{hmol}, ($\times 10^{-15}$ m^2/Vs)	2–3
Trap energy for electron, E_{Te}, (eV)	0.7–1
Trap energy for hole, E_{Th}, (eV)	0.7–1
Trap density for electron, NT_e, ($1/m^3$)	6.25×10^{20}
Trap density for hole, NT_h, ($1/m^3$)	6.25×10^{20}
Ramping rate of the voltage, k_{ramp}, (kV/s)	−0.5

Assuming that the applied electric field is high enough, the potential barrier for charge de-trapping can be reduced to zero, hence the charge can migrate freely in the material [41]. Such a special condition is herein referred to as a surpassing Poole-Frenkel (SPF) conduction. The breakdown is considered to take place when charges localized by the deepest trap overcome the potential barrier following the SPF conduction. The corresponding electric field needed to reach the SPF could be regarded as the threshold electric field F_c. When the reduction of potential barrier caused by the Poole-Frenkel effect is equal to the deepest trap center, F_c can be calculated by the following expression:

$$F_c = {E_t}^2 \pi\varepsilon_0\varepsilon_r/{q_e}^3 \quad (15)$$

where E_t is the deepest trap center in the polymer in eV.

The model of charge transport in this study was implemented using the finite element software COMSOL Multiphysics. The partial differential Equations (3)–(6) were solved by selecting suitable physics modules available in the software. The coupling between different modules was carried out so that the charge densities obtained as solutions of equations (3) and (4) were updated in Poisson's Equation (6) for calculating the electric field distribution.

4. Simulation Results and Discussion

4.1. Space Charge Accumulation

Figure 3a shows the distribution of space charge in the sample with random filler orientation at different positions varying as a function of time. The cathode and anode are at the positions of x = 0 and x = 200 μm, respectively. There is little charge injection and the position of accumulated space charges is near the electrode at the initial time. As the time lapses, more charges are injected into the material and the space charge accumulation tends to distribute in the whole volume of the composite. This suggests that the accumulation of space charge becomes more serious over time. It is possible that most of the injected charges can only transport to the vicinity of the electrode at the beginning while they can migrate towards the bulk of the sample over time. Comparisons of space charge distribution as a function of time in the samples with various filler orientations are depicted in Figure 3b. As regards to samples with parallel filler orientation, at the position of 195 μm, the space charge density is 65.0 C/m^3 at 20 s, 86.8 C/m^3 at 40 s, 126.6 C/m^3 at 60 s and 174.3 C/m^3 at 80 s. It is obvious that the space charge accumulation becomes more serious with time, and this feature exists in other samples. It should be kept in mind that the polymer with perpendicular orientation of the filler is assumed to have the deepest trap energy, the material with filler distributed randomly follows, and the trap energy of sample having filler parallel orientation is the shallowest. It can be observed that the amount of accumulated charge decreases with the increase of trap energy. For example, the maximum space charge density is 530.3 C/m^3 at 80 s in the sample with parallel orientation of filler whereas that is 249.2 C/m^3 when the filler orientation is perpendicular to the normal vector of the material. When the micro-filler is distributed randomly in the composite, the space charge density reaches its peak at 320.7 C/m^3. Another interesting feature that can be derived from Figure 3b is that the amount of accumulated holes near the anode is larger than that of electrons near the cathode in all samples. Such phenomena should be attributed to the lower mobility of holes.

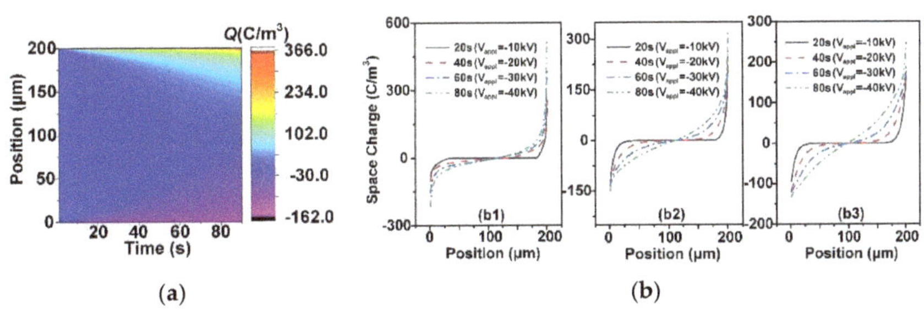

Figure 3. Space charge distribution in the sample with random orientation (**a**) and comparisons of space charge distribution as a function of time in the polymer with various orientations: parallel orientation (**b1**) random orientation (**b2**) perpendicular orientation (**b3**).

The comparisons of the density of free charge and trapped charge in different samples are illustrated in Figure 4a,b, respectively. The sample with perpendicular filler orientation has been taken as an example. At x = 3 μm, the density of free charge is −8.3 C/m^3 at 20 s, −9.9 C/m^3 at 40 s, 13.2 C/m3 at 60 s and −22.8 C/m^3 at 80 s. As a comparison, the corresponding values are −54.2, −90.6, −98.6 and −99.7 C/m^3, respectively, for the trapped charge. The density of charge increases over time for both free charge and trapped charge, which is similar to the features observed from Figure 4(a1,b1) and Figure 4(a2,b2). It can be seen from Figure 4 that the density of free charge and trapped charge are affected by the trap energy. For example, at x = 5 μm nearby the cathode, the density of free and trapped electrons in the sample with parallel filler orientation are −75.9 and −41.5 C/m^3 at t = 80 s, whereas they are −23.4 and −90.6 C/m^3 in the sample with randomly distributed filler, respectively. As for the material with perpendicular filler orientation, the densities of free and trapped charges are −19.2 and −99.3 C/m^3. As mentioned above, the probability of charge de-trapping is inversely proportional to the trap energy in the polymer [20], thus the charges captured by the deeper trap are more difficult to de-trap while the charges localized in the shallower trap are easier to de-trap. Therefore, with the growth of trap energy, the density of trapped charge increases, whereas the density of free charge decreases.

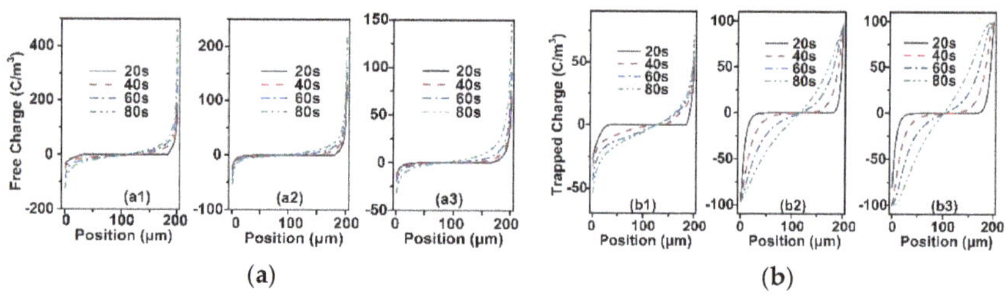

Figure 4. Comparisons of free charge (**a**) and trapped charge (**b**) in the polymer with various orientations: parallel orientation (**a1**,**b1**) random orientation (**a2**,**b2**) perpendicular orientation (**a3**,**b3**).

4.2. Electric Field Distribution

It has been generally accepted that the electric field in the polymer can be distorted due to the space charge accumulation [27]. The distorted electric field in the material is demonstrated in Figure 5a, taking the sample with filler random orientation as an example. At the initial moment, a uniform electric field distribution can be observed while the electric field distorts as time elapses. It can be also noticed from Figure 5a that, at the middle of the composite, the electric field is 268.6 kV/mm at 80 s whereas the values of electric field are 19.5 and 14.8 kV/mm at anode and cathode, respectively. It can be concluded that the electric field in the sample bulk is higher than that at the electrode. Similar results can be also noticed in the other two samples, as depicted in Figure 5b. Since homocharges are accumulated near the electrode as discussed in Figure 3 and the equivalent electric field formed by the charges is opposite to the applied field, the field at the electrode is weakened. As mentioned above, more positive charges accumulate near the anode than negative charges near the cathode due to their lower carrier mobility. According to Poisson's equation (6), the derivative of the electric field with respect to position is proportional to the density of accumulated charges. Therefore, the position at which the maximum electric field appears is biased toward the anode as shown in Figure 5b. Another interesting feature that could be derived from Figure 5b is that the maximum superimposed electric field in the composite increases with time.

In order to compare the electric field distortion, the electric field distribution factor f is calculated, which is defined as the radio of the superimposed electric field to the applied electric field calculated only from the applied voltage and the sample thickness. Figure 6a depicts the factor of electric field distortion at different times as a function of position in the sample with random filler orientation. At the initial moment, the internal electric field is almost determined by the applied voltage, resulting in a uniform distribution in the bulk with the distortion of electric field only near the sample electrode. With the lapse of the time for voltage application, the electric field in the bulk distributes unevenly. Comparisons of f in various samples as a function of position in different times are exhibited in Figure 6b. For example, at 80 s, the electric field distortion factor of the material with parallel filler orientation reaches its peak at 1.28, while the maximum of f is 1.35 in the sample having randomly oriented filler. In addition, f maximizes at 1.38 in the composite with perpendicular filler orientation. It can be concluded from Figure 6b that the distortion of electric field becomes more serious with the growth of trap energy as time goes by.

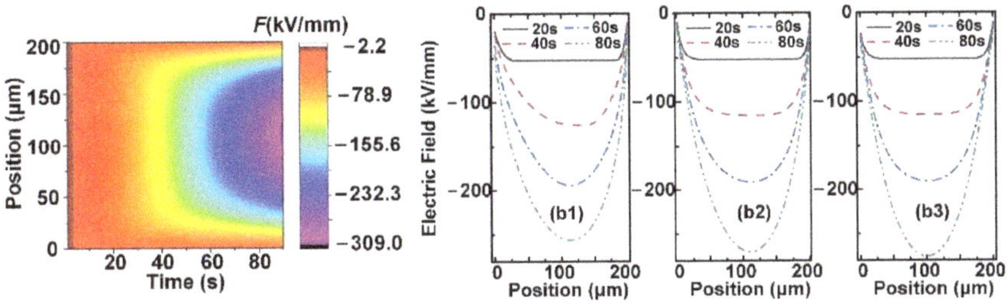

Figure 5. Electric field distribution in the sample with random orientation (**a**) and comparisons of electric field distribution as a function of position in the polymer with various orientations (**b**): parallel orientation (**b1**) random orientation (**b2**) perpendicular orientation (**b3**).

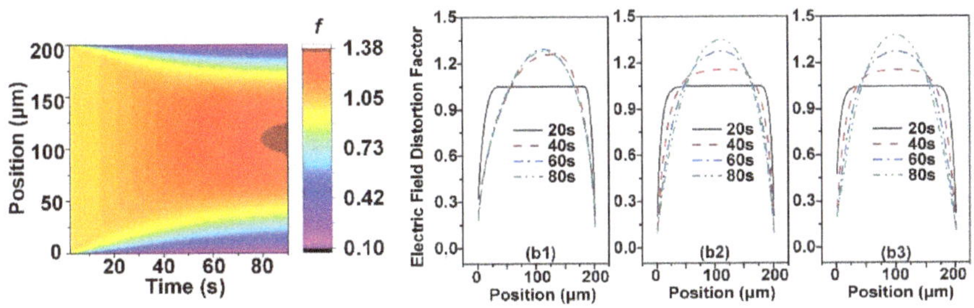

Figure 6. Electric field distribution factor in the sample with random orientation (**a**) and comparisons of electric field distribution factor as a function of position in the polymer with various orientations (**b**): parallel orientation (**b1**) random orientation (**b2**) perpendicular orientation (**b3**).

Figure 7 shows comparisons of the absolute value of maximum electric field in the various samples as a function of time. The values of the assumed threshold electric field F_c calculated by Equation (15) are 250, 302 and 383.9 kV/mm. for the sample with parallel, random and perpendicular filler orientation, respectively. The breakdown of the sample with parallel filler orientation takes place firstly at 81 s while the applied electric field is 202.5 kV/mm. The breakdown is most difficult to occur for the material with perpendicular filler distribution at 106 s when the electric field with a value of 265.0 kV/mm is applied to the composite. As h-BN is distributed randomly in the sample, the time of breakdown occurred is 90 s and the corresponding electric field is 225.0 kV/mm. The possible charge transport behavior in the composite with perpendicular and parallel filler orientations is shown in Figure 8a,b. The doped particles are distributed mainly in the amorphous region. Figure 8a shows that, when the h-BN is perpendicular to the normal vector of the sample, the charge injection is suppressed and the amount of charge recombination is lower compared with the sample with parallel filler orientation. As a comparison, Figure 8b demonstrates that more charges are injected from the electrode into the material while the density of charge recombination is higher in the sample with parallel filler orientation. Meanwhile the density of trapped charge is largest in the sample with perpendicular filler orientation, whereas that is smallest when the filler is parallel to the normal vector of the sample as mentioned above and shown in Figure 8. Such behavior should be attributed to the deeper trap energy in the composite with perpendicular orientation. It was generally accepted that the addition of nano-sized particles would introduce deep traps at the interface between the polymer matrix and the particle [42] whereas the trap energy could become shallower due to the defects introduced by micro-particles [43]. In this paper, a synergistic effect between the micro- and nano-sized particles is considered in that the de-trapped charge could be captured again quickly by the trap, by which the overall time for the carrier remaining in the trap becomes longer and hence the trap depth would be deepened. Such an effect is strong for the sample with perpendicular filler orientation, but is weak while the h-BN is parallel to the normal vector of the material. Due to the different filler orientations, the synergistic effect varies in the composite, which results in differenced in trap depth in the composite, thus affecting the charge transport and breakdown behavior of the sample. It was reported that the deep trap level was beneficial for the breakdown strength [44], and such a phenomenon was also verified in this paper via numerical simulation method. It should be noted that the trend of the simulation results in this paper was consistent with the experimental results reported in [42,44], and the order of magnitude was also identical to the results described in [36], thus verifying the reasonability of our simulation results. In short, the perpendicular micro-filler orientation in the composite is suggested to suppress the charge injection and hinder the charge transport dynamics, due to the formation of deep traps originated from long-time charge capture. As a result, micro/nano-composites with perpendicular filler orientation have higher DC breakdown strength as compared with the other two cases.

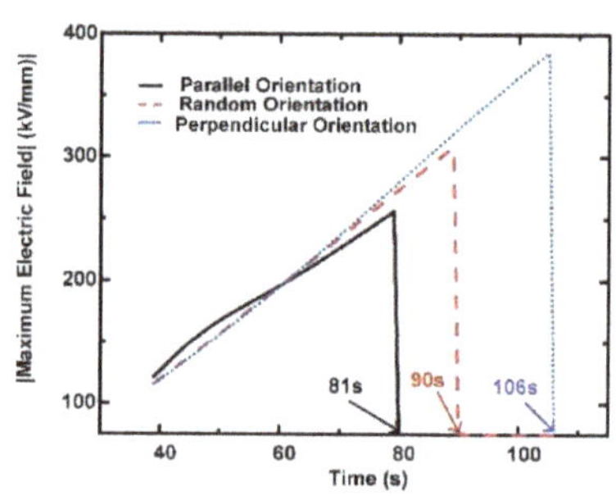

Figure 7. Comparisons of maximum electric field in the polymer with various orientations as a function of time.

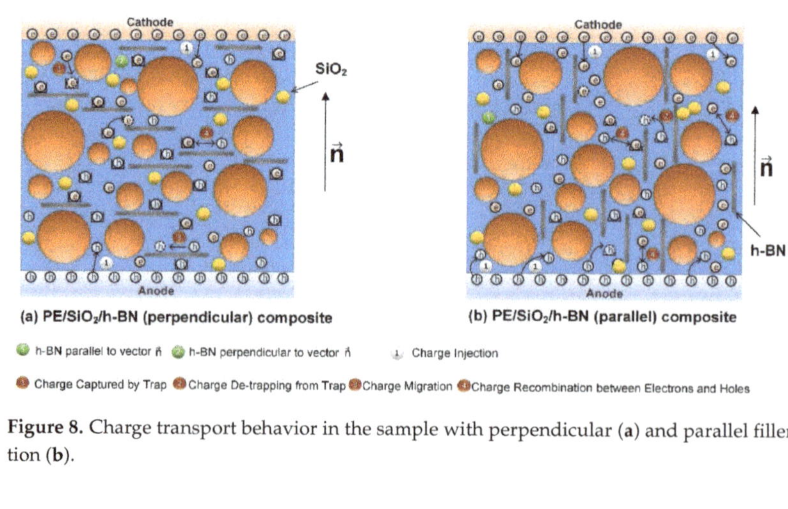

Figure 8. Charge transport behavior in the sample with perpendicular (**a**) and parallel filler orientation (**b**).

5. Conclusions

The charge transport and DC breakdown behaviors of polyethylene-based micro/nano-composites with various orientations of the h-BN filler have been investigated by numerical simulation based on the BCT model in this paper. The main conclusions may be summarized as follows:

(1) The density of accumulated space charge is the highest in the sample with parallel h-BN orientation, the material having filler distributed randomly follows and the lowest charge density appears as the h-BN is perpendicular to the normal vector of the composite surface.
(2) The electric field in the sample is distorted due to the space charge accumulation. The factor of electric field distortion is the smallest when the h-BN has a parallel orientation, whereas the distortion of electric field is most serious in the sample with the perpendicular orientation of h-BN.
(3) The breakdown of the sample most easily occurs when the h-BN has the parallel orientation, whereas the sample that has the highest breakdown strength as the h-BN with a perpendicular orientation.

Author Contributions: Conceptualization, X.X. and Y.G.; methodology, X.X. and Y.G.; software, X.X.; data curation, J.L.; formal analysis, X.X. and Z.S.; investigation, H.Z. and T.H.; writing—original draft preparation, X.X.; writing—review and editing, Y.G.; project administration, Y.G.; funding acquisition, Y.G. All authors have read and agreed to the published version of the manuscript.

Funding: This work is financially supported by State Grid Tianjin Electric Power Company (5500-202099277A-0-0-00).

Institutional Review Board Statement: Not applicable.

Informed Consent Statement: Not applicable.

Data Availability Statement: Not applicable.

Conflicts of Interest: The authors declare no conflict of interest.

References

1. Chen, G.; Hao, M.; Xu, Z.; Vaughan, A.; Cao, J.; Wang, H. Review of high voltage direct current cables. *CSEE J. Power Energy Syst.* **2015**, *1*, 9–21. [CrossRef]
2. Zhou, Y.; Peng, S.; Hu, J.; He, J. Polymeric insulation materials for HVDC cables: Development, challenges and future perspective. *IEEE Trans. Dielectr. Electr. Insul.* **2017**, *24*, 1308–1318. [CrossRef]
3. Hanley, T.L.; Burford, R.P.; Fleming, R.J.; Barber, K.W. A general review of polymeric insulation for use in HVDC cables. *IEEE Electr. Insul. Mag.* **2003**, *19*, 13–24. [CrossRef]
4. Huang, X.; Jiang, P.; Tanaka, T. A review of dielectric polymer composites with high thermal conductivity. *IEEE Electr. Insul. Mag.* **2011**, *27*, 8–16. [CrossRef]
5. Huang, X.; Zhi, C.; Jiang, P.; Colberg, D.; Bando, Y.; Tanaka, T. Polyhedral oligosilsesquioxane-modified boron nitride nanotube based epoxy nanocomposites: An ideal dielectric material with high thermal conductivity. *Adv. Funct. Mater.* **2013**, *23*, 1824–1831. [CrossRef]
6. Zhou, W.; Zuo, J.; Zhou, A.; Zhang, X. Thermal, electrical, and mechanical properties of hexagonal boron nitride–reinforced epoxy composites. *J. Compos. Mater.* **2013**, *48*, 2517–2526. [CrossRef]
7. Zeng, X.; Yu, S.; Sun, R. Thermal behavior and dielectric property analysis of boron nitride-filled bismaleimide-triazine resin composites. *J. Appl. Polym. Sci.* **2013**, *128*, 1353–1359. [CrossRef]
8. Wang, Z.; Iizuka, T.; Kozako, M.; Ohki, Y.; Tanaka, T. Development of epoxy/bn composites with high thermal conductivity and sufficient dielectric breakdown strength part I- sample preparations and thermal conductivity. *IEEE Trans. Dielectr. Electr. Insul.* **2012**, *18*, 1963–1972. [CrossRef]
9. Sato, K.; Horibe, H.; Shirai, T.; Hotta, Y.; Nakano, H.; Nagai, H. Thermally conductive composite films of hexagonal boron nitride and polyimide with affinity-enhanced interfaces. *J. Mater. Chem.* **2010**, *20*, 2749–2752. [CrossRef]
10. Gaska, K.; Kadar, R.; Rybak, A.; Siwek, A.; Gubanski, S. Gas barrier, thermal, mechanical and rheological properties of highly aligned graphene-ldpe nanocomposites. *Polymers* **2017**, *9*, 294.
11. Tanimoto, M.; Yamagata, T.; Miyata, K.; Ando, S. Anisotropic Thermal Diffusivity of Hexagonal Boron Nitride-Filled Polyimide Films: Effects of Filler Particle Size, Aggregation, Orientation, and Polymer Chain Rigidity. *ACS Appl. Mater. Interfaces* **2013**, *5*, 4374–4382. [CrossRef] [PubMed]

12. Tsekmes, I.A.; Morshuis PH, F.; Smit, J.J.; Kochetov, R. Enhancing the thermal and electrical performance of epoxy microcomposites with the addition of nanofillers. *IEEE Electr. Insul. Mag.* **2015**, *31*, 32–42. [CrossRef]
13. Donnay, M.; Tzavalas, S.; Logakis, E. Boron nitride filled epoxy with improved thermal conductivity and dielectric breakdown strength. *Compos. Sci. Technol.* **2015**, *110*, 152–158. [CrossRef]
14. Wang, Z.; Iizuka, T.; Kozako, M.; Ohki, Y.; Tanaka, T. Development of epoxy/bn composites with high thermal conductivity and sufficient dielectric breakdown strength partII- breakdown strength. *IEEE Trans. Dielectr. Electr. Insul.* **2011**, *18*, 1973–1983. [CrossRef]
15. Pourrahimi, A.M.; Olsson, R.T.; Hedenqvist, M.S. Nanocomposites: The role of interfaces in polyethylene/metal-oxide nanocomposites for ultrahigh-voltage insulating materials. *Adv. Mater.* **2018**, *30*, 1703624. [CrossRef]
16. Pourrahimi, A.M.; Kumara, S.; Palmieri, F.; Yu, L.; Lund, A.; Hammarstr, T.; Hagstrand, P.; Scheblykin, L.; Fabiani, D.; Xu, X.; et al. Repurposing Poly(3-hexylthiophene) as a Conductivity-Reducing Additive for Polyethylene-Based High-Voltage Insulation. *Adv. Mater.* **2021**, *33*, 2100714. [CrossRef]
17. Gaska, K.; Xu, X.; Gubanski, S.; Kadar, R. Electrical, mechanical, and thermal properties of ldpe graphene nanoplatelets composites produced by means of melt extrusion process. *Polymers* **2017**, *9*, 11.
18. Koerner, H.; Jacobs, J.; Tomlin, D.; Busbee, J.; Vaia, R. Tuning polymer nanocomposite morphology: AC electric field manipulation of epoxy–montmorillonite (clay) suspensions. *Adv. Mater.* **2004**, *16*, 297–302. [CrossRef]
19. Yan, H.; Tang, Y.; Long, Y.; Li, Y. Enhanced thermal conductivity in polymer composites with aligned graphene nanosheets. *J. Mater. Sci.* **2004**, *49*, 5256–5264. [CrossRef]
20. Agarwal, S.; Khan, M.; Gupta, R. Thermal conductivity of polymer nanocomposites made with carbon nanofibers. *Polym. Eng. Sci.* **2008**, *48*, 2474–2481. [CrossRef]
21. Shen, H.; Guo, J.; Wang, H.; Zhao, N.; Xu, J. Bioinspired modification of h-BN for high thermal conductive composite films with aligned structure. *ACS Appl. Mater. Interfaces* **2016**, *7*, 5701–5708. [CrossRef]
22. Nan, W.; Su, J.; Wang, Y.; Chen, D. Fabrication of Ni@SiC composite nanofibers by electrospinning and autocatalytic electroless plating techniques. *Results Phys.* **2018**, *12*, 853–858.
23. Xu, X.; Gao, Y.; Li, J.; Song, Z.; Peng, J.; Du, B. Charge Transport and DC Breakdown in Polymer Based Micro/Nano-Composite with Different Filler Orientations. In Proceedings of the 2020 IEEE International Conference on High Voltage Engineering and Application (ICHVE), Beijing, China, 6–10 September 2020; pp. 1–4.
24. Min, D.; Li, S. Simulation on the influence of bipolar charge injection and trapping on surface potential decay of polyethylene. *IEEE Trans. Dielectr. Electr. Insul.* **2014**, *21*, 1627–1636. [CrossRef]
25. Teyssedre, G.; Laurent, C. Charge transport modeling in insulating polymers: From molecular to macroscopic scale. *IEEE Trans. Dielectr. Electr. Insul.* **2005**, *12*, 857–875. [CrossRef]
26. Laurent, C.; Teyssedre, G.; Roy, S.; Baudoin, F. Charge dynamics and its energetic features in polymeric materials. *IEEE Trans. Dielectr. Electr. Insul.* **2013**, *20*, 357–381. [CrossRef]
27. Li, S.; Zhu, Y.; Min, D.; Chen, G. Space charge modulated electrical breakdown. *Sci. Rep.* **2016**, *6*, 32588. [CrossRef] [PubMed]
28. Roy, S.; Teyssedre, G.; Laurent, C.; Montanari, G.; Palmieri, F. Description of charge transport in polyethylene using a fluid model with a constant mobility: Fitting model and experiments. *J. Phys. D Appl. Phys.* **2006**, *39*, 1427. [CrossRef]
29. Chen, G.; Zhao, J.; Li, S.; Zhong, L. Origin of thickness dependent DC electrical breakdown in dielectrics. *Appl. Phys. Lett.* **2012**, *100*, 2135. [CrossRef]
30. Li, S.; Min, D.; Wang, W.; Chen, G. Modelling of dielectric breakdown through charge dynamics for polymer nanocomposites. *IEEE Trans. Dielectr. Electr. Insul.* **2016**, *23*, 3476–3485. [CrossRef]
31. Hoang, A.T.; Serdyuk, Y.V.; Gubanski, S.M. Charge transport in LDPE nanocomposites Part II—Computational approach. *Polymers* **2016**, *8*, 103.
32. Kuik, M.; Koster, L.; Dijkstra, A.; Wetzelaer, G.; Blom, P. Non-radiative recombination losses in polymer light-emitting diodes. *Org. Electron.* **2012**, *13*, 969–974. [CrossRef]
33. Shockley, W.; Read, W. Statistics of the recombinations of holes and electrons. *Phys. Rev.* **1952**, *87*, 835–842. [CrossRef]
34. Gao, Y.; Xu, B.; Wang, X.; Jia, T. Charge transport in low density polyethylene based micro/nano-composite with improved thermal conductivity. *J. Phys. D Appl. Phys.* **2019**, *52*, 285302. [CrossRef]
35. Min, D.; Li, Y.; Yan, C.; Xie, D.; Li, S.; Wu, Q.; Zhao, Z. Thickness-Dependent DC electrical breakdown of polyimide modulated by charge transport and molecular displacement. *Polymers* **2018**, *10*, 1012. [CrossRef]
36. Min, D.; Cui, H.; Wang, W.; Wu, Q.; Zhao, Z.; Li, S. The coupling effect of interfacial traps and molecular motion on the electrical breakdown in polyethylene nanocomposites. *Compos. Sci. Technol.* **2019**, *184*, 107873. [CrossRef]
37. Xie, D.; Min, D.; Yang, L.; Li, S. Temperature- and thickness-dependent electrical breakdown modulated by a coupling model of charge transport and molecular chain dynamics. *J. Phys. D Appl. Phys.* **2019**, *52*, 365305. [CrossRef]
38. Min, D.; Li, S.; Ohki, Y. Numerical simulation on molecular displacement and dc breakdown of LDPE. *IEEE Trans. Dielectr. Electr. Insul.* **2016**, *23*, 507–516. [CrossRef]
39. Bhutta, M.; Yang, L.; Ma, Z.; Nazir, M.; Akram, S.; Mehmood, M.; Faiz1, N. Simulation of thickness controlled dc breakdown of XLPE regulated by space charge & molecular chain movement. *IEEE Trans. Dielectr. Electr. Insul.* **2020**, *27*, 1143–1151.
40. Frenkel, J. On pre-breakdown phenomena in insulators and electronic semi-conductors. *Phys. Rev.* **1938**, *54*, 647–648. [CrossRef]

41. Wu, K.; Dissado, L.A.; Okamoto, T. Percolation model for electrical breakdown in insulating polymers. *Appl. Phys. Lett.* **2004**, *85*, 4454–4456. [CrossRef]
42. Wang, W.; Li, S. Correlation between trap parameters and breakdown strength of polyethylene/alumina nanocomposites. In Proceedings of the 2014 International Symposium on Electrical Insulating Materials (ISEIM), Niigata, Japan, 1–5 June 2014; pp. 73–76.
43. Tanaka, T.; Wang, Z.; Iizuka, T.; Kozako, M.; Ohki, Y. High thermal conductivity epoxy/BN composites with sufficient dielectric breakdown strength. In Proceedings of the 2011 Annual Report Conference on Electrical Insulation and Dielectric Phenomena (CEIDP), Cancun, Mexico, 16–19 October 2011; pp. 691–694.
44. Wang, W.; Li, S.; Min, D. Understanding the conduction and breakdown properties of polyethylene nanodielectrics: Effect of deep traps. *IEEE Trans. Dielectr. Electr. Insul.* **2016**, *23*, 564–572. [CrossRef]

energies

MDPI

Article

Autonomous Analysis of Infrared Images for Condition Diagnosis of HV Cable Accessories †

Lixiao Mu [1], Xiaobing Xu [2], Zhanran Xia [1], Bin Yang [1], Haoran Guo [1], Wenjun Zhou [2] and Chengke Zhou [2,3,*]

1 State Grid Hubei Electric Power Company, Wuhan Power Supply Company, Wuhan 430072, China; mulx@hb.sgcc.com.cn (L.M.); xiazr2@hb.sgcc.com.cn (Z.X.); Yangb59@hb.sgcc.com.cn (B.Y.); pengh62@hb.sgcc.com.cn (H.G.)
2 School of Electrical Engineering and Automation, Wuhan University, Wuhan 430072, China; xuxiaobing@whu.edu.cn (X.X.); wjzhou@whu.edu.cn (W.Z.)
3 School of Computing, Engineering and Built Environment, Glasgow Caledonian University, Glasgow G4 0BA, UK
* Correspondence: c.zhou@gcu.ac.uk
† This paper is an extended version of our paper published in 2020 IEEE International Conference on High Voltage Engineering and Application (ICHVE), Beijing, China, 6–10 September 2020.

Abstract: Infrared thermography has been used as a key means for the identification of overheating defects in power cable accessories. At present, analysis of thermal imaging pictures relies on human visual inspections, which is time-consuming and laborious and requires engineering expertise. In order to realize intelligent, autonomous recognition of infrared images taken from electrical equipment, previous studies reported preliminary work in preprocessing of infrared images and in the extraction of key feature parameters, which were then used to train neural networks. However, the key features required manual selection, and previous reports showed no practical implementations. In this contribution, an autonomous diagnosis method, which is based on the Faster RCNN network and the Mean-Shift algorithm, is proposed. Firstly, the Faster RCNN network is trained to implement the autonomous identification and positioning of the objects to be diagnosed in the infrared images. Then, the Mean-Shift algorithm is used for image segmentation to extract the area of overheating. Next, the parameters determining the temperature of the overheating parts of cable accessories are calculated, based on which the diagnosis are then made by following the relevant cable condition assessment criteria. Case studies are carried out in the paper, and results show that the cable accessories and their overheating regions can be located and assessed at different camera angles and under various background conditions via the autonomous processing and diagnosis methods proposed in the paper.

Keywords: cable accessories; infrared image processing; Faster RCNN; Mean-Shift algorithm; smart condition diagnosis

Citation: Mu, L.; Xu, X.; Xia, Z.; Yang, B.; Guo, H.; Zhou, W.; Zhou, C. Autonomous Analysis of Infrared Images for Condition Diagnosis of HV Cable Accessories. *Energies* **2021**, *14*, 4316. https://doi.org/10.3390/en14144316

Academic Editors: Issouf Fofana and Bo Zhang

Received: 24 June 2021
Accepted: 14 July 2021
Published: 17 July 2021

1. Introduction

Power cables have been widely used in urban power systems, and their safe operation is key to the reliability of the power grid [1]. Cable accessories, which are used to connect cables with other electrical equipment or different sections of cables, are the weak links of cable systems [2]. Reference [3,4] indicated that the manufacturing of fault-free cable accessories is almost impossible, and poor workmanship during installations and design defects may result in cable faults. Previous studies showed that when there exist defects in cable accessories, such as poor contact of metal connectors, misalignment of stress cones, damage of insulation layer or impurities, and bubbles in the internal medium, the temperature of defective cable accessories was usually higher than those fault free cable accessories [5]. Because the use of infrared thermography to identify abnormal temperatures has many advantages, such as no physical contact, non-intrusive, high efficiency, and so on, it is widely used in inspection and maintenance activities of power cable circuits [6]. At present, the processing and analysis of infrared images taken during inspections mainly require visual inspection. This is time-consuming and laborious on the one hand, and on the other hand, it relies too much on expert experience and is prone to erroneous diagnosis. Therefore, the realization of autonomous condition diagnosis in cable accessories would greatly benefit the practitioners involved in cable maintenance and inspections.

In previously published researches, the image features related to the temperature gradients of equipment were used as the input of neural networks for the autonomous diagnosis of electrical equipment. In order to analyze the temperature-related information, Rahmani et al. used the Zernike moment as an image feature of fuse bases [7]; Huda et al. extracted the first-order histogram and gray level co-occurrence matrix of infrared images captured from main switchboards [8–10]; Jaffery et al. extracted the RGB color moment of images of fuse cabinets [11]. In the above studies, the key features needed to be selected manually, which was, in fact, a heuristic process. The quality of the selected features was largely dependent on human expertise.

Thus far, there have been few studies on the recognition of infrared images for condition diagnosis of cable accessories. Previously reported research came from the same research team where a number of techniques were investigated for infrared image processing of cable terminations, including the adaptive denoising method based on layer-by-layer optimal basic wavelet and Bayesian estimation [12], the denoising method based on interscale correlation of the wavelet coefficients and the bivariate shrinkage function [13], the improved hybrid Fourier-wavelet denoising method [14] and the identification method based on the Radon transform and the Fourier–Mellin transform [15]. Although the above methods were proven to be effective in preliminary image processing, the published work failed to carry out condition diagnosis of real-world infrared images.

This paper proposes a method for autonomous diagnosis of overheating defects in cable accessories based on a Faster RCNN network and Mean-Shift algorithm. Firstly, the collected infrared images of cable accessories during routine inspection activities are used as samples to complete the training of the Faster RCNN network thus as to identify and locate the objects to be diagnosed. Then, the Mean-Shift clustering algorithm is used to segment the images. This helps to extract the overheating area quickly and accurately. Finally, the temperature characteristic parameters are calculated, thus the condition of cable accessories can be diagnosed according to pre-set diagnostic criteria.

2. Object Localization Based on Faster RCNN Network

Cable accessories of interest in this paper include cable terminations and cable grounding boxes. For more information, readers may refer to [16,17]. For the FLIR T630 thermal imager, a handheld camera, which has been applied in work presented in this paper, does not have a specified minimum and maximum shooting distance. When we applied it, the distance between the camera and the target was usually between 1 and 10 m. It was around 1 m when pictures were taken of cable joints and was 3–10 m for cable terminations. As a result, the infrared images under analysis may contain the targeted cable accessories and may also contain other undesired background objects.

The Faster RCNN network, which can identify and locate the desired objects contained in given images, is one of the most advanced algorithms for target detection. Figure 1 shows the flowchart of the Faster RCNN network. Firstly, in order to produce the feature map, it extracts the features of the detected image through a Convolutional Neural Network (CNN). Then, the Region Proposal Network (RPN) is used to propose the possible regions. Next, on the basis of the feature map and proposed regions, the Region of Interest (RoI) pooling layer is applied to extract the proposal feature maps, which are sent to the subsequent network. Finally, the autonomous recognition and positioning of the objects are realized through the object detection layer [18].

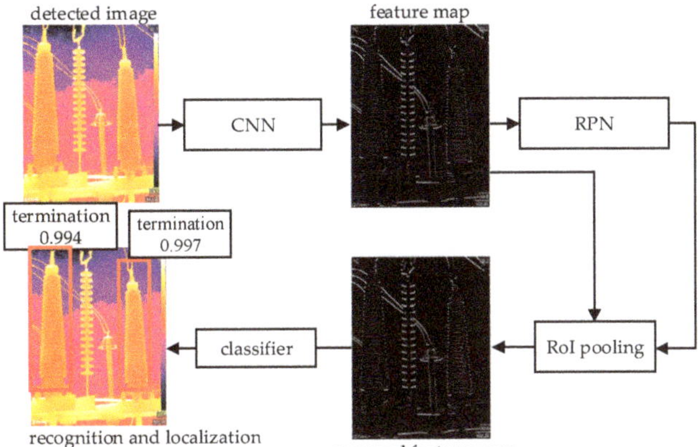

Figure 1. Model of Faster RCNN network.

2.1. *Sturcture of Faster RCNN Network*

2.1.1. Convolutional Neural Network

Convolutional Neural Network is used to extract feature maps of input images, and the feature maps are shared with the subsequent RPN network and RoI pooling layer. In this paper, the VGG16 Convolutional Neural Network, which includes 13 Convolutional (Conv) layers, 13 Rectified Linear Units (ReLU) layers, and 4 Pooling layers, shown in Figure 2, is adopted.

Figure 2. Structure of VGG16 convolutional neural network.

Convolutional layers are applied to detect features of images. It is composed of several convolutional kernels (equivalent to weight matrix k) and additive bias b. Each convolutional kernel can be regarded as a kind of feature detector, which filters the whole image by sliding on the image to capture the corresponding features. From the mathematical point of view, the corresponding features are obtained through the convolutional operation. Suppose the mth layer is a Convolutional layer, then its output vector is as follows:

$$\boldsymbol{x}_j^m = \sum_{\boldsymbol{x}_i \in \boldsymbol{M}_j} \boldsymbol{x}_i^{m-1} * \boldsymbol{k}_{ij}^m + \boldsymbol{b}_j^m \quad (1)$$

where $\boldsymbol{x}_j^m$ is the jth output of this layer; $\boldsymbol{M}_j$ is the set of input vectors; $\boldsymbol{k}_{ij}^m$ is a convolutional kernel; $\boldsymbol{b}_j^m$ is the additive bias, and * represents the convolutional operation.

The size of the output image is as follows:

$$s_{\text{output}} = \frac{s_{\text{input}} - s_{\text{kernel}} + 2 * p}{d} + 1 \quad (2)$$

where s_{output} represents the size of the output image; s_{input} represents the size of the input image; s_{kernel} represents the size of the convolutional kernel; p represents the number of pixels to be filled, and d represents the step length of the convolutional kernel sliding on the image.

After each convolutional layer, the ReLU layers are applied to enhance the nonlinear characteristics, and the ReLU function is given in (3). Between the Convolutional layers, the Pooling layers are inserted periodically to reduce the dimensionality of features.

$$f(x) = max(0, x) \tag{3}$$

where $f(x)$ stands for the ReLU function, and x represents the characteristic parameters of the output of the convolutional layer.

2.1.2. Region Proposal Network

The Region Proposal Network is used to complete the preliminary positioning of the objects. As is shown in Figure 3, the Region Proposal Network firstly generates a set of rectangular bounding boxes in the detected image. These regions are represented by four-dimensional vectors (x, y, w, h), where x and y denote the region's center coordinates, while w and h denote the width and height. Then, the classification layer is applied to obtain the object score of each proposed region, based on which the Softmax classifier is used to identify the regions that include the diagnostic objects by calculating the probability, using the formula given in (4). On the other hand, to make the positioning more accurate, the regression layer is applied to realize the bounding box regression, based on the formulae given in (5) and (6).

$$P = \frac{e^{z_1}}{e^{z_1} + e^{z_2}} \tag{4}$$

where P represents the probability that the proposed region contains diagnostic objects; z_1 and z_2 represent the foreground score and background score of proposed regions, respectively.

$$\begin{cases} G'_x = A_w d_x(A) + A_x \\ G'_y = A_h d_y(A) + A_y \end{cases} \tag{5}$$

$$\begin{cases} G'_w = A_w \exp(d_w(A)) \\ G'_h = A_h \exp(d_h(A)) \end{cases} \tag{6}$$

where (A_x, A_y, A_w, A_h) denote the unadjusted coordinates of the bounding box. (G'_x, G'_y, G'_w, G'_h) denote the adjusted coordinates. $d_x(A)$ and $d_y(A)$ denote the translation parameters, $d_w(A)$ and $d_h(A)$ denote the scale parameters.

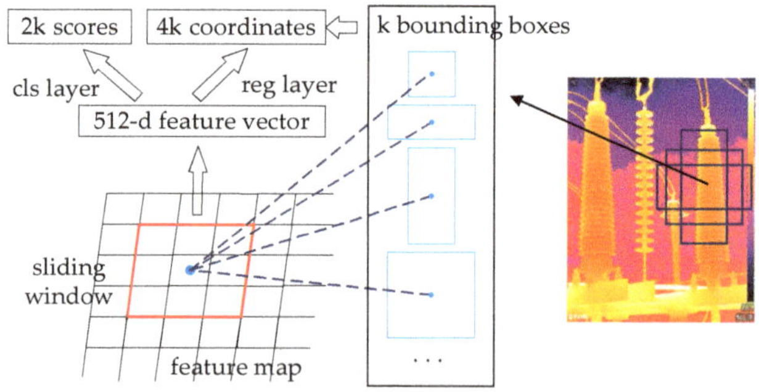

Figure 3. Structure of RPN network.

2.1.3. Region of Interest Pooling Layer

The RoI Pooling layer can obtain the fixed-length feature vectors by analyzing input data of different sizes. The operating principle is shown in Figure 4. Assuming that the size of input feature map is 8 × 8, and the box calibration region is the proposed feature.

According to the transformation factors p_w and p_h, the proposed feature is divided into $p_w \times p_h$ blocks ($p_w = p_h = 2$ in this example). Then the maximum pooling is applied for each block in order to the maximum value of each block.

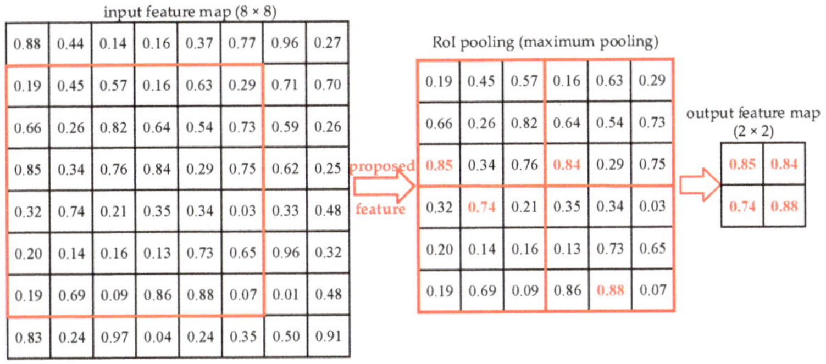

Figure 4. Operating principle of RoI pooling layer.

In the Faster RCNN network, the pooling process of the RoI Pooling layer is shown in Figure 5. Firstly, the regions proposed by the RPN are mapped to the feature map obtained by the CNN, to allow the proposal feature maps of different sizes to be extracted. Next, according to p_w and p_h ($p_w = p_h = 7$ in this paper), maximum pooling is applied to convert the proposal features into feature maps with the fixed spatial extent of 7×7. Finally, the fixed-length proposal feature maps are sent to the subsequent network.

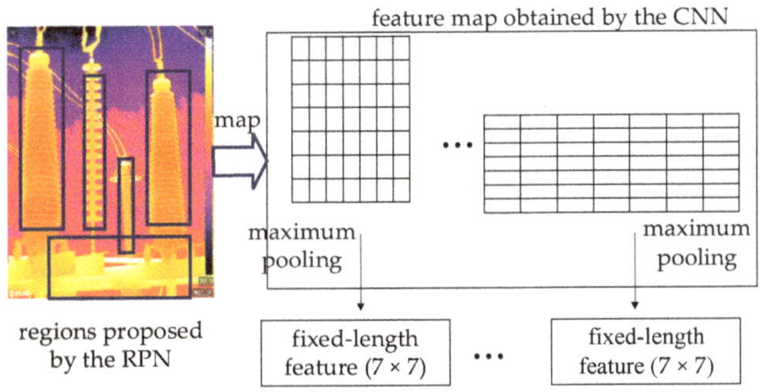

Figure 5. Pooling process of RoI pooling layer.

2.1.4. Object Detection Layer

As is shown in Figure 6, based on the proposal feature maps, the Softmax classifier is used to achieve object identification (objects to be identified include cable terminations and grounding boxes in this paper). On the other hand, the regression layer is applied to complete the second bounding box regression, which makes the localization more precise.

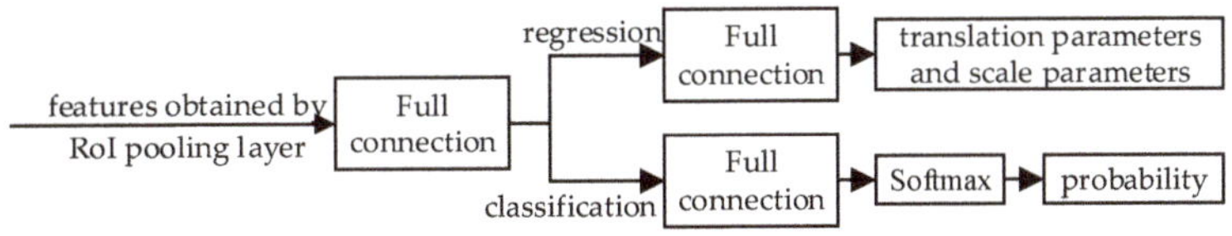

Figure 6. Operating principle of object detection layer.

2.2. Autonomous Detection Results of Faster RCNN Network

The infrared images captured by the infrared thermal imagers (FLIR T630) were used as samples to train the Faster RCNN network. Table 1 shows the specifications of the FLIR T630 handheld thermal cameras. When the infrared images of power cable accessories were taken, the emissivity coefficients of imagers were set to 0.9.

Table 1. Specifications of the FLIR T630 handheld thermal cameras.

Specifications	
Detector resolution	640 × 480
Accuracy	±2 °C or ±2% of reading
Thermal sensitivity	<30 mK @ 30 °C
Operating temperature range	−15 °C to 50 °C
Operating humidity	95% relative humidity
Object temperature range	−40 °C to 650 °C
Atmosphere transmission correction	automatic
External optics and widows correction	automatic
Optics transmission correction	automatic
Reflected apparent temperature correction	automatic

The abnormal heating phenomenon usually occurs at connection fittings and sleeves of the cable terminations or the connections of the grounding boxes. Therefore, when constructing the training samples, if the diagnostic objects are the terminations, the labeled target should be the connection fittings and sleeves. If the objects under analysis are grounding boxes, the marked target should be the connections. After completing the training, the Faster RCNN network can realize the autonomous identification and positioning of the cable terminations and grounding boxes in images.

Taking Figures 7 and 8 as examples, where Figures 7a and 8a, respectively, show the original infrared images of grounding box and cable termination taken during routine inspection activities. Figures 7b and 8b show the recognition and positioning results of the trained Faster RCNN network. To eliminate the influence of the interference information, the image contents inside the proposal regions were kept, while other contents were eliminated by setting the pixels' components of red (R), green (G), and blue (B) to zero. The extracted connections of the grounding box are shown in Figure 7c. The extracted connection fitting is shown in Figure 8c, and the extracted sleeve in Figure 8d. The results show that based on the Faster RCNN network, the diagnostic objects were accurately extracted, and the interference of complex background and foreground, which my hamper subsequent image processing, were eliminated.

(a) (b) (c)

Figure 7. Result of eliminating the interference information in the image of the grounding box: (**a**) original infrared image; (**b**) detecting result of Faster RCNN network; (**c**) extracting results of connections.

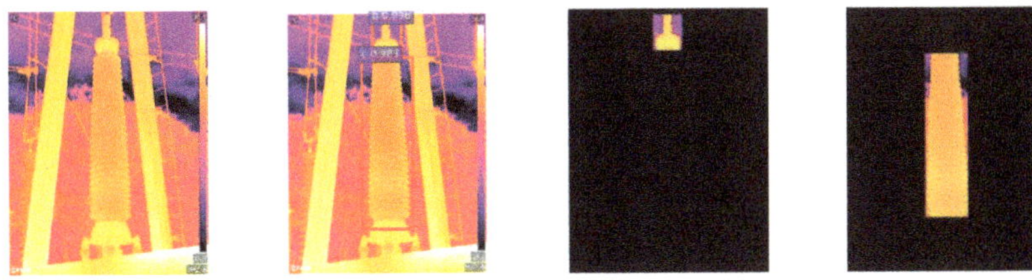

(**a**) (**b**) (**c**) (**d**)

Figure 8. Results of eliminating the interference information in the image of termination: (**a**) original infrared image; (**b**) detecting results of Faster RCNN network; (**c**) extracting result of the connection fitting; (**d**) extracting result of the sleeve.

3. Extraction of Suspected Abnormal Heating Regions Based on Mean-Shift Algorithm

After extracting the diagnostic objects, it is necessary to extract the suspected abnormal heating regions among the objects. In subsequent processing, their temperature distribution is the basis for condition diagnosis.

The gray information of the infrared images reflects the temperature distribution. The greater the gray value of the pixel, the higher the corresponding temperature. Therefore, the gray value is to be extracted from the infrared images. The graying formula is as follows [19]:

$$I = 0.299R + 0.587G + 0.114B \tag{7}$$

where, I is the gray value of the pixel; R is the red component; G is the green component, and B is the blue component.

The Mean-Shift algorithm, which has been widely used in clustering, is essentially an iterative search algorithm [20]. In this paper, the gray values of pixels were used as the data samples, and the Mean-Shift algorithm was applied to cluster the pixels. The clustering process is shown in Figure 9. Firstly, a pixel was randomly selected to be the clustering center, and other pixels, of which the grayscale difference with the center less than the bandwidth, were placed in the same class. The bandwidth h_r realized adaptive selection based on the asymptotic mean integrated square error (AMISE) is as shown in (8) and (9) [21]. Then, the Mean-Shift vector was calculated, and the original center moved the vector to obtain the new center [22]. The clustering center was updated until the convergence condition was satisfied. The above steps were repeated until all the pixels were traversed.

$$h_r = \left(\frac{4}{d+2}\right)^{1/(d+4)} n^{-1/(d+4)}\sigma \tag{8}$$

$$\sigma = \sqrt{\frac{1}{n-1}\sum_{i=1}^{n}(x_i - \overline{x})^2} \tag{9}$$

where, h_r represents the bandwidth; d the dimension of the feature space; n the number of samples; σ the standard deviation; x the sample, and $\overline{x}$ represents the average value of the sample data.

In order to describe the Mean-Shift clustering process more intuitively, this paper takes the cable grounding box as an example, as is shown in Figure 10. Figure 10a shows the result of autonomous positioning and identification of the diagnostic objects (connections) in infrared images by trained Faster RCNN network. Figure 10b shows the corresponding grayscale image. Figure 10c shows the three-dimensional visualization result of the pixels' gray information, where each scatter point corresponds to a pixel in the image, (x, y) represents the position information of the corresponding pixel in the original image, and z represents the gray value of the pixel. Figure 10d shows the result of clustering by the

Mean-Shift algorithm. In order to reflect the clustering result more intuitively, the pixels belonging to the same category were marked with the same color.

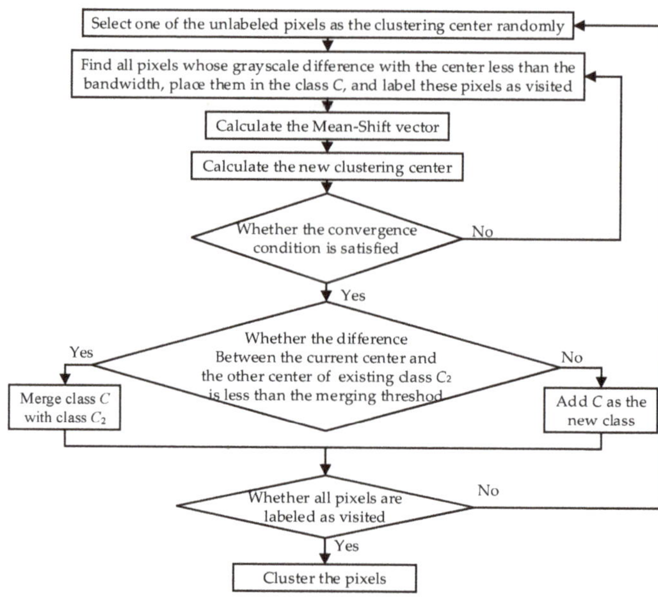

Figure 9. Clustering process of Mean-Shift algorithm.

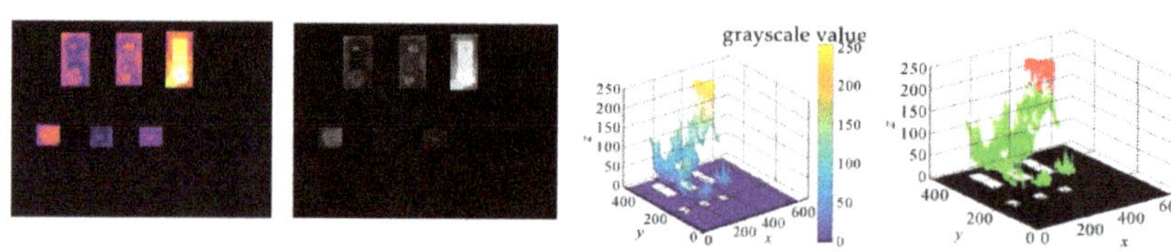

Figure 10. Clustering result of Mean-Shift algorithm: (**a**) detecting results of Faster RCNN network; (**b**) grayscale image; (**c**) three-dimensional visualization result of the pixel gray information; (**d**) clustering result.

As can be seen in Figure 10d, the pixels in the example were adaptively divided into three categories. The pixels marked in black correspond to the background area in the original image. The pixels marked in green correspond to the regions under normal conditions, and the pixels marked in red correspond to the suspected overheating regions. Therefore, after clustering based on the Mean-Shift algorithm, this paper retained the category with the center, where the gray value was the greatest, as the suspected overheating area.

Taking the infrared images of the HV cable accessories captured during an inspection as the testing objects. In order to achieve the best results, after realizing the autonomous identification and positioning of the diagnostic objects, this paper, respectively, applied the Maximum Between-Class Variance (OTSU) algorithm, the K-Means algorithm, and the Mean-Shift algorithm to segment the images. The OTSU algorithm can adaptively calculate the segmentation threshold, and the pixels of which the gray values are lower than the threshold were categorized into the background, while other pixels, of which the gray values are higher than the threshold, were categorized into the foreground [23]. The K-Means algorithm randomly selected k pixels as clustering centers according to the given k, and then classified the remaining pixels to the most similar center before it updated the clustering centers to the mean value of each category. The above steps were repeated until the convergence condition is satisfied [24,25].

Based on the clustering results, the suspected abnormal heating regions were extracted, as is shown in Figure 11. Figure 11a shows the original infrared images of grounding box I, grounding box II, termination I, and termination II. Figure 11b shows the detecting results of diagnostic objects by the Faster RCNN network. Figure 11c, Figure 11d,e show the extracting results of suspected abnormal heating regions by different clustering algorithms.

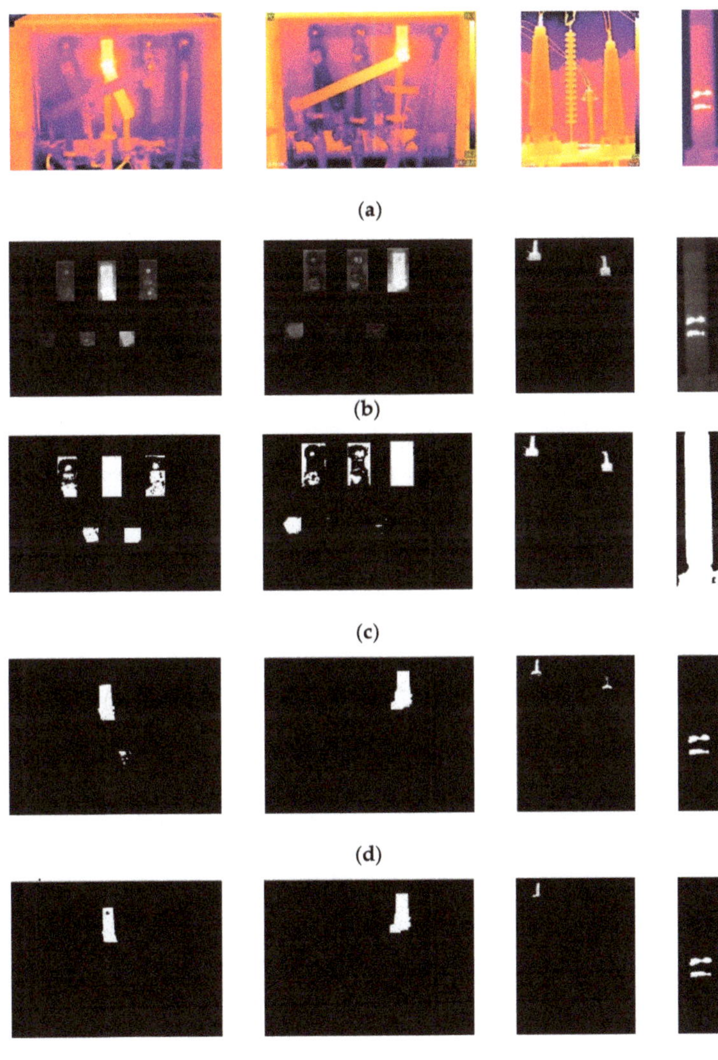

Figure 11. Results of extracting the suspected abnormal heating regions: (**a**) original infrared images, (**b**) detecting results of diagnostic objects by Faster RCNN network, (**c**) extracting results of the OTSU algorithm, (**d**) extracting results of the K-Means algorithm, (**e**) extracting results of the Mean-Shift algorithm.

As shown in Figure 11, the adaptive segmentation threshold calculated by the OTSU algorithm was often too low to distinguish the abnormal heating regions from the area under normal conditions. Thus, in the final segmentation results, the abnormal heating area was almost submerged in the diagnostic objects. The K-Means algorithm can specify the number of categories manually, which solves the disadvantage of the OTSU algorithm that it can only achieve dichotomy. Therefore, compared with the OTSU algorithm, the clustering results by K-Means algorithm represented an improvement. However, for different images, the most suitable number of clustering categories is also different. Thus, some regions under normal conditions were still identified as abnormal heating regions wrongly by the K-Means algorithm. The Mean-Shift algorithm can adaptively select the most appropriate number of clustering categories according to the gray information of the image, which can overcome the disadvantage of the K-Means algorithm. The testing results showed that the Mean-Shift algorithm could extract the abnormal heating regions accurately, and its performance was better than the OTSU algorithm and the K-Means algorithm.

4. Positioning of Reference Regions

In order to accurately realize condition diagnosis after extracting the abnormal heating regions, the reference regions should be localized in the reference phases, which do not have abnormal heating regions (assuming that at least one phase is in normal status in this paper). This article applied different methods to locate the positions of reference regions for different HV cable accessories.

4.1. Grounding Boxes

If the diagnostic object was a grounding box, the Faster RCNN network would have directly positioned the connections (including the upper side connections and the lower side connections). The phase (one of the three phases), which contained the overheating area, was noted as the suspected overheating phase, and the remaining phases, which always had the same height, were regarded as the reference phases. The whole reference phases can be deemed as the reference regions.

Figure 12 shows the positioning results of reference regions of the grounding boxes. To describe the positioning process more intuitively, the four-dimensional vectors (x_1, y_1, x_2, y_2) were used to represent the position of each phase, where x_1 and x_2 denote the horizontal coordinates of top-left corner and bottom-right corner of each phase, while y_1 and y_2 denote the vertical coordinates. The position information of the suspected overheating phase of the grounding box I was (268, 65, 339, 184), while the position information of other phases was (148, 65, 219, 184), (390, 65, 461, 184), (100, 290, 142, 338), (214, 290, 256, 338), and (339, 290, 381, 338). Thus, the connections corresponding to the first two data sets were regarded as the reference regions because they had the same vertical coordinates with the overheating phase. The positioning method of reference regions of grounding box II was the same.

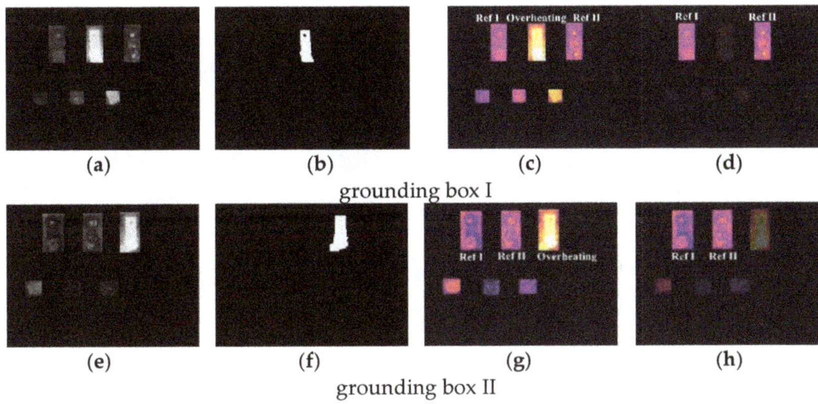

Figure 12. Reference regions locating results of grounding boxes: (**a**,**e**) grayscale image, (**b**,**f**) extracting results of suspected overheating regions, (**c**,**g**) suspected overheating phase and reference phases, (**d**,**h**) locating results of reference regions.

4.2. Cable Terminations

When the object to be diagnosed was a cable termination, the Faster RCNN network was firstly applied to locate the connection fittings and sleeves. The phase that included abnormal heating regions were regarded as the suspected abnormal heating phase, and the remaining phases were considered as the reference phases. Considering the similarity of the structure among the three phases of the termination, the reference regions can be positioned according to the abnormal heating area's position information and the size ratio of different phases of terminations identified in the image. Taking Figure 13 as an example, (10) and (11) were used to calculate the coordinate information of the reference regions' pixels. The highlighted areas in Figure 13d,h show the positioning results of reference regions.

$$\begin{cases} \mathbf{X}_{\text{hot}} = [x_1, x_2, x_3, \dots, x_n] \\ \mathbf{Y}_{\text{hot}} = [y_1, y_2, y_3, \dots, y_n] \end{cases} \tag{10}$$

$$\begin{cases} \mathbf{X}_{\text{ref}} = (\mathbf{X}_{\text{hot}} - x_{\text{min1}})/(x_{\text{max1}} - x_{\text{min1}}) * (x_{\text{max2}} - x_{\text{min2}}) + x_{\text{min2}} \\ \mathbf{Y}_{\text{ref}} = (\mathbf{Y}_{\text{hot}} - y_{\text{min1}})/(y_{\text{max1}} - y_{\text{min1}}) * (y_{\text{max2}} - y_{\text{min2}}) + y_{\text{min2}} \end{cases} \tag{11}$$

where $\mathbf{X}_{\text{hot}}$ and $\mathbf{Y}_{\text{hot}}$, respectively, denote the horizontal coordinates and vertical coordinates of the abnormal heating regions' pixels; $\mathbf{X}_{\text{ref}}$ and $\mathbf{Y}_{\text{ref}}$, respectively, denote the horizontal coordinates and vertical coordinates of the reference regions' pixels. n denotes the number of pixels in the abnormal heating regions; $(x_{\text{min1}}, x_{\text{max1}})$ and $(y_{\text{min1}}, y_{\text{max1}})$ the horizontal coordinates and vertical coordinates of the top-left corner and bottom-right corner of the suspected abnormal heating phase. $(x_{\text{min2}}, x_{\text{max2}})$ and $(y_{\text{min2}}, y_{\text{max2}})$ the vertical coordinates of the top-left corner and bottom-right corner of the reference phase.

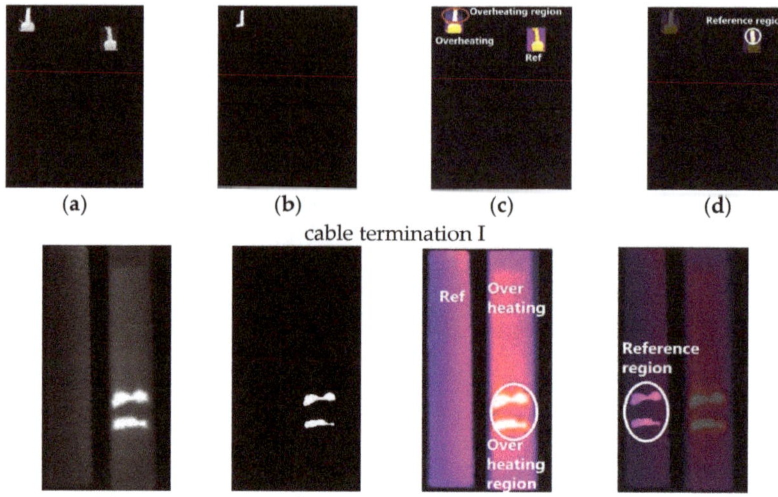

(a) (b) (c) (d)
cable termination I

(e) (f) (g) (h)
cable termination II

Figure 13. Location results of reference regions of cable terminations: (**a**,**e**) grayscale image, (**b**,**f**) extracting results of suspected abnormal heating regions, (**c**,**g**) suspected abnormal heating phase and reference phases, (**d**,**h**) locating results of reference regions.

5. Calculation of Temperature Parameters and Condition Diagnosis

In order to diagnose the severity of the overheating defect, T_1 and T_2, which denote the maximum temperature of the identified abnormal heating area and reference area, respectively, were extracted. Combined with the ambient temperature T_0, the temperature characteristic parameters can be calculated according to (12)~(14).

$$T_r = T_1 - T_0 \quad (12)$$

$$T_d = T_1 - T_2 \quad (13)$$

$$\delta = (T_1 - T_2)/(T_1 - T_0) \quad (14)$$

where T_r is the value of temperature rise; T_d is the value of temperature difference, and δ represents the value of relative temperature difference.

According to the corresponding diagnostic criteria, the autonomous diagnosis of overheating defects in cable accessories can be achieved based on the calculated temperature characteristic parameters. Table 2 shows the diagnostic criteria from the "Guidelines for defective grading standards of power transmission equipment of Guangdong Power Grid Company (version 2018)". Table 3 shows the calculated temperature parameters and diagnosis results of Figures 12 and 13. The temperature difference of grounding box I in Figure 12 was 6.4 °C. Because it was between 5 °C and 15 °C, the status of grounding box I was regarded as the general defect. The temperature difference of grounding box II was only 2.9 °C, which was lower than 5 °C. In addition, the temperature rise was only 4.6 °C, which was less than 15 °C, the condition of grounding box II was diagnosed to be normal. Similarly, the temperature difference of connection fittings of cable terminations I in Figure 13 was 50.6 °C, which was higher than 40 °C, thus the termination I was judged to have a defect needing urgent action. The temperature difference of sleeves of cable terminations II was 11.2 °C, thus the condition of this termination was determined as having a major defect.

Table 2. Diagnostic standard for HV cable accessories.

Defective Part	Defective Appearance	Severity Level
Connection fittings of cable terminations or connections of grounding boxes	Temperature of heating spot >130 °C; Temperature difference >40 °C; Relative temperature difference >95% and temperature rise >15 °C	Urgent
	Temperature of heating spot: 90~130 °C; Temperature difference: 15~40 °C; Relative temperature difference: 80%~95% and temperature rise >15 °C	Major
	Temperature difference: 5~15 °C; Relative temperature difference: 35%~80% and temperature rise >15 °C	General
sleeves of cable terminations	Temperature difference >4 °C	Major
	Temperature difference: 2~4 °C	General

Table 3. Temperature calculation and diagnostic results of cable accessories.

Cable Accessories	Temperature/°C Heating	Ref I	Ref II	Environment	T_r [1]/°C	Max T_d [2]/°C	Max σ [3]/%	Diagnostic Results
Grounding box I	31.4	25	26.4	23	8.4	6.4	76.2	General defect
Grounding box II	29.6	26.7	27.1	25	4.6	2.9	63	Normal
Termination I	86.3	35.7	\	32	54.3	50.6	93.18	Urgent defect
Termination II	47.3	36.1	\	22	25.3	11.2	44.27	Major defect

[1] T_r means the temperature rise, [2] max T_d means the maximum temperature difference, [3] max σ means the maximum relative temperature difference.

The proposed method has been applied to test against actual infrared images, including 50 images of cable terminations and 50 images of grounding boxes. The testing results are shown in Table 4.

Table 4. Testing results of inspection images.

Cable Accessories	Actual Condition		Autonomous Diagnosis Results	
Cable terminations	Normal	32	Normal	32
	General defect	2	General defect	2
	Major defect	7	Major defect	7
	Urgent defect	9	Urgent defect	9
Grounding boxes	Normal	40	Normal	39
	General defect	7	General defect	8
	Major defect	1	Major defect	1
	Urgent defect	2	Urgent defect	2

6. Conclusions

This paper is an extension of a conference paper, which the authors previously published [25]. It proposed an autonomous method to analyze infrared images for the diagnosis of insulation conditions of cable accessories. The approach included the positioning and identification of diagnostic objects, the extraction of key regions, and the calculation of temperature parameters for condition assessment of cable accessories. The conclusions are as follows:

(1) The method for autonomous positioning and identification of cable accessories in infrared images based on Faster RCNN network was proposed. This method was applied to test against actual infrared images, and results showed that the autonomous location and

recognition could be achieved at different shooting angles and under various background conditions.

(2) The method for extracting suspected abnormal heating regions based on Mean-Shift algorithm was proposed. Case studies showed that this method, which was superior to other alternatives, can realize image adaptive segmentation. It can extract the abnormal heating regions of grounding boxes and cable terminations accurately.

(3) The proposed method may potentially be productive as it helps reduce the dependence on human efforts and expertise and helps improve the practice of intelligent condition monitoring.

Author Contributions: Conceptualization, L.M., Z.X., C.Z., X.X.; methodology, L.M., Z.X., C.Z., W.Z.; software, B.Y., H.G.; validation, L.M., Z.X., C.Z., W.Z.; formal analysis, L.M., Z.X., C.Z., X.X.; investigation, L.M., Z.X., C.Z., X.X.; resources, B.Y., H.G.; data curation, B.Y., H.G.; writing—original draft preparation, X.X., L.M., Z.X.; writing—review and editing, C.Z., W.Z.; visualization, X.X., L.M., Z.X.; supervision, L.M., Z.X., C.Z., W.Z.; project administration, L.M., Z.X., C.Z., W.Z.; funding acquisition, L.M., Z.X., B.Y., H.G. All authors have read and agreed to the published version of the manuscript.

Funding: This research was funded by Science and Technology Project of State Grid Hubei Electric Power Company, grant number SGHBWH00YJJS2001955.

Institutional Review Board Statement: Non applicable.

Informed Consent Statement: Non applicable.

Data Availability Statement: Non applicable.

Conflicts of Interest: The authors declare no conflict of interest.

References

1. Lowczowski, K.; Lorenc, J.; Tomczewski, A.; Nadolny, Z.; Zawodniak, J. Monitoring of MV cable screens, cable joints and earthing systems using cable screen current measurements. *Energies* **2020**, *13*, 3438. [CrossRef]
2. Zhou, L.J.; Zhu, S.B.; Bai, L.L.; Liu, Y.; Zhu, L.; Guo, L. Influence of stress tubes interface on partial discharge of vehicle cable terminal at low temperatures. *High Voltage Eng.* **2019**, *45*, 1266–1273.
3. Reda, A.M.; Al-Yafei, A.M.S.; Howard, I.M.; Forbes, G.L.; McKee, K.K. Simulated in-line deployment of offshore rigid field joint—A testing concept. *Ocean Eng.* **2016**, *112*, 153–172. [CrossRef]
4. Reda, A.; Abu-Siada, A.; Howard, I.M.; McKee, K.K. A testing platform for subsea power cable deployment. *Eng. Fail. Anal.* **2019**, *96*, 142–157. [CrossRef]
5. Xiong, L.; Chen, Y.H.; Jiao, Y.; Wang, J.; Hu, X. Study on the effect of cable group laying mode on temperature field distribution and cable ampacity. *Energies* **2019**, *12*, 3397. [CrossRef]
6. Ullah, I.; Khan, R.U.; Yang, F.; Wuttisittikulkij, L. Deep learning image-based defect detection in high voltage electrical equipment. *Energies* **2020**, *13*, 392. [CrossRef]
7. Rahmani, A.; Haddadnia, J.; Seryasat, O.L. Intelligent fault detection of electrical equipment in ground substations using thermos vision technique. In Proceedings of the 2010 2nd International Asia Conference on Mechanical and Electronics Engineering (ICMEE), Kyoto, Japan, 1–3 August 2010.
8. Huda, A.S.N.; Taib, S.; Jadin, M.S.; Ishak, D. A semi-automatic approach for thermographic inspection of electrical installations within buildings. *Energy Build* **2012**, *55*, 585–591. [CrossRef]
9. Huda, A.S.N.; Taib, S.; Ghazali, K.H.; Jadin, M.S. A new thermographic NDT for condition monitoring of electrical components using ANN with confidence level analysis. *ISA Trans.* **2014**, *53*, 717–724. [CrossRef]
10. Ullah, I.; Yang, F.; Khan, R.; Liu, L.; Yang, H.S.; Gao, B.; Sun, K. Predictive maintenance of power substation equipment by infrared thermography using a machine-learning approach. *Energies* **2017**, *10*, 1987. [CrossRef]
11. Jaffery, Z.A.; Dubey, A.K. Design of early fault detection technique for electrical assets using infrared thermograms. *Int. J. Electr. Power Energy Syst.* **2014**, *63*, 753–759.
12. Wu, J.Z.; Niu, H.Q.; Xu, J. Adaptive de-noising method for infrared image of porcelain bushing cable terminal based on layer optimal basic wavelet and Bayes estimation. *Electr. Meas. Instrum.* **2016**, *53*, 83–89.
13. Niu, H.Q.; Wu, J.Z.; Xu, J.; Wu, Q.; Gao, Z.J.; Zheng, W.J. Denoising of infrared images of porcelain sleeve cable terminal considering inter-scale correlation. *J. South China Univ. Technol. (Natl. Sci. Ed.)* **2017**, *45*, 15–21.
14. Wu, J.Z.; Niu, H.Q.; Zhang, H.; Xu, J. A hybrid Fourier-wavelet de-noising method for infrared image of porcelain sleeve cable terminal using GSM model for wavelet coefficients. *Electr. Meas. Instrum.* **2018**, *55*, 113–117.
15. Niu, H.Q.; Wu, J.Z.; Xu, J.; Zheng, W.Q. Infrared image recognition of cable terminal based on Randon and Fourier-Mellin Transform. *J. South China Univ. Technol. (Natl. Sci. Ed.)* **2016**, *44*, 47–52.

16. Wan, L.; Zhou, K.; Li, X.T.; Wu, K. Toward understanding development mechanism of partial discharge in air gap defects in cable termination by analysis of electric field characteristics. *High Voltage Eng.* **2014**, *40*, 3709–3716.
17. Yang, J.; Zhu, X.L.; Dong, X.; Lu, Y.; Li, N. On-line monitoring and diagnosis of HV cable faults based on sheath currents. *High Voltage Eng.* **2016**, *42*, 3616–3625.
18. Ren, S.Q.; He, K.M.; Girshick, R.; Sun, J. Faster R-CNN: towards real-time object detection with region proposal networks. In Proceedings of the 2016 Conference and Workshop on Neural Information Processing Systems (NIPS), Barcelona, Spain, 9–10 December 2016.
19. Yang, Y. Infrared image analysis and processing of power equipments. Master's Thesis, Beijing Jiaotong University, Beijing, China, 2015.
20. Kim, N.; Park, S.; Lee, J.; Choi, J.K. Load profile extraction by Mean-Shift clustering with Sample Pearson Correlation Coefficient distance. *Energies* **2018**, *11*, 2397. [CrossRef]
21. Hong, Y.P.; Yi, J.Q.; Zhao, D.B. Improved mean shift segmentation approach for natural images. *Appl. Math. Comput.* **2007**, *185*, 940–952. [CrossRef]
22. Zhou, Z.Q.; Feng, Z.X.; Zhou, D.G.; Xu, X.L.; Gu, K.K. Fault region extraction of electrical equipment in infrared image by using an extended Mean Shift method. *Infrared Technol.* **2019**, *41*, 78–83.
23. Otsu, N. A threshold selection method from gray-level histograms. *IEEE Trans. Syst. Man Cybern.* **1979**, *9*, 62–66. [CrossRef]
24. Zhang, Y.J.; Yang, L.F.; Ge, S.Y.; Zhou, H.X. Short-term photovoltaic power forecasting based on Kmeans algorithm and support vector machine. *Power Syst. Prot. Control* **2018**, *46*, 118–124.
25. Xu, X.B.; Zhou, C.K.; Zhou, W.J.; Liao, Y.Q.; Wei, Y.L.; Yuan, J. Autonomous diagnosis of overheating defects in cable accessories based on Faster RCNN and Mean-Shift algorithm. In Proceedings of the 2020 IEEE International Conference on High Voltage Engineering (ICHVE 2020), Beijing, China, 6–10 September 2020.

energies

MDPI

Article

DC Aging Mechanism of Co_2O_3-Doped ZnO Varistors †

Xia Zhao [1], Haibin Shen [1], Men Guo [2,*], Ziming He [1], Yupeng Li [3] and Ran Wen [4]

1 China Electric Power Research Institute, Beijing 100192, China; zhaoxia@epri.sgcc.com.cn (X.Z.); shenhb@epri.sgcc.com.cn (H.S.); hezm@epri.sgcc.com.cn (Z.H.)
2 State Key Laboratory of Electrical Insulation and Power Equipment, Xi'an Jiaotong University, Xi'an 710049, China
3 State Key Laboratory of Power Transmission Equipment & System Security and New Technology, Chongqing University, Chongqing 400044, China; 201911021005@cqu.edu.cn
4 School of Electronics and Information, Xian Polytechnic University, Xi'an 710048, China; wenran1994@126.com
* Correspondence: guomen@stu.xjtu.edu.cn

† This paper is an extended version of our paper published in the 2020 IEEE International Conference on High Voltage Engineering and Applications (ICHVE), Beijing, China, 6–10 September 2020.

Abstract: Lowered power loss and asymmetrically electrical parameters are reported in the DC aging of Co_2O_3-doped ZnO varistors in this paper. Based on the frequency domain dielectric responses of the pristine and degraded samples, the present study explores the roles of point defects in the aging process via dielectric relaxations and their parameters. It is found that breakdown field increased more in the positive direction than the negative direction. Nonlinearity increased in the positive direction, whereas it decreased in the negative direction, and leakage current density increased more in the negative direction than the positive direction. Given the lowest migration energy of Zinc interstitial (Zn_i, 0.57 eV) and high oxygen ion conductivity in Bi_2O_3-rich phase, it is speculated that Zni and adsorbed oxygen (O_{ad}) migrate under DC bias, and then change the defect structure and the double Schottky barrier (DSB) at grain boundaries. As a result, the forward-biased barrier height gradually decreases more than the reverse-biased one.

Keywords: ZnO varistors; DC aging; double Schottky barrier; point defects

Citation: Zhao, X.; Shen, H.; Guo, M.; He, Z.; Li, Y.; Wen, R. DC Aging Mechanism of Co_2O_3-Doped ZnO Varistors . *Energies* **2021**, *14*, 4011. https://doi.org/10.3390/en14134011

Academic Editor: Issouf Fofana

Received: 18 June 2021
Accepted: 1 July 2021
Published: 3 July 2021

1. Introduction

ZnO varistor, one of the polycrystalline semiconductors, manifests excellently non-ohmic current–voltage (*I-V*) characteristics so that they are vastly used for overvoltage protection in the electrical system [1]. Recently, as high voltage direct current power transmission develops rapidly, ZnO varistors are available to a very high voltage level of ±1100 kV on the transmission line and at the converter station. In these applications, the ZnO blocks are continuously stressed by pure DC or a dominant DC component with superimposed continuous (high frequency) transients, and ion migrations in ZnO varistors are more serious. Therefore, the aging under DC loading was significantly different compared to the AC world. Most of the available blocks were stable under AC but not necessarily stable under a DC stress. For low-voltage ZnO varistors, simulations are developing to describe the DC current-voltage (*I-V*) behavior and to obtain extended DC *I-V* characteristics in the time domain [2,3].

The DSB is generally accepted to explain the non-ohmic characteristics of ZnO varistor, which is also employed to analyze its aging phenomena [4–7]. At an early stage, Eda and Matsuoka studied the influence of DC-biased stress on ZnO varistors by thermally stimulated current, where they suggested that the aging arose from migration of ions in the depletion layer [8]. Gupta and Carlson further pointed out that the migration was relevant to zinc interstitial (Zn_i) under electrical-thermal driving force [9]. Via positron annihilation spectroscopy, Ramanachalam et al. [10] unveiled that positron trap density decreased with

DC stress increasing and associated positron lifetime increased, which was attributed to recombination of zinc interstitial migrating to grain boundary with zinc vacancy, leading to lower barrier height. Besides DC operation voltage, conductivity at contact points between varistor granules may also be increased due to accumulated damage from Joule heating by multiple high voltage strokes [11].

Besides ion migration, some other interesting experimental results were also reported. Ohbuchi et al. [12] analyzed the impact of AC aging on the interface states of ZnO varistors by deep level transient spectroscopy and found that aging phenomena were highly referring to the emission rate of the interface states, which they claimed arose from migration of adsorbed oxygen ion (O_{ad}). Takahashi et al. [13] found that oxygen atmosphere could inhibit the leakage current during the aging process and resume aged samples. Ramirez et al. argued that it was due to the adsorption of oxygen influenced by β/δ-Bi_2O_3 at grain boundaries and then the control of barrier [14]. Mielcarek et al. [15] also investigated the influence of β/δ-Bi_2O_3 phases on aging phenomena.

Cobalt oxide is one of the commonly used dopants to modify the DSBs at grain boundaries. Cobalt ions in ZnO grains exit as Co^{2+}, which replaces Zn at the lattice sites in the manner of tetrahedral coordination [16]. Co_2O_3 can increase the oxygen partial pressure in the sintering processing and decrease the density of donor, which decreases the conductivity of the ZnO grains [17]. Co was conducive to forming interface states between the conduction band and the valence band of ZnO [18]. Except for changing multi-scale defect structure, the phase of Bi_2O_3 can also be influenced by Co_2O_3 [19]. In a word, studies on Co_2O_3-doped ZnO varistors have been mainly focused on reducing grain boundary resistivity, enhancing the nonlinearity, preventing Bi_2O_3 evaporation, and improving stability against AC aging. However, the effect of Co_2O_3 dopant on the stability of ZnO varistors against DC aging has not been systematically presented yet. It is also helpful to understand DC aging mechanism based on Co_2O_3-doped ZnO varistors due to its obvious modification effect on the defect structure.

In this paper, the *I-V* characteristics of Co_2O_3-doped ZnO varistors in low-current (pre-breakdown) region were measured before and after DC aging. Power loss was continuously recorded during DC aging. Broadband dielectric spectroscopy was utilized to analyze relaxation processes, by which the variation of defects due to aging was compared. Finally, dynamics of ion migrations during DC aging was deduced, of which the effect on the DSB was also analyzed.

2. Experiment

A series of ZnO varistor block samples were prepared via the solid-state reaction method. Besides (95−x) mol% ZnO, raw materials contained approximately 1 mol% Bi_2O_3, 1 mol% Sb_2O_3, 0.5 mol% $MnCO_3$, 1 mol% NiO, and 1.5 mol% SiO_2 (mainly for inhibiting grain growth). The content of Co_2O_3 was chosen as x = 0, 0.28, 0.55, 0.83, 1.1, and 1.38 mol%. The raw materials were mixed by a 12 h ball-milling with deionized water and an appropriate amount of polyvinyl alcohol (PVA). Next, the slurries were processed into particles with a size of 80~120 μm via spray granulation. The particles were added 3 wt% water to be stale for 24 h and then were pressed into pellets under 100 MPa for 10 s. The pellets were pre-sintered at 600 °C to discharge PVA. Finally, the pellets were sintered in air at 1150 °C for 2 h, and naturally cooled. All samples were designated as S1, S2, S3, S4, S5, and S6 for short, with increasing Co_2O_3 content. Based on Archimedes principle of water displacement, densities and porosities were measured for all samples.

After polishing both sides of the samples, Al electrodes were sputtered for electrical measurements. The nonlinear coefficient $\alpha = 1/\lg(E_B/E_A)$ was calculated from the current density-electric field (*J-E*) curves, where E_A and E_B (called breakdown field) is an electric field under 0.1 $mAcm^{-2}$ and 1 $mAcm^{-2}$, respectively. The leakage current density (J_L) was acquired at $0.75E_B$. These varistors were heated at 135 °C and applied an electric field of $0.9E_B$ for more than 160 h. The direction, the same as that of the electric field during aging, was referred to as the positive direction, and the opposite direction is the negative direction. After aging J-E curves of both directions were taken into consideration. Dielectric properties were measured by an impedance analyzer (Novocontrol, Concept 80, Montabaur, Germany) in a frequency range of 10^{-1}~10^6 Hz and a temperature range of −110~200 °C.

3. Results and Discussion

3.1. Macro Electrical Properties

Phase compositions and microstructures of all samples have been presented elsewhere [19]. α-Bi_2O_3 phase was detected in samples S1–S4, whereas δ-Bi_2O_3 phase was detected in samples S5 and S6. δ-Bi_2O_3 phase trended to appear in those samples with higher Co_2O_3 content. Spine phase $Zn(Zn_{4/3}Sb_{2/3})O_4$ was identified in S1–S3, and spine phase $Co(Co_{4/3}Sb_{2/3})O_4$ was identified in S4–S6. Average grain sizes (λ), densities (D), and porosities (H) are shown in Table 1. λ changed little with increasing Co_2O_3 content. D decreased at low contents, and then increased with increasing Co_2O_3 content. Changes of H were opposite to D. It was found that microstructures changed little after doping Co_2O_3.

Table 1. Average grain sizes (λ), densities (D), and porosities (P) of Co_2O_3-doped ZnO varistors.

No.	λ (μm)	D (gcm^{-3})	H (%)
S1	9.98	5.32	0.016
S2	10.27	5.25	0.018
S3	10.85	5.28	0.024
S4	10.19	5.23	0.051
S5	10.24	5.23	0.020
S6	10.71	5.42	0.021

Electrical properties of these as-sintered varistors were reported elsewhere. During DC aging, power loss (P) was continuously recorded, as shown in Figure 1. P_0 is a maximum of P at the beginning of aging. P rapidly decreases in the initial time. Ion migrations were most intense when DC voltage was applied. With increasing aging time, equilibrium was established between migrations and diffusions. Thus, P gradually decreased. Approximately after 20 h, P of S5 gradually increases, whereas those of others decrease with lower rates than before. This indicates that these samples except S5 are stable under the DC aging test. As to S5, the increase of P is due to δ-Bi_2O_3, which is a phase of Bi_2O_3 with the highest conductivity of oxygen ion. During aging, adsorbed oxygen can desorb and easily escape from the varistor through δ-Bi_2O_3 [20]. As a result, the interface states and Schottky barrier height decrease, and P gradually increases due to increasing leakage current. Leakage current density (J_L) vs. aging time (t) satisfies

$$J_L = J_{L0} + K_T\sqrt{t}, \quad (1)$$

where J_{L0} is the leakage current density at t = 0; K_T is the degradation rate. J_L of S5 was extrapolated via Equation (1), where it would be equal to the initial value after 1489 h, with J_{L0} = 53.04 μA and K_T = 1.45 μA·cm^{-2}·h$^{0.5}$.

As an example, J-E curves of S1 are shown in Figure 2. After DC aging, J-E curves became asymmetric in both positive direction and negative direction. Leakage currents increased in both directions, where the increase is more in the negative direction. Breakdown field also increased in both directions, where the increase is more in the positive direction. All electrical parameters of S1–S6 before and after DC aging are shown in Figure 3. Breakdown field increased more in the positive direction than the negative direction, nonlinearity increased in the positive direction, whereas it decreased in the negative direction. Leakage current density increased more in the negative direction than the positive direction. It is found that after DC aging, electrical properties degraded more seriously in the negative direction.

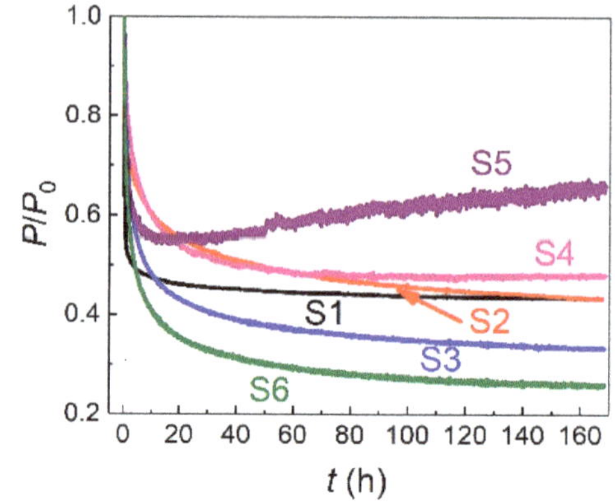

Figure 1. Power losses of S1–S6 during DC aging. Aging condition: Heated at 135 °C and applied an electric field of $0.9E_B$ for more than 160 h, where E_B is the electric field when current density is 1 $mAcm^{-2}$ and is called breakdown field.

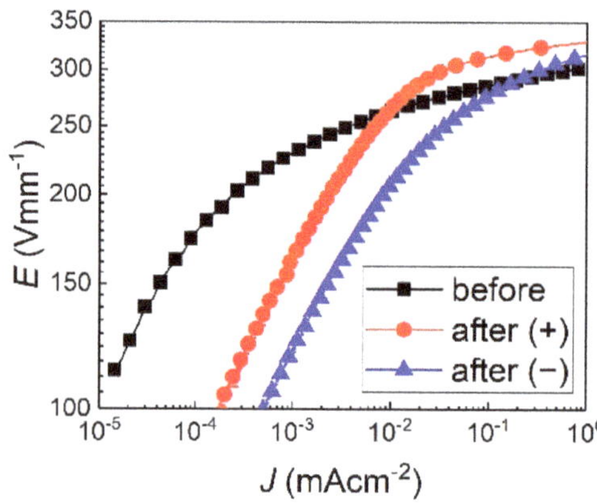

Figure 2. *J-E* characteristics of S1 before and after DC aging.

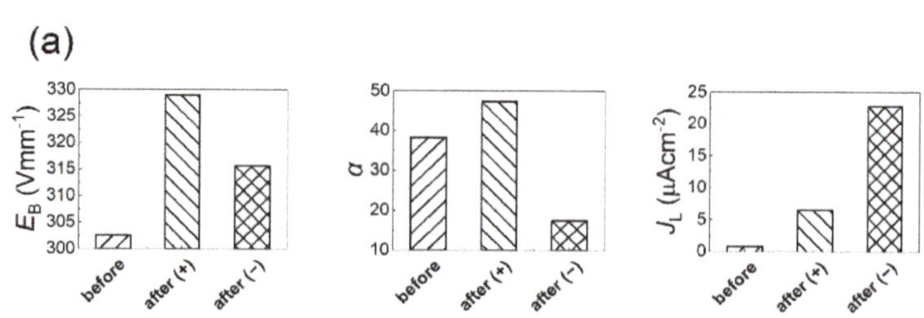

Figure 3. *Cont.*

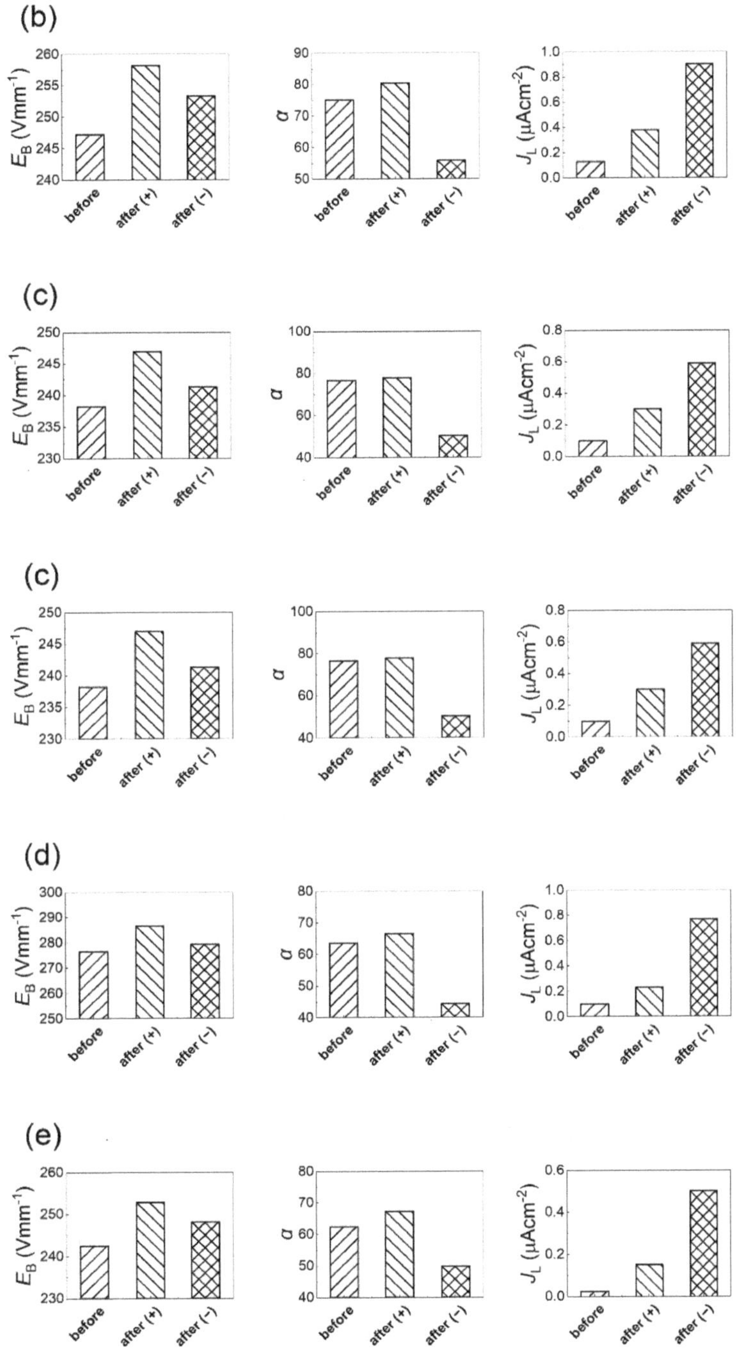

Figure 3. *Cont.*

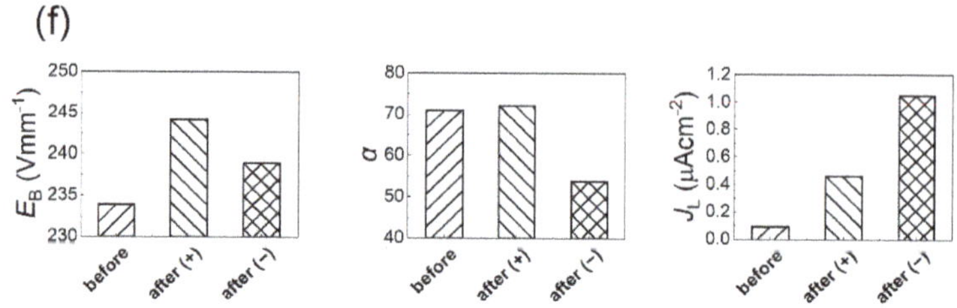

Figure 3. Electrical parameters of S1–S6 before and after DC aging (**a**–**f**) S1–S6.

The nonlinear coefficient $\alpha = 1/\lg(E_B/E_A)$ was calculated from the current density-electric field (*J*-*E*) curves, where E_A and E_B (called breakdown field) are electric fields under 0.1 mAcm^{-2} and 1 mAcm^{-2}, respectively. The leakage current density (J_L) was acquired at $0.75E_B$.

3.2. Dielectric Response

Relaxations of intrinsic point defects of Zn_i and oxygen vacancy (V_O) can be detected at lower temperatures. Zn_i is the most mobile ion in ZnO varistors and plays a decisive role in aging. Relaxations of extrinsic defects respond at higher temperatures. Extrinsic defects, such as grain boundaries and intergranular phases, may be changed due to temperature rise, for example, partial melting arose from intensive heat generated in the varistors when high currents flow [21]. These defects decide the property of DSB.

As an example, dielectric responses of S1 at −100 °C and 200 °C before DC aging are presented to analyze these defects, which is shown in Figure 4. Two obvious loss peaks were manifested in ε″ spectroscopy at −100 °C in Figure 3a, while only a vague one at 200 °C in Figure 3b was detected. Frequency has effects on the electrical characteristics of varistors [22]. The reason is that DC conductivity exponentially increased at higher temperatures, which masked loss peak(s) at low frequencies. Here, frequencies lower than 10 Hz are referred to as low frequencies. Compared to Figure 2, the dielectric responses for the pristine sample had little difference. Note that some relaxation peaks were hidden by DC conductivity at high temperatures, a derivative method was applied to characterize relaxation processes [23,24]. It was found that five relaxations exist in a temperature range of −110~200 °C and a frequency range of 10^{-1}~10^6 Hz. At low temperatures (−110~−60 °C), two loss peaks were detectable, which were designated as *Peak A* and *Peak B*. At high temperatures (150~200 °C), two loss peaks were also characterized that were designated as *Peak C* and *Peak D*. Because *Peak E* had not completely manifested in the spectra, it is not discussed in the following content.

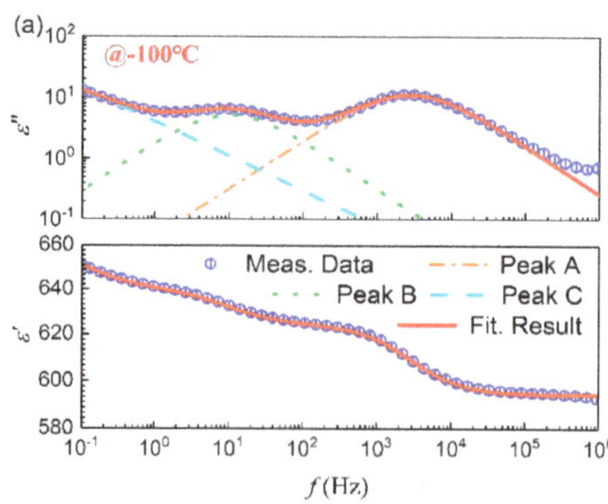

Figure 4. *Cont.*

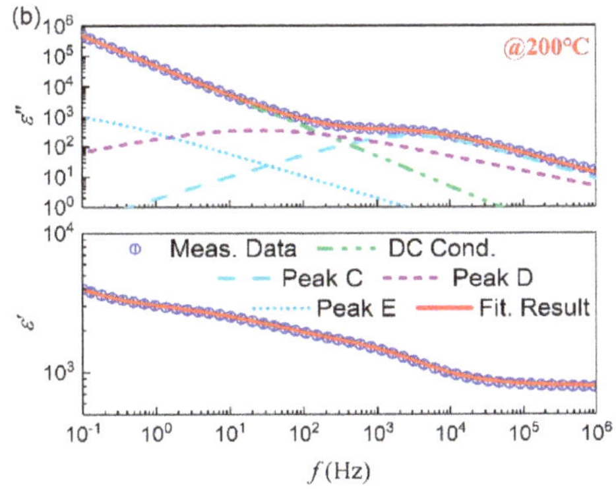

Figure 4. Dielectric responses of S1 before DC aging (**a**) −100 °C (**b**) 200 °C.

To obtain parameters of all loss peaks and DC conductivity, Cole–Cole model was applied to fit the complex permittivity [25]:

$$\varepsilon^* = \varepsilon_\infty + \frac{\sigma_{dc}}{j\omega\varepsilon_0} + \sum_i \frac{\Delta\varepsilon_i}{1 + (j\omega\tau_i)^{1-\alpha_i}} \tag{2}$$

In this expression, $\varepsilon^* = \varepsilon' - j\varepsilon''$ is the complex permittivity, ε_∞ the limiting high-frequency permittivity, ω the angular frequency, σ_{dc} the dc conductivity, ε_0 the vacuum permittivity, $\Delta\varepsilon_i$ and τ_i the relaxation amplitude and the relaxation time for the *i*th relaxation, respectively, α_i the shape (or relaxation time distribution) parameters.

The genetic algorithm was used to fit measurement results. For results at lower temperatures, all parameters were obtained based on ε'', and examined over ε'. For results at higher temperatures, σ_{dc} was firstly calculated via ε'' at the low frequencies. Relaxation parameters were obtained from ε'. Then, all parameters were examined by ε''. Frequency ranges of 'steps' in ε' spectra or loss peaks in ε' spectra were obtained in advance for globally optimal solutions. All parameters of loss peaks of S1–S6 before and after DC aging are shown in Figure 5. *Peak A* and *Peak B* were ascribed to the relaxation of $Zn_i^{\bullet\bullet}$ and $V_O^{\bullet}$ in depletion layers [26]. After DC aging, $\Delta\varepsilon_i$ and α_i of the two peaks changed a bit. This is because of the change of defect distribution in depletions. τ_i of *Peak A* and *Peak B* almost did not change due to the intrinsic nature of the two defects. *Peak C* and *Peak D* were related to extrinsic defects, where *Peak C* was caused by inter-granular phase and *Peak D* is speculated due to the interface states [27,28]. Zhao et al. recently found that *Peak C* was related to both Bi_2O_3-rich phase and the spinel phase, and was therefore, due to relaxation of interface between the two phases [24]. Because of the extrinsic nature of the two defects, all parameters of *Peak C* and *Peak D* change to some extent.

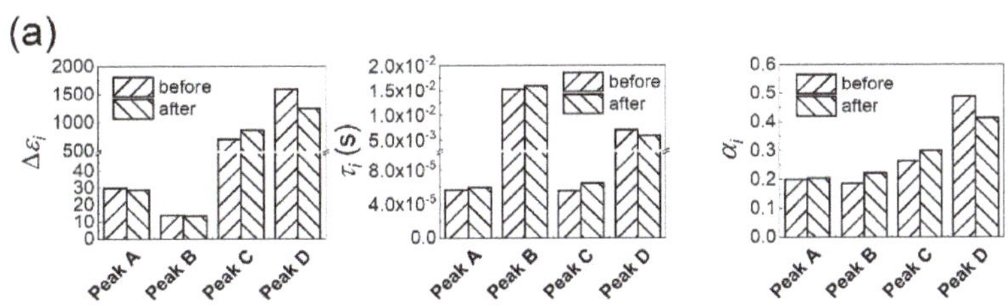

Figure 5. *Cont.*

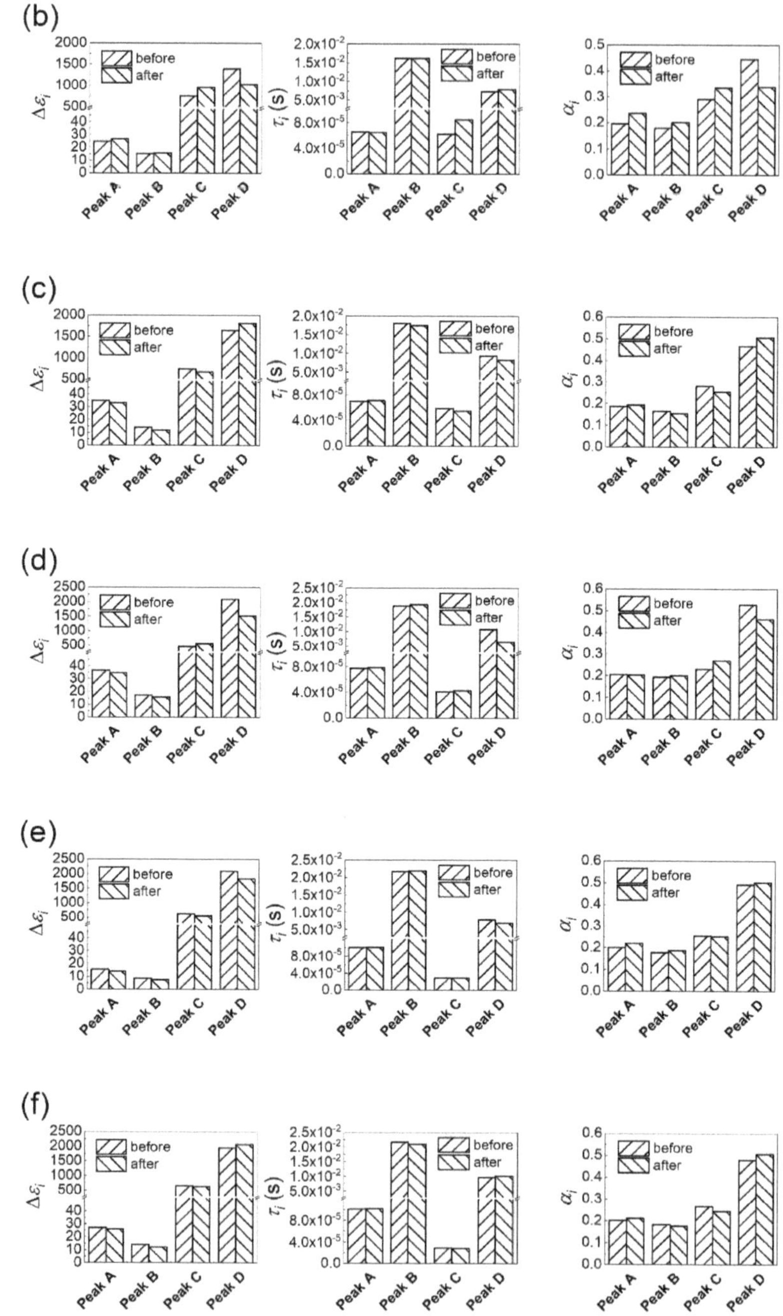

Figure 5. Parameters of all loss peaks before and after DC aging (**a–f**) S1–S6.

DC conductivities (σ_{dc}) were obtained from σ' spectra at 0.1 Hz, which was also calculated via dielectric responses at higher temperatures. σ_{dc} of S1–S6 at 200 °C before and after DC aging are shown in Figure 6a. After DC aging, σ_{dc} increases for every sample. As σ_{dc} was obtained from frequency-domain conductivity at the lowest measurement frequency, it reflects the property of the DSB [29,30]. Based on the Arrhenius equation, activation energies (E_{dc}) of σ_{dc} were also calculated, as shown in Figure 6b. It is found that E_{dc} decreases for all samples after DC aging. Therefore, the reason why σ_{dc} increases is mainly because E_{dc} decreases, and the decrease of E_{dc} is due to the aging of the DSB.

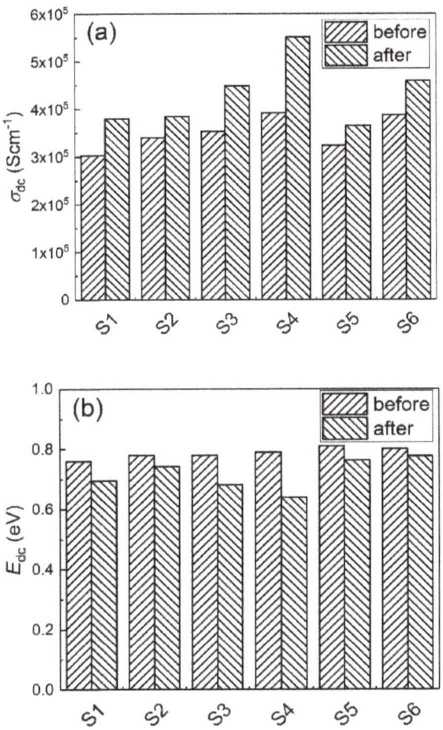

Figure 6. DC conductivities (**a**) and their activation energies (**b**) of S1–S6 before and after DC aging.

3.3. Discussion

DSB model was utilized to explained DC aging mechanism, as shown in Figure 7. *Peak A* derives from electron relaxations between donor levels of Zn_i and the conduction band in the depletion layers. *Peak B* derives from similar processes of V_O. *Peak C* and *Peak D* have higher activation energies, and are related to electron emissions from different kinds of interfaces (boundaries). As leakage current density increased more in the negative direction than the positive direction, one could deduce that the forward-biased barrier height (Φ_{bF}) decreased more and the reverse-biased barrier (Φ_{bR}) decreased a bit during aging, i.e., $\Phi_{bF} < \Phi_{bR} < \Phi_{b0}$. In ZnO semiconductor, Zn_i were too mobile to be stable at room temperature, with the migration energy about 0.57 eV [31]. Oxygen, which could be adsorbed at the grain boundary and form localized acceptor states, played a key role in forming a potential barrier [32,33].

When the DC stress was applied, the depletion width (w_F) of the forward-biased barrier became narrow, while the reverse-biased barrier became wide, i.e., $w_F < w_0 < w_R$. As aging time increased, metastable cations, which were mainly Zn_i, migrated into grains for the forward-biased barrier. As to the reverse-biased barrier, metastable cations migrated to the grain boundary, some of which recombined with part of the localized states. At the grain boundary, metastable anions migrated from the forward-biased barrier to the reverse-biased one, which replenished the interface states in the reverse-biased barrier and was postulated being O_{ad} or its aliovalent species [34]. Finally, the barrier widths got narrower and satisfied $w_R < w_F < w_0$. Therefore, all parameters of loss peaks and DC conductivity due to the change of defect structure.

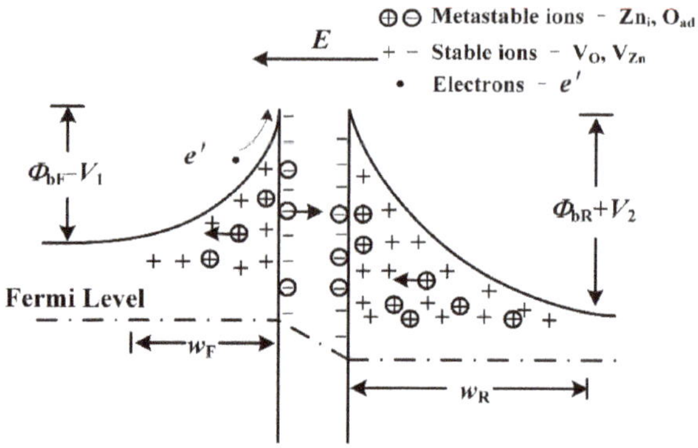

Figure 7. The schematic for ion migration based on the double Schottky barrier model.

The relation among the Schottky barrier height (Φ_b), the density of the interface states (N_i) and the donor (N_d), and the depletion width (w) satisfies the following expression [35]:

$$\Phi_b = \frac{eN_d w^2}{2\varepsilon_r\varepsilon_0} = \frac{eN_i{}^2}{2\varepsilon_r\varepsilon_0 N_d}, \quad (3)$$

where e is the charge of an electron, ε_r the relative permittivity of ZnO. Hence, the previous conclusion for the deformation of the DSB was deduced. Meanwhile, the complex defects related to *Peak C* and *Peak D* also changed due to the migration of oxygen ion in the Bi_2O_3-rich phases. The specific defects relative to *Peak D* need further characterization. Besides, asymmetric conduction of ZnO varistors is also associated with nonuniform distribution of additives and grain sizes, especially due to ion redistributions after DC aging [36,37]. Conductivity, or say DSB, is a function of both voltage and temperature [38]. Ion distributions led to changes of the functions in both the positive and the negative directions. Establishing such a function is still one of the remaining challenges in the field [39].

4. Conclusions

DC aging mechanism was investigated based on Co_2O_3-doped ZnO varistors. During the aging process, the power loss decreased continuously except S5 with δ-Bi_2O_3. After DC aging, breakdown field increased more in the positive direction than the negative direction, nonlinearity increased in the positive direction, whereas it decreased in the negative direction, and leakage current density increased more in the negative direction than the positive direction.

After DC aging, parameters of four relaxations were compared with the initial value, where the amplitudes of all relaxations changed. It was deduced that the migrations of Zn_i and O_{ad} changed the species of the interface states and then deformed the DSB. As a result, the forward-biased barrier height decreased more than the reverse-biased one.

Author Contributions: Conceptualization, X.Z. and H.S.; methodology, X.Z.; software, Z.H.; validation, X.Z., M.G. and Y.L.; formal analysis, Z.H.; investigation, X.Z.; resources, Z.H.; data curation, M.G.; writing—original draft preparation, M.G.; writing—review and editing, X.Z. and R.W.; visualization, Z.H.; supervision and project administration, H.S. and M.G. All authors have read and agreed to the published version of the manuscript.

Funding: This research was supported by The National Key R&D Program of China (2018YFC0809400).

Data Availability Statement: Data sharing not applicable. No new data were created or analyzed in this study. Data sharing is not applicable to this article.

Conflicts of Interest: The authors declare no conflict of interest.

References

1. He, J.L. *Metal Oxide Varistors;* TsingHua University Press: Beijing, China, 2019; pp. 1–4.
2. Kaufmann, B.; Billovits, T.; Kratzer, M.; Teichert, C.; Supancic, P. A modelling approach to describe the DC current-voltage behaviour of low-voltage zinc oxide varistors. *Open Ceram.* **2021**, *6*, 100113. [CrossRef]
3. Tsovilis, T.E.; Topcagic, Z. DC Overload Behavior of Low-Voltage Varistor-Based Surge Protective Devices. *IEEE Trans. Power Deliv.* **2020**, *35*, 2541–2543. [CrossRef]
4. Blatter, G.; Greuter, F. Carrier transport through grain boundaries in semiconductors. *Phys. Rev. B* **1986**, *33*, 3952–3966. [CrossRef] [PubMed]
5. Blatter, G.; Greuter, F. Electrical breakdown at semiconductor grain boundaries. *Phys. Rev. B* **1986**, *34*, 8555–8572. [CrossRef]
6. Gupta, T.K. Application of Zinc Oxide Varistors. *J. Am. Ceram. Soc.* **1990**, *73*, 1817–1840. [CrossRef]
7. Gupta, T.K.; Carlson, W.G. A grain-boundary defect model for instability/stability of a ZnO varistor. *J. Mater. Sci.* **1985**, *20*, 3487–3500. [CrossRef]
8. Eda, K.; Iga, A.; Matsuoka, M. Degradation mechanism of non-ohmic zinc oxide ceramics. *J. Appl. Phys.* **1980**, *51*, 2678–2684. [CrossRef]
9. Gupta, T.K.; Carlson, W.G. Barrier voltage and its effect on stability of ZnO varistor. *J. Appl. Phys.* **1982**, *53*, 7401–7409. [CrossRef]
10. Ramanachalam, M.S.; Rohatgi, A.; Schaffer, J.P.; Gupta, T.K. Characterization of ZnO varistor degradation using lifetime positron-annihilation spectroscopy. *J. Appl. Phys.* **1991**, *69*, 8380–8386. [CrossRef]
11. Yang, P.; Gomez, C.A.; Andrews, S.; Sorenson, J.D.; Chen, K.S. High-voltage surge protection by a varistor-filled air gap. *J. Am. Ceram. Soc.* **2021**, *104*, 3247–3259. [CrossRef]
12. Ohbuchi, Y.; Kawahara, T.; Okamoto, Y.; Morimoto, J. Characterization of interface states in degraded ZnO varistors. *Jpn. J. Appl. Phys.* **2002**, *41*, 190–196. [CrossRef]
13. Takahashi, K.; Miyoshi, T.; Maeda, K. Degradation of Zinc Oxide Varistors. *MRS Proc.* **1981**, *5*, 399. [CrossRef]
14. Ramirez, M.A.; Simoes, A.Z.; Bueno, P.R.; Márquez, M.A.; Orlandi, M.O.; Varela, J.A. Importance of oxygen atmosphere to recover the ZnO-based varistors properties. *J. Mater. Sci.* **2006**, *41*, 6221–6227. [CrossRef]
15. Mielcarek, W.; Gubanski, A.; Paściak, G.; Prociów, K.; Warycha, J.; Wrobel, J.M. The effect of bismuth oxide polymorph forms on degradation processes in ZnO varistors. *Ceram. Int.* **2013**, *39*, 8219–8226. [CrossRef]
16. Anderson, R.S. Lattice-vibration effects in the spectra of ZnO:Ni and ZnO:Co. *Phys. Rev.* **1967**, *164*, 398–405. [CrossRef]
17. Kim, E.D.; Kim, C.H.; Oh, M.H. Role and effect of Co_2O_3 additive on the upturn characteristics of ZnO varistors. *J. Appl. Phys.* **1985**, *58*, 3231–3235. [CrossRef]
18. Yano, Y.; Takai, Y.; Morooka, H. Interface states in ZnO varistor with Mn, Co and Cu impurities. *J. Mater. Res.* **1994**, *9*, 112–118. [CrossRef]
19. Guo, M.; Wang, Y.; Wu, K.N.; Zhang, L.; Zhao, X.; Lin, Y.; Li, J.Y. Revisiting the effects of Co_2O_3 on multiscale defect structures and relevant electrical properties in ZnO varistors. *High Volt.* **2020**, *5*, 241–248. [CrossRef]
20. Shuk, P.; Wiemhofer, H.D.; Guth, U.; Göpel, W.; Greenblatt, M. Oxide ion conducting solid electrolytes based on Bi_2O_3. *Solid State Ion.* **1996**, *89*, 179–196. [CrossRef]
21. Tanaka, T.; Baba, Y.; Tsujimoto, Y.; Tsukamoto, N. FDTD Electromagnetic and Thermal Simulation of a Metal Oxide Varistor Element Considering the Temperature Dependence of Its Resistivity. *Electricity* **2021**, *2*, 158–167. [CrossRef]
22. Staikos, E.T.; Tsovilis, T.E. Low-frequency response of low-voltage metal-oxide varistors used for telecommunication systems protection. In Proceedings of the 2020 IEEE International Conference on High Voltage Engineering and Application (ICHVE), Beijing, China, 6–10 September 2020; pp. 1–4.
23. Huang, Y.; Wu, K.; Xing, Z.; Zhang, C.; Hu, X.; Guo, P.; Li, J.Y. Understanding the validity of impedance and modulus spectroscopy on exploring electrical heterogeneity in dielectric ceramics. *J. Appl. Phys.* **2019**, *125*, 084103. [CrossRef]
24. Zhao, X.; Shi, W.D.; Zhang, B.Y.; Guo, M.; Shen, H.B. Relaxation peaks of ZnO varistors and their correlations with electrical properties. *Jpn. J. Appl. Phys.* **2021**, *60*, 031006. [CrossRef]
25. Cole, K.S.; Cole, R.H. Dispersion and absorption in dielectrics I. alternating current characteristics. *J. Chem. Phys.* **1941**, *9*, 341–351. [CrossRef]
26. Cheng, P.; Li, S.; Zhang, L.; Li, J.Y. Characterization of intrinsic donor defects in ZnO ceramics by dielectric spectroscopy. *Appl. Phys. Lett.* **2008**, *93*, 012902. [CrossRef]
27. Zhao, X.; Li, J.Y.; Li, H.; Li, S.T. Intrinsic and extrinsic defect relaxation behavior of ZnO ceramics. *J. Appl. Phys.* **2012**, *111*, 6132. [CrossRef]
28. West, A.R.; Andres-Verges, M. Impedance and Modulus Spectroscopy of ZnO Varistors. *J. Electroceramics* **1997**, *1*, 125–132. [CrossRef]
29. Huang, Y.; Wu, K.; Tang, Z.; Xin, L.; Zhang, L.; Li, J. Investigation of electrical inhomogeneity in ZnO varistor ceramics based on electronic relaxations. *Ceram. Int.* **2019**, *45*, 1110–1114. [CrossRef]

30. He, J.; Li, S.; Lin, J.; Zhang, L.; Feng, K.; Zhang, L.; Li, J. Reverse manipulation of intrinsic point defects in ZnO-based varistor ceramics through Zr-stabilized high ionic conducting β_{III}-Bi_2O_3 intergranular phase. *J. Eur. Ceram. Soc.* **2018**, *38*, 1614–1620. [CrossRef]
31. Janotti, A.; Van de Walle, C.G. Fundamentals of zinc oxide as a semiconductor. *Rep. Prog. Phys.* **2009**, *72*, 126501. [CrossRef]
32. Stucki, F.; Greuter, F. Key role of oxygen at zinc oxide varistor grain boundaries. *Appl. Phys. Lett.* **1990**, *57*, 446–448. [CrossRef]
33. Bueno, P.R.; Leite, E.R.; Oliveira, M.M.; Orlandi, M.O.; Longo, E. Role of oxygen at the grain boundary of metal oxide varistors: A potential barrier formation mechanism. *Appl. Phys. Lett.* **2001**, *79*, 48–50. [CrossRef]
34. Eda, K. Zinc oxide varistors. *IEEE Electr. Insul. Mag.* **1989**, *5*, 28–30. [CrossRef]
35. Emtage, P.R. The physics of zinc oxide varistors. *J. Appl. Phys.* **1977**, *48*, 4372–4384. [CrossRef]
36. Chiang, Y.; Kingery, W.D.; Levinson, L.M. Compositional changes adjacent to grain boundaries during electrical degradation of a ZnO varistor. *J. Appl. Phys.* **1982**, *53*, 1765–1768. [CrossRef]
37. Topcagic, Z.; Tsovilis, T.E. Varistor Electrical Properties: Microstructural Effects. In *Encyclopedia of Materials: Technical Ceramics and Glasses*; Pomeroy, E.M., Ed.; Elsevier: Oxford, UK, 2021; pp. 254–271.
38. Zhou, Q.; Yang, H.; Huang, X.; Wang, M.; Ren, X. Numerical modelling of MOV with Voronoi network and finite element method. *IET High Volt.* **2021**, 1–8. [CrossRef]
39. Greuter, F. ZnO Varistors: From Grain Boundaries to Power Applications. In *Oxide Electronics*; Wiley: Hoboken, NJ, USA, 2021.

energies

MDPI

Article

Research on the Non-Contact Pollution Monitoring Method of Composite Insulator Based on Space Electric Field †

Dongdong Zhang [1,2,*], Hong Xu [1], Jin Liu [3], Chengshun Yang [1], Xiaoning Huang [1], Zhijin Zhang [2] and Xingliang Jiang [2]

1 School of Electrical Engineering, Nanjing Institute of Technology, Nanjing 210000, China; xuhong0126@163.com (H.X.); yangchengshun@126.com (C.Y.); liuguofu0@126.com (X.H.)
2 State Key Laboratory of Power Transmission Equipment & System Security and New Technology, Chongqing University, Chongqing 400000, China; zhangzhijing@cqu.edu.cn (Z.Z.); xljiang@cqu.edu.cn (X.J.)
3 State Grid Zhejiang Ninghai County Power Supply Company, Ningbo 315000, China; liujin_njit@163.com
* Correspondence: zhangdd@njit.edu.cn
† This paper is an extended version of our paper published in 2020 IEEE International Conference on High Voltage Engineering and Application (ICHVE), Beijing, China, 6–10 September 2020.

Abstract: Through spatial electric field monitoring, it is expected to realize insulator pollution condition monitoring and contamination flashover warning in a non-contact way. Therefore, in this paper, the spatial electric field distribution characteristics of 110 kV composite insulators are simulated, where the effects of different surface states and their discharge levels on the spatial electric field of insulators are analyzed. On this basis, a non-contact monitoring method for composite insulator pollution based on the spatial electric field is proposed. The results show that there are significant differences in the spatial electric field of the composite insulator among three conditions, namely cleaning, pollution layer wetting, and dry band arcing. Increases of pollution layer wetting and dry band arcing would lead to an increase of the amplitude of the spatial electric field of the insulator. Verification experiments well indicated that it is feasible to identify the degree of pollution layer wetting as well as dry band arcing of the insulator string by fixed-point monitoring, the spatial electric field signal at the cross-strand of d = 0.5 m and directly opposite the last three positions. Research results can provide references for the online monitoring of overhead line polluted insulators and its flashover warning.

Keywords: finite element; composite insulator; electric field strength; operating state; pollution flashover

Citation: Zhang, D.; Xu, H.; Liu, J.; Yang, C.; Huang, X.; Zhang, Z.; Jiang, X. Research on the Non-Contact Pollution Monitoring Method of Composite Insulator Based on Space Electric Field . *Energies* **2021**, *14*, 2116. https://doi.org/10.3390/en14082116

Academic Editor: Issouf Fofana

Received: 10 March 2021
Accepted: 8 April 2021
Published: 10 April 2021

1. Introduction

Compared with the traditional string of porcelain and glass insulators, composite insulators, with excellent electrical and mechanical properties, are widely accepted in high-voltage power transmission systems. However, aging and deterioration of composite insulators are particularly serious in some typical heavily soiled, coastal, high-temperature, and high-humidity areas, so that their water-repellent performance and anti-fouling flash performance are significantly weakened, which has led to several pollution flashover accidents [1–4]. In 2016, the insulators of the ±600kV DC transmission line in Dezhou, Shandong were discharged due to heavy pollution in the shed. In 2018, the post-insulators of the Henan substation were polluted by heavy industrial dust, resulting in multiple pollution flashovers throughout the year.

Therefore, effective and stable online monitoring and early warning of pollution flashover for the pollution of external insulation in harsh environments still have practical engineering significance. Some scholars have conducted a large number of studies on the online monitoring of dirty insulators. In the current research, the leakage current method is widely adopted, but it still has some problems [5–9]. The literature [9] points out that the method of measuring the leakage current using a centrifugal structural current sensor or a microcurrent sensor has the disadvantages of single feature and consuming excessive transmission load. At present, relevant scientific research institutions are still carrying out tests. As a matter of fact, this method still has certain shortcomings. There is no unified conclusion about the selection of characteristic parameters of current signal leakage, pollution degree evaluation algorithms, and the corresponding early warning threshold settings. In addition, the sensors for leakage current monitoring require contact installation with the tower, which is not easy for post-maintenance.

Furthermore, other scholars have proposed some advanced online monitoring methods, such as thermal imaging, ultrasound, UV imaging, etc. [10–13]. However, the literature [14] points out that thermal imaging is influenced and limited by climatic conditions, while the ultrasound method has the problems of coupling and attenuation, and it is also limited by the performance of the ultrasonic converter. For the UV imaging-based external insulation detection technology for power transmission and distribution, there have been many years of research experience, but there is still no unified opinion on the selection of characteristic parameters such as spot area and photon number, and the recognition algorithm is complicated [15]. In addition, the actual operating insulators in the line generally only have severe corona discharge at the high-voltage end, so there are certain defects in the applicability of ultraviolet detection.

As one of the important parameters of electrical equipment, the electric field strength can well reflect the operating state of the external insulation of power transmission and transformation, and it can be measured in a non-contact manner [16,17]. Existing studies have shown that contamination of the surface of an insulator affects its surrounding electric field distribution. For example, the literature [18,19] pointed out that the electric field distribution around the insulator under different contaminated conditions is significantly different, and when the contaminated composite insulator discharges along the surface, the electric field distribution of the local arc head significant changes from a weak vertical component to a strong vertical component structure. The authors in [20] simulated the microscopic development process of the local arc of the pollution flashover and proposed the influence of different pollution distribution conditions and the salt density conductivity of different pollution layers on the local potential and electric field distribution of composite insulators. They [20] also proved that the insulator pollution flashover process is directly associated with the variation of its surrounding electric field strength.

However, the above research did not compare the electric field distribution characteristics of the insulators under different surface conditions in order to obtain the relationship between the changes of insulator spatial electric field and its surface pollution conditions, which can provide theoretical support for non-contact pollution insulator condition monitoring and the provision of a pollution flashover warning. In addition, other monitoring and detection methods have their own defects, as were aforementioned. In summary, to propose a novel and non-contact condition monitoring method of a polluted insulator based on the spatial electric field, this paper takes 110 kV composite insulator string as the object and studies the spatial electric field variation laws of insulator strings under different pollution conditions through finite element simulation. The influence of observation position and distance on the spatial electric field distribution of the insulator string is compared and analyzed. Furthermore, the correlation relationship between the spatial electric field amplitude and pollution surface conditions is obtained and experimentally verified. The research results can provide a reference for non-contact monitoring and pollution flashover early warning of the external pollution insulation of the power system.

2. Finite Element Simulation Model for Pollution Insulator

2.1. Basic Simulation Settings

The basic setup of the simulation calculation domain is as follows: (1) the domain is solved with single-phase loading voltage [21]; (2) the three-dimensional axisymmetric structure of the insulator string is simplified to a two-dimensional model for analysis; (3) the pollution layer, water film, etc., are set as uniform media; (4) the simulation mainly considers the changes in electric field distribution caused by the boundary conditions and properties of the medium, while ignoring corona discharge, water droplet flow, and other factors.

The conductivity of the insulator surface in a clean state is very small, and its electric field is a capacitively distributed axisymmetric electric field. The electrostatic field method can be used to establish a two-dimensional cross-sectional model for solution. Once the contamination on the surface of the contaminated insulator is moistened, the conductivity increases sharply, and the surface leakage current increases, so the surface electric field of the insulator becomes a non-static field distribution. Considering the power frequency alternating current, the electric field distribution on the surface of the shed exhibits resistance–capacitance properties. Therefore, the electrostatic field method alone cannot solve the electric field distribution of contaminated insulators. In this paper, the calculation domain of the electric field distribution of the dirty insulator is set as the quasi-current field, and the complex domain is applied. At this time, the insulator potential changes from a scalar to a complex vector, and the calculation domain potential distribution equation is

$$\frac{1}{r}\frac{\partial}{\partial r}\left\{r(\sigma + jw\varepsilon)\frac{\partial \phi}{\partial r}\right\} + \frac{\partial}{\partial z}\left\{(\sigma + jw\varepsilon)\frac{\partial \phi}{\partial z}\right\} = 0 \tag{1}$$

Boundary conditions for composite insulators are as follows:

$$\begin{cases} \varphi|_{l_0} = U \\ \varphi|_{l_1} = 0 \end{cases} \tag{2}$$

Since the electric field strength is very small at a place far away from the insulator, it has no effect on the simulation results. Therefore, on the boundary of the model, we set the bottom surface as the ground and the other five surfaces as the artificial cut surface. The voltage on the artificially cut surface is zero.

The formula for the boundaries of different media are as follows:

$$\begin{cases} \phi_1 = \phi_2 \\ (\sigma_1 + jw\varepsilon_1)\frac{\partial \phi_1}{\partial n} = (\sigma_2 + jw\varepsilon_2)\frac{\partial \phi_2}{\partial n} \end{cases} \tag{3}$$

The finite element solution of the variable partition equation is as follows:

$$F(\phi) = \frac{1}{2}\int_{\Omega}(\sigma + j\omega\varepsilon)(\nabla\phi)^2 d\Omega + \int_{\Gamma = L_1}(\sigma + j\omega\varepsilon)\left(\frac{1}{2}f_1\phi^2 - f_2\phi^2\right)dT \tag{4}$$

The function F is the intermediate variable for solving the electric potential distribution based on the variational method, and its minimum value is the solution of the differential equation, where ε and σ are the conductivity and dielectric constant of the material, respectively; r and z are the coordinates in the model; l_0 is the high-voltage electrode; l_1 is the ground electrode; Φ is the potential; Ω is the solving field; T is the curve surrounding the solving field; and W is the power angle frequency.

2.2. Simulation Model

This paper selected the FXBW$_4$-110/120 composite insulator as the simulation sample. As shown in Figure 1, the minimum nominal creepage distance of the insulator is 3600 mm, the structure height is 1440 mm, the insulation distance is 1200 mm, and the dielectric material properties are as shown in Table 1 [22–24]. The conductivity of the water was set at 0.03 S/m to simulate the pollution severity level I. The insulator shed number was set from No. 1 to No. 27 with the high voltage end as the starting point.

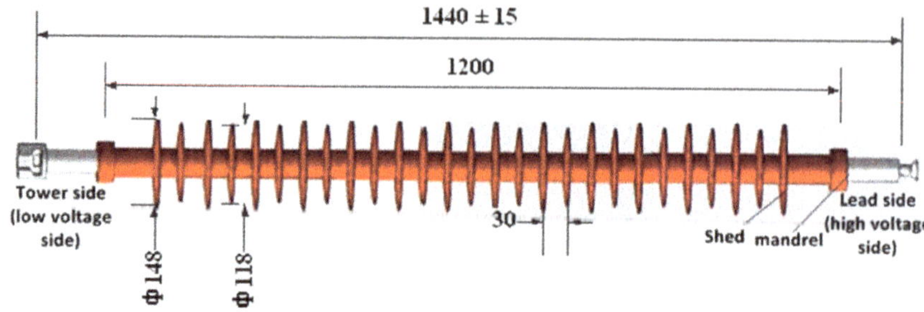

Figure 1. Diagram of FXBW$_4$-110/120 composite insulator structure diagram (unit: mm).

Table 1. Properties of each dielectric material.

Type of Material	Relative Dielectric Constant ε	Conductivity (S/m)
Silicone rubber	3.5	1×10^{-12}
Goldsmith	1000	5.998×10^{7}
Mandrel	6	1×10^{-12}
Water droplet/film	81	0.03
Air	1.02	1×10^{-22}

In this paper, the unbounded domain was processed by the manual truncation method, and the structure height of the composite insulator was set as 3 times the calculation domain, as shown in Figure 2a. The spatial electric field distribution on a straight line parallel to the insulator string was studied at different distances d outside the shed, as shown in Figure 2b. The high voltage end of the composite insulator was the highest amplitude-phase voltage, and the low voltage end and the side of the gold were grounded, of which the potential was 0 kV.

2.3. Pollution Conditions Setting Method

In this paper, simulation models of composite insulators with different wet pollution states and different arc starting stages were established. The specific settings were as follows:

(1) The water bead morphology and water film coverage area on the surface of the shed are defined by the water spray classification method. As shown in Figure 2c, the water bead formation model corresponds to HC1. As shown in Figure 2d, the intermittent water band model corresponds to HC4.

(2) In actual situations, the high-voltage end and the low-voltage end are most likely to produce local arcs. As shown in Figure 2e, this paper simulates two different arc starting stages, namely high voltage end arcing (stage 1), and high and low voltage two-end arcing (stage 2). The local arc is simulated by setting the potential difference on the surface boundary of the insulator shed surface, and local arc bridging regions are simulated with alternating potentials. In order to obtain the value of this potential difference, we need to calculate the voltage of the arc in the simulation using Equation (5). The arc volt–ampere characteristic equation is as follows:

$$U_a = AI^{-n_a}L_a \tag{5}$$

where U_a is the peak value of the local arc voltage; I is the peak value of leakage current; L_a is the arc length; and A and n_a are arc coefficients, which are associated with environmental parameters. When the composite insulator string starts arcing, the leakage current is not high. According to the recommended value in the literature [25], in this paper, I was set as 10 mA, L_a was set as 60 mm, A was set as 60, and n_a was set as 0.35. Under this case, the arc potential was about 1500 V, and this value was used in the computation model to set the boundary potential.

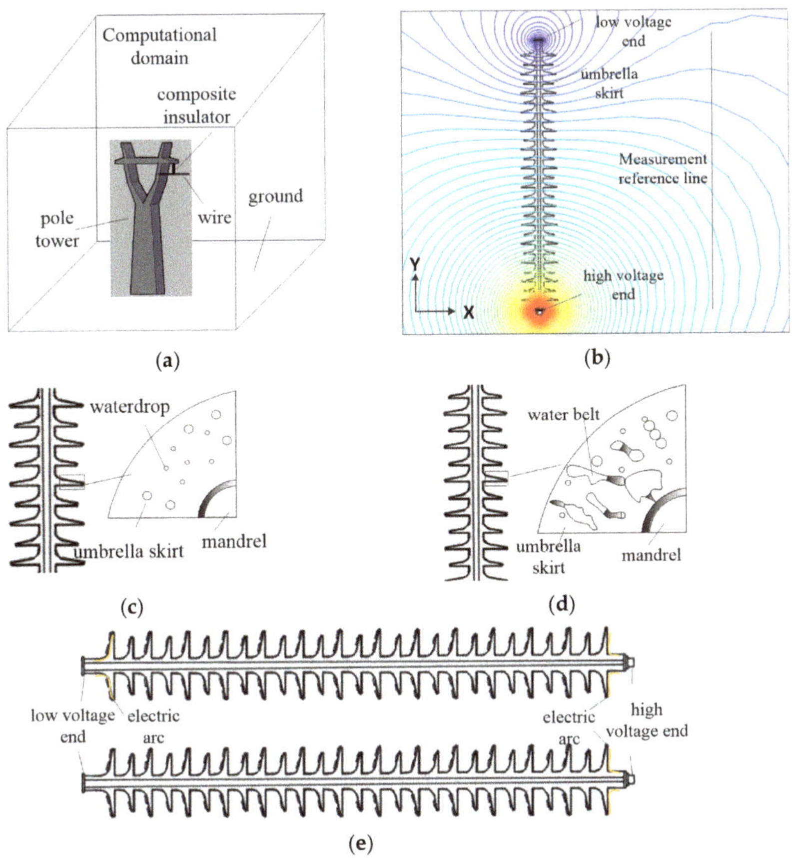

Figure 2. Simulation diagrams of composite insulator in different pollution conditions: (**a**) composite insulator string model, (**b**) spatial electric field simulation measurement, (**c**) water bead formation model, (**d**) water belt formation model, (**e**) local arcing model.

3. Space Electric Field Distribution Simulation Results and Analysis

3.1. Electric Field Distribution of Clean Insulator Strings

Figure 3 shows the electric field magnitude distribution of clean composite insulator strings at different radial distances. As shown in Figure 3, an increase of d would lead to a decrease of the spatial electric field magnitude. When the radial distance d increases from 0.2 m to 0.5 m, the insulator string spatial electric field magnitude attenuation can reach 80%. When d is greater than 0.5 m, the distribution curve no longer has a saddle shape. Therefore, in this paper, d = 0.3 m and 0.5 m were selected to study the spatial electric field distribution characteristics of insulator strings under different pollution conditions. The electric field change rate ΔE was set as follows:

$$\begin{cases} \Delta E_{i,k} = (|E_k - E_i|)/E_i \\ \Delta E_{i,m} = (|E_m - E_i|)/E_i \end{cases} \quad (6)$$

where i represents the clean composite insulator string, k represents the pollution layer wetting condition, and m represents the dry band arcing condition.

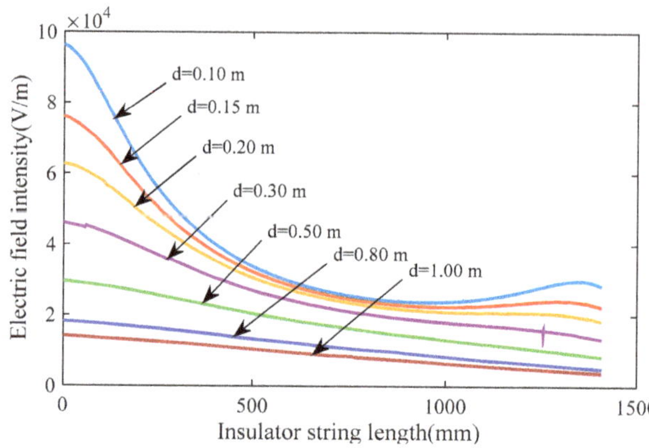

Figure 3. Electric field distribution of clean composite insulator string.

3.2. Electric Field Distribution of Insulator Strings under Surface Pollution Wetting Condition

The spatial electric field distribution of the insulator string under different pollution wetting conditions is shown in Figure 4. As shown in Figure 4, when a water film formed on the insulator pollution surface, the spatial electric field showed obvious fluctuations with the change of the distance of the high voltage end. This may be due to the change of the resistance–capacitance distribution characteristics of the insulator string caused by the formation of conductive water bands. Compared to the clean condition, the spatial electric field values of the insulator under the pollution wetting condition changed significantly. The maxima of $\Delta E_{i,k}$ were 3.2% and 53.7% for the droplet formation condition and water film formation condition, respectively, when d = 0.3 m. When d = 0.5 m, the maxima of $\Delta E_{i,k}$ were 3.2% and 44.2% under the two respective conditions. The maximum values of $\Delta E_{i,k}$ occurred at around 1000 mm and 1400 mm from the high voltage end.

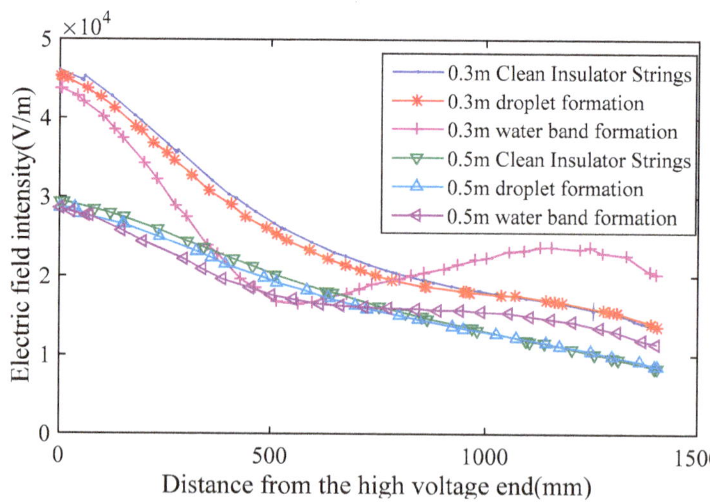

Figure 4. Effects of wet pollution condition on space electric field of composite insulator.

3.3. Electric Field Distribution of Insulator Strings under Dry Band Arcing Conditions

Figure 5 shows the electric field distribution of the composite insulator string under dry band arcing conditions. As shown in Figure 5, in the case of a local arc, the spatial electric field distribution of the insulator string no longer exhibited a saddle shape, because the local arc and the conductive water film greatly changed the potential distribution on the surface of the insulator. Compared to the clean condition, the spatial electric field values of the insulator under dry band arcing conditions changed significantly. The maximum values of $\Delta E_{i,k}$ were 103.6% and 151.9% for the arcing stage 1 and arcing stage 2, respectively, when d = 0.3 m. When d = 0.5 m, the maximum values of $\Delta E_{i,k}$ were 86.7% and 131.5% under the two respective arcing stages. The maximum values of $\Delta E_{i,m}$ also appeared at about 1000–1400 mm from the high voltage end.

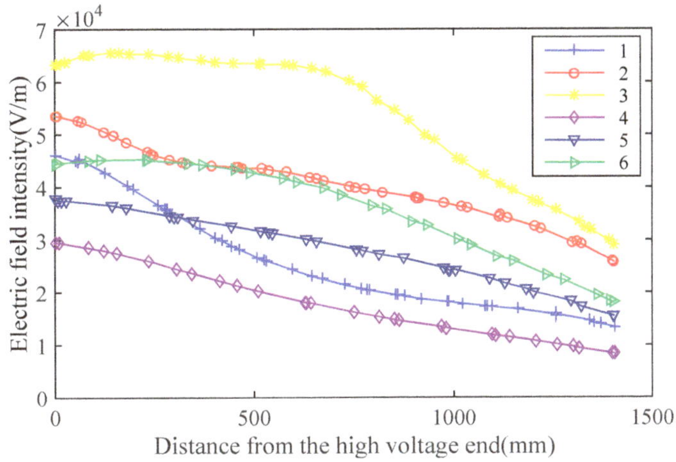

Figure 5. Effect of local arc on space field of composite insulator (1—d = 0.3 m clean insulator string, 2—d = 0.3 m arcing stage 1, 3—d = 0.3 m arcing stage 2, 4—d = 0.5 m clean insulator string, 5—d = 0.5 m arcing stage 1, 6—d = 0.5 m arcing stage 2).

4. Non-Contact Monitoring Method for Contaminated Insulators

Previous research has shown that the amplitude of the insulator's spatial electric field changes most obviously under different surface contamination states. Therefore, the simulation results of the spatial electric field amplitude of the insulators under all the above-mentioned typical pollution conditions were selected and drawn into a broken line diagram, as shown in Figure 6. As shown in Figure 6, the amplitude of the spatial electric field near the low-voltage end of the insulator changed obviously with the variation of pollution conditions, under both d = 0.3 m and d = 0.5 m. The maximum rate of change of the spatial electric field amplitude of the insulator was similar, and they all appeared at a distance of about 1000–1400 mm from the high voltage end, that is, the low voltage end.

Taking into account the convenience of on-site installation and subsequent maintenance, the safety net distance d of the monitoring sensor should be as large as possible; therefore, in this paper, d = 0.5 m was selected as the representative position for monitoring the spatial electric field distribution characteristics of the insulator string under different surface pollution states.

As was mentioned, it is expected to be close to the low-voltage end, and the greater the d, the easier it is to install and use the space electric field probe. Therefore, d = 0.5 m was selected, and the spatial electric field amplitudes were compared at the corresponding positions of the three insulators of No. 25, No. 26, and No. 27 under different pollution layer conditions, which are recorded as A, B, and C, respectively (as shown in Figure 7), and the simulation results were drawn as a line chart, as shown in Figure 8.

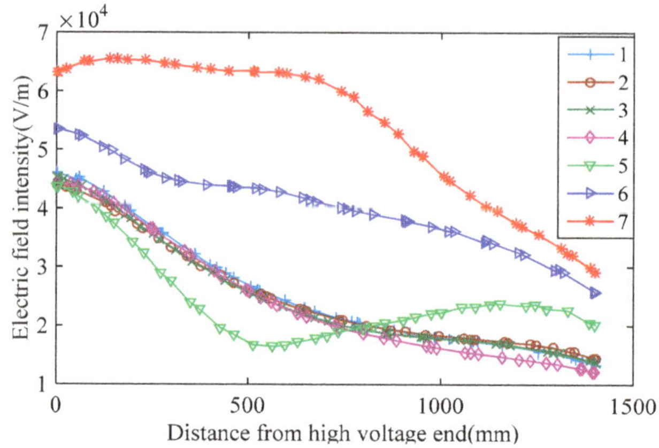

(**a**) d = 0.3 m

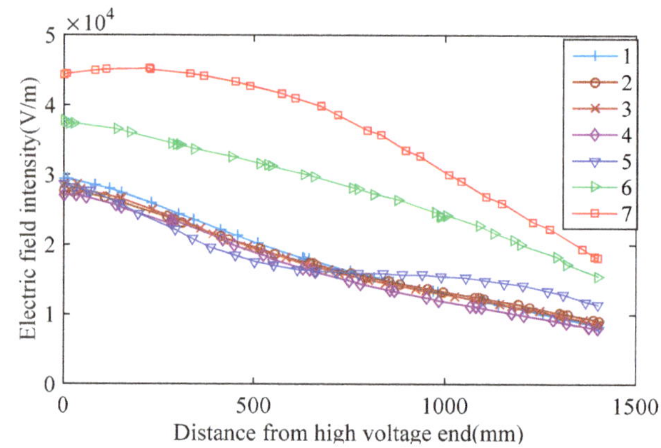

(**b**) d = 0.5 m

Figure 6. Effect of different pollution conditions on the space electric field of composite insulator. (1—clean insulator strings, 2—dry state, 3—a few drops of water, 4—many water drops, 5—water band formation, 6—localized arc phase I, 7—localized arc phase II).

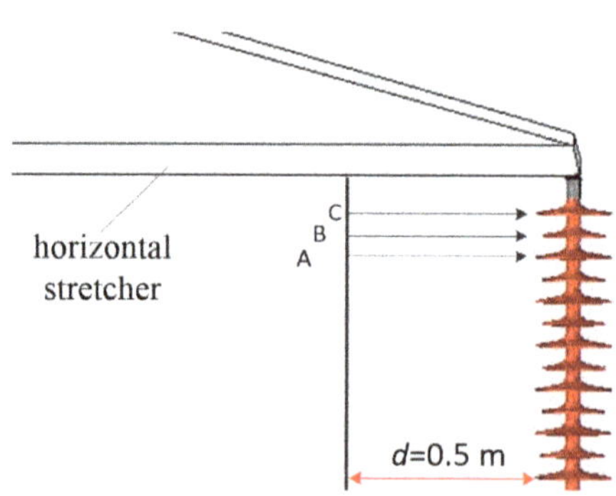

Figure 7. Monitoring position diagram of insulator unit No. 25, No. 26, No. 27.

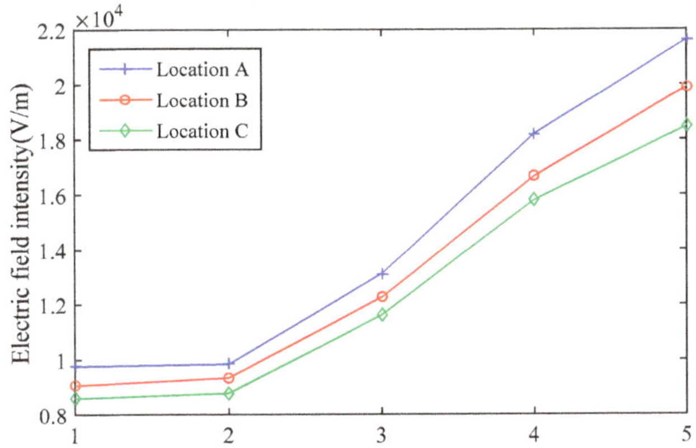

Figure 8. Changes in the amplitude of the electric field at three locations with different pollution states (1—clean insulator strings, 2—droplet formation, 3—water band formation, 4—localized arc phase I, 5—localized arc phase II).

It can be seen from Figure 8 that under the condition of surface pollution, when the insulator developed from a surface pollution wetting state to a stage where local arcs were generated, the amplitude of the space electric field of the insulator at three locations gradually increased, and the closer to the pollution flashover, the greater the increase. Therefore, the above results indicate that it is feasible to identify the degree of contamination and the degree of arc development by monitoring the space electric field signal at a point of d = 0.5 m at 110 kV composite insulator cross-arm, directly facing the designated patch position. At that location, the electric field value is in the scale of 8 kV/m to 22 kV/m under different pollution layer conditions, and after calculation, when the insulator pollution surface is wetting, the space electric field amplitude may increase by 3–43.1%. When a local arc occurs, the amplitude of the space electric field may increase by 83.6–119.4%.

5. Artificial Verification Test

The research results in this paper preliminarily show that there is a clear correlation between the changing rate of the spatial electric field of the polluted composite insulator string and its surface pollution condition. Based on the simulation, the spatial electric field waveform during insulator flashover was monitored at a fixed point in laboratory to propose a non-contact method for pollution insulator surface state identification and pollution flashover warning. Due to the limitation of test power supply capacity, this paper selected a 35 kV short composite insulator as the sample for the artificial pollution test. The structure parameters of the sample are shown in Table 2.

Table 2. Structure parameters of the sample (unit: mm).

Type	Height	Shed Diameter	Creepage Distance	Structure Diagram
FXBW-35/70	620	129/89	1280	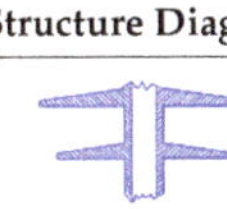

The test layout and circuit diagram are shown in Figure 9. The test power supply was a 100 kVA test transformer with a maximum AC output voltage of 50 kV and a rated current of 2.0 A. A plexiglass jar with an inner diameter of 80 cm and a height of 120 cm was used to simulate the environment of the fog chamber. The sample was suspended vertically in the fog chamber, and the ultrasonic fog generator was passed into the fog chamber from the top to wet the surface of the sample. In order to fully wet the surface of the entire series of test samples, four fans were installed under the glass jar to distribute the fog evenly and fill the entire fog chamber. A d-dot sensor [26] was used to measure the electric field strength in the space around the insulator, which could convert the electric field strength signal into a millivolt–volt voltage signal for oscilloscope acquisition. It was hung on the cross arm of the insulator during the test, and the distance from the insulator was 0.5 m. In order to increase the contrast of the spatial electric field signal, the leakage current of the contaminated insulator string was measured by a 1 ohm non-inductive sampling resistor in this test, and the signal was also recorded by an oscilloscope.

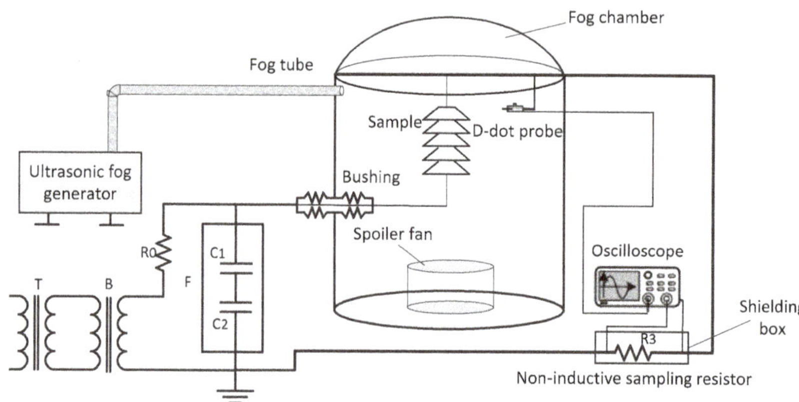

(**a**) Test circuit: T is the regulator, B is the transformer, R_0 is the protecting resistor, F is the voltage divider with its two capacitor C_1 and C_2, R_3 is the 1 ohm sampling resistor.

(**b**) Test equipment

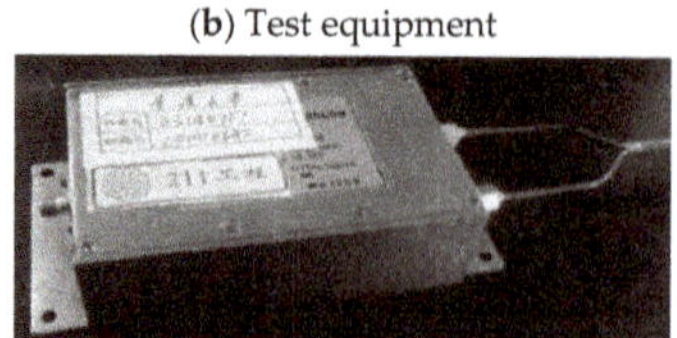

(**c**) D-dot probe

Figure 9. Laboratory test arrangement and circuit diagram.

In this paper, the withstand voltage method was used to simulate the pollution flashover process of the insulator under the operating lines, that is, applying the operating voltage first and then opening the fog generator. The corresponding electrical measurement signals during the withstand process were recorded. The test procedure is as follows:

(1) Cleaning. The sample was cleaned and allowed to dry naturally.
(2) Polluting. The equivalent salt deposit density, which was set as 0.05 mg/cm^2, was simulated by sodium chloride, and the non-soluble deposit density was simulated by kaolin. Their mas ratio was set as 1:6. During polluting of the insulator sample, the surface was firstly covered with a thin layer of kaolin, and then the surface was uniformly brushed to make it evenly coated with a pollution layer, and then the polluted sample was dried in the shade for 24 h.
(3) Evaluating. The sample was hung, the wiring circuit was arranged, the voltage was evenly increased to the rated operating voltage of the insulator, and the ultrasonic fog generator and the fans were turned on. In order to fully wet the surface of the entire series of the sample, the wind direction of the fans went up vertically, and the speed of wind was 2–3 m/s. The arc development on the surface of the sample and the waveform of the oscilloscope were recorded in real time, including electric field signals and leakage current signals. If the flashover did not occur for one hour, the test was stopped. During the test, the arc development was observed and the signal changes recorded by the oscilloscope, and a single rising edge trigger mode was set to intercept the spatial electric field as well as the leakage current waveform at the moment of local arc generation.

The withstand voltage time of this test was set as 1 h. During the test, local arcs were found on the surface of the insulator many times, but no flashover occurred till the end of the withstand process. This may be due to low contamination degree, cold fog, and the long withstand time, which caused the loss of the pollution layer. In addition, due to the good hydrophobicity of the composite insulating material and the insufficient pollution degree, a stable power frequency arc was not generated during the test, so the recorded leakage current waveform during pollution flashover of the sample showed pulse characteristics. The typical waveform obtained during the test is shown in Figure 10.

It can be seen from the figure that the spatial electric field signal at the characteristic position of the insulator under dry conditions was basically a sinusoidal waveform with no obvious distortion. After the dirty layer was wet for 12 min, the dirty layer on the surface of the insulator showed obvious conductive characteristics, and the resistive component of the leakage current increased. Coupled with the influence of corona discharge interference, the phase of the space electric field signal was shifted and the waveform was distorted. After 35 min of pressure, the dirt layer was fully wet, and the local arcs were alternately extinguished, causing multiple pulses of leakage current and space electric field signals. The distortion of the spatial electric field signal was more obvious, especially at its peak.

Comparing the leakage current and the signal waveform of the space electric field at different stages, it can be seen from Figure 10 that for the leakage current in the three states of the pollution flashover discharge of the composite insulator, the surface was dry, the pollution layer was wet, and the local arc started. The change of the current average value was not very obvious, but the amplitude change was obvious, and the largest pulse amplitude reached 500 mA, while the smallest pulse amplitude was about 10 mA. However, the spatial electric field signal waveform not only had obvious amplitude changes, but also had obvious pulse signals. Moreover, the effective value of the spatial electric field waveform also had an obvious upward trend in the three states, which was consistent with the previous simulation results, thus verifying the previously proposed non-contact monitoring method of insulator pollution flashover based on the spatial electric field signal. In addition, for the spatial electric field signal, there are obvious distortion characteristics in the state where the dirt layer is wet and there is local arcing. In summary, compared with the leakage current signal, the space electric field can not only be measured in a non-contact manner, but it also has richer change characteristics if serving as a basis for early warning of pollution flashover.

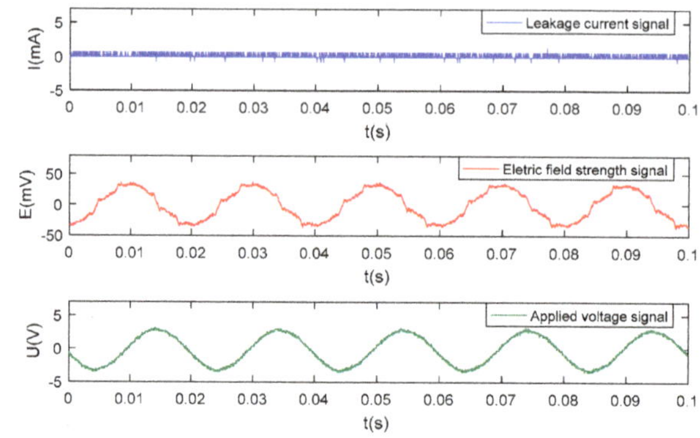

(a) T = 0 min, dry condition

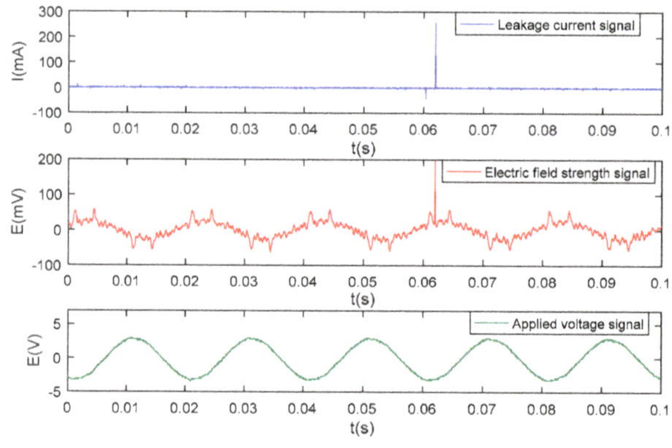

(b) T = 12 min, pollution layer wetting

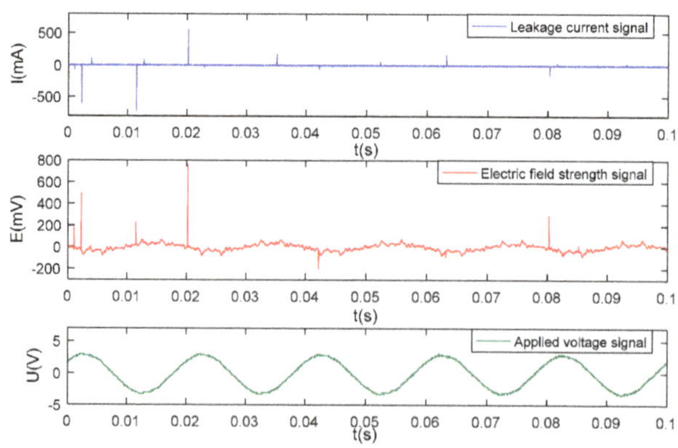

(c) T = 28 min, surface wetting and dry band arcing

Figure 10. Typical waveform during the artificial tests.

After calculation, the comparison of the characteristic parameters of the leakage current signal and the spatial electric field signal under different contaminated insulator surface conditions is shown in Table 3.

Table 3. Comparison of various characteristic parameters of leakage current signal and spatial electric field signal under different contaminated insulator surface conditions.

Signal Type	Spatial Electric Field Signal				Leakage Current Signal	
	Fundamental Wave Amplitude (mV)	Pulse Amplitude (mV)	Variation by 1 Wave Amplitude	Simulation Results	Fundamental Wave Amplitude (mA)	Pulse Amplitude (mA)
Dry	20	30	/	/	1.4	2
Pollution layer wetting	30	56	33.3%	3–43.1%	0.8	10–44
Wetting with arcs	41	190–680	90.5%	83.6–119.4%	3.1	16–560

It can be seen from the data in the table that under different contaminated insulator surface conditions, the change characteristics of the spatial electric field signal were more obvious than the leakage current. The test results also confirmed the validity of the previous simulation conclusions for 110 kV insulators. It is feasible to realize pollution flashover warning by monitoring the spatial electric field signal through a non-contact manner. In the follow-up, based on the artificial pollution test, the relationship between the characteristic parameters of the space electric field signal and the salt density of the insulator surface and the degree of wetness and discharge will be established, and corresponding pollution flashover warning measures are to be proposed.

When recommending the adoption of this methodology on live HVAC lines, we would like to supplement that compared with the leakage current monitoring device, which needs to arrange a current-collecting-ring on the surface of the insulator shed, the non-contact electric field monitoring device in this paper is hung under a cross-arm and 0.5 m away from the string, so we are sure that it has no influence on the insulation distance and thus will not impact the withstand performance on a live tower structure application. Secondly, compared to the insulator string, the electric field sensor is very small and away from the string, so it has little effect on the electric field distribution of the whole string, especially the electric field changing properties under different surface conditions. Moreover, the other phases' rated voltages will have less effects on the proposed non-contact method in a live system, not only because the other phase voltage source is far away from the electric field monitoring device compared with its own phase, but also because the grounded cross-arms and steel tower architecture of the other phase itself will play the role of electromagnetic shielding. Normally, since the same tower corresponds to the same pollution level, it is recommended to install only one non-contact monitoring device to realize flashover warning. However, for a wider tower structure with multiple phases, two or three devices may be better.

6. Conclusions

This paper studies the spatial electric field variation laws of insulator string under different pollution conditions through simulations and artificial verification tests and proposes a non-contact condition monitoring method of polluted insulators based on the spatial electric field. The conclusions are as follows:

(1) In the simulation, with the aggravation of the degree of pollution layer wetting and dry band arcing, the insulator spatial electric field amplitude near the low voltage end has an obvious increasing trend.

(2) In the simulation, for the 110 kV composite insulator, it is feasible to identify its pollution layer wetting condition and local arc development through fixed-point monitoring of the spatial electric field signal at a point of $d = 0.5$ m at the cross-arm, directly facing the last three units of the string.

(3) For the above monitoring locations, the simulation results show that the electric field value is in the scale of 8 kV/m to 22 kV/m under different pollution layer conditions. Compared with clean conditions, the space electric field amplitude increases by 3–43.1% under pollution layer wetting conditions and increases by 83.6–119.4% under local arcing conditions.

(4) The artificial test verified well the proposed method by simulation. Test results show that under different contaminated insulator surface conditions, the change characteristics of the spatial electric field signal are more obvious than the leakage current, which are waveform distortion, fundamental wave amplitude increase, and pulse amplitude increase. Thus, it is feasible to realize pollution flashover warning by monitoring the spatial electric field signal through a non-contact manner.

Author Contributions: Conceptualization, D.Z., H.X., Z.Z., X.J.; methodology, D.Z., H.X., J.L.; software, D.Z., J.L.; validation, H.X., J.L.; formal analysis, C.Y.; in-vestigation, H.X., J.L.; resources, D.Z., X.H.; data curation, H.X., J.L.; writing—original draft preparation, D.Z., H.X., J.L.; writing—review and editing, H.X., J.L.; visualization, H.X., J.L.; supervision, D.Z.; project administration, D.Z.; funding acquisition, D.Z., X.H. All authors have read and agreed to the published version of the manuscript.

Funding: This resea rch was funded by Open Research Fund of Jiangsu Collaborative Innovation Center for Smart Distribution Network, Nanjing Institute of Technology, grant number XTCX201806; Jiangsu Natural Science Foundation Project, grant number BK20181021; High-level Talent Introduction Scientific Research Foundation of Nanjing Institute of Technology, grant number YKJ201819; Nanjing Institute of Technology Graduate Science and Technology and Innovation Foundation Project, grant number TB202117042. The APC was funded by YKJ201819.

Conflicts of Interest: The authors declare no conflict of interest.

References

1. Yang, Z.; Fan, P.; Wang, Q.F.; Wang, S.W.; Chen, J.; Huang, W.; Wu, B.Q.; Zhou, D. Electric field analysis of internal gaps in composite insulators for high-voltage transmission lines. *Power Grid Clean Energy* **2019**, *35*, 1–8.
2. Zhang, Z.J.; Yang, S.H.; Jiang, X.L.; Qiao, X.H.; Xiang, Y.Z.; Zhang, D.D. DC flashover dynamic model of post insulator under non-uniform pollution between windward and leeward sides. *Energies* **2019**, *12*, 2345. [CrossRef]
3. Majid Hussain, M.; Farokhi, S.; McMeekin, S.G.; Farzaneh, M. Risk Assessment of Failure of Outdoor High Voltage Polluted Insulators under Combined Stresses Near Shoreline. *Energies* **2017**, *10*, 1661. [CrossRef]
4. Majid Hussain, M.; Farokhi, S.; McMeekin, S.G.; Farzaneh, M. Effect of uneven wetting on E-field distribution along composite insulators. In Proceedings of the 2016 IEEE Electrical Insulation Conference (EIC), Montreal, QC, Canada, 19–22 June 2016.
5. Hao, Y.; Liao, Y.; Kuang, Z.; Sun, Y.; Shang, G.; Zhang, W.; Mao, G.; Yang, L.; Zhang, F.; Li, L. Experimental Investigation on Influence of Shed Parameters on Surface Rainwater Characteristics of Large-Diameter Composite Post Insulators under Rain Conditions. *Energies* **2020**, *13*, 5011. [CrossRef]
6. Wang, S.H.; Tan, G.L. Fuzzy evaluation of insulator pollution flashover based on improved EMD de-noising and entropy weight method. *IEEE J. Inf. Comput. Sci.* **2015**, *12*, 5687–5696. [CrossRef]
7. Ghosh, R.; Chatterjee, B.; Chakravorti, S. A Novel Leakage Current Index for the Field Monitoring of Overhead Insulators Under Harmonic Voltage. *IEEE Trans. Ind. Electron.* **2018**, *65*, 1568–1576. [CrossRef]
8. Majid Hussain, M.; Farokhi, S.; McMeekin, S.G.; Farzaneh, M. Effect of cold fog on leakage current characteristics of polluted insulators. In Proceedings of the 2015 International Conference on Condition Assessment Techniques in Electrical Systems (CATCON), Bangalore, India, 10–12 December 2015.
9. Chandrasekar, S.; Kalaivanan, C.; Montanari, G.C.; Cavallini, A. Partial discharge detection as a tool to infer pollution severity of polymeric insulators. *IEEE Trans. Dielectr. Electr. Insul.* **2010**, *17*, 181–188. [CrossRef]
10. Ma, Z.B. Research and Design of Online Monitoring and Management System of Insulator Pollution in Distribution Networks in Mining Areas. Master's Thesis, China University of Mining and Technology, Xuzhou, China, 2019.
11. Jiang, T.; Wang, J.; He, W.; Yang, F. Non-contact on-line insulators monitoring based on UV-pulse detecting. In Proceedings of the World Automation Congress, Hawaii, HI, USA, 28 September–2 October 2008.
12. Albano, M.; Waters, R.T.; Charalampidis, P.; Griffiths, H.; Haddad, A. Infrared analysis of dry-band flashover of silicone rubber insulators. *IEEE Trans. Dielectr. Electr. Insul.* **2016**, *23*, 304–310. [CrossRef]
13. Prasad, P.S.; Rao, B.P. Condition monitoring of 11 kV overhead power distribution line insulators using combined wavelet and LBP-HF features. *IET Gener. Transm. Distrib.* **2017**, *11*, 1144–1153. [CrossRef]
14. Jin, L.; Tian, Z.; Ai, J.; Zhang, Y.; Gao, K. Condition evaluation of the contaminated insulators by visible light images assisted with infrared information. *IEEE Trans. Instrum. Meas.* **2018**, *67*, 1349–1358. [CrossRef]

15. Wei, J.C. Research on the Method of Ultraviolet Image Diagnosis of Faulty Insulators and Its Influencing Factors. Master's Thesis, Shandong University, Shandong, China, 2018.
16. Zhang, D.; Chen, S. Intelligent Recognition of Insulator Contamination Grade Based on the Deep Learning of Ultraviolet Discharge Image Information. *Energies* **2020**, *13*, 5221. [CrossRef]
17. Qi, B.; Zhao, X.L.; Zhang, S.Q.; Huang, M.; Li, C.R. Measurement of the electric field strength in transformer oil under impulse voltage. *IEEE Trans. Dielectr. Electr. Insul.* **2017**, *24*, 1256–1262. [CrossRef]
18. Arshad; Nekahi, A.; McMeekin, S.G.; Farzaneh, M. Effect of Pollution Severity on Electric Field Distribution along a Polymeric Insulator. In Proceedings of the 2015 IEEE 11th International Conference on the Properties and Applications of Dielectric Materials (ICPADM), Sydney, NSW, Australia, 19–22 July 2015.
19. Majid Hussain, M.; Farokhi, S.; McMeekin, S.G.; Farzaneh, M. The effects of salt contamination deposition on HV insulators under environmental stresses. In Proceedings of the 2015 IEEE 11th International Conference on the Properties and Applications of Dielectric Materials (ICPADM), Sydney, NSW, Australia, 19–22 July 2015.
20. Kong, D.S. Experimental Study on Surface Discharge of Silicone Rubber Insulators. Master's Thesis, Huazhong University of Science and Technology, Wuhan, China, 2016.
21. Rasara, H.L.; Wong, K.L. Study of electrical field and micro arcing in a wooden supporting structure using Finite Element Method. In Proceedings of the 2012 International Conference on High Voltage Engineering and Application, Shanghai, China, 17–20 September 2012.
22. Qiao, X.H.; Zhang, Z.J.; Jiang, X.L.; Sundararajan, R.J.; You, J.W. DC pollution flashover performance of HVDC composite insulator under different non-uniform pollution conditions. *Electr. Power Syst. Res.* **2020**, *185*. [CrossRef]
23. Jamshid, M.; Mohammad, M.; Shayegani, A.; Abbas, A. Effect of accumulated surface charges on DC flashover of SiR insulators under pollution and aging conditions. *Electr. Eng.* **2020**, *102*, 2123–2133. [CrossRef]
24. Bhavani, J.; Phani, K.C. Finite Element Modeling of Voltage and Electric Field Distribution along the Insulators. In Proceedings of the 2019 4th International Conference on Recent Trends on Electronics, Information, Communication & Technology, Bangalore, India, 17–18 May 2019.
25. Jiang, X.L.; Shu, L.C.; Sun, C.X. *Power System Pollution and Icing Insulation*, 1st ed.; China Electric Power Press: Beijing, China, 2009; pp. 59–124.
26. Huang, Q.J. Study on Partial Disrcharge On-line Mornitoring of High Voltage Switchgears Based on Ultral-High-Frequency Method. Master's Thesis, Chongqing University, Chongqing, China, 2013.

energies

MDPI

Article

Research on the Time-Domain Dielectric Response of Multiple Impulse Voltage Aging Oil-Film Dielectrics †

Chenmeng Zhang [1,*], Kailin Zhao [1,2], Shijun Xie [1], Can Hu [3], Yu Zhang [1] and Nanxi Jiang [1]

1 State Grid Sichuan Electric Power Research Institute, Chengdu 610041, China; 201811131098@cqu.edu.cn (K.Z.); sj-xie@163.com (S.X.); zy863129@163.com (Y.Z.); zcmwhu@163.com (N.J.)
2 State Key Laboratory of Power Transmission Equipment & System Security and New Technology, Chongqing University, Chongqing 400044, China
3 State Grid Sichuan Electric Power Company, Chengdu 610041, China; bobozzz123@126.com
* Correspondence: zcm@whu.edu.cn; Tel.: +86-028-6999-5620

† This paper is an extended version of our paper published in International Conference on High Voltage Engineering and Application 2020, Beijing, China, 6–10 September 2020.

Citation: Zhang, C.; Zhao, K.; Xie, S.; Hu, C.; Zhang, Y.; Jiang, N. Research on the Time-Domain Dielectric Response of Multiple Impulse Voltage Aging Oil-Film Dielectrics. *Energies* **2021**, *14*, 1948. https://doi.org/10.3390/en14071948

Academic Editors: Issouf Fofana and Bo Zhang

Received: 6 March 2021
Accepted: 30 March 2021
Published: 1 April 2021

Abstract: Power capacitors suffer multiple impulse voltages during their lifetime. With the multiple impulse voltage aging, the internal insulation, oil-film dielectric may deteriorate and even fail in the early stage, which is called accumulative effect. Hence, the time-domain dielectric response of oil-film dielectric with multiple impulse voltage aging is studied in this paper. At first, the procedure of the preparation of the tested samples were introduced. Secondly, an aging platform, impulse voltage generator was built to test the accumulative effect of capacitor under multiple impulse voltage. Then, a device was used to test the time-domain dielectric response (polarization depolarization current, PDC) of oil-film dielectric in different aging states. And finally, according to the PDC data, extended Debye model and characteristic parameters were obtained by matrix pencil algorithm identification. The results indicated that with the increase of impulse voltage times, the time-domain dielectric response of oil-film dielectric changed accordingly. The polarization current curve moved up gradually, the insulation resistance decreased when subjected to the repeated impulses. In frequency domain, the frequency spectrum of tan δ changed along with the impulse accumulation aging, especially at low frequency. At last, combined with the aging mechanism of oil-film dielectric under multiple impulse voltage, the test results were discussed.

Keywords: oil-film dielectric; accumulative effect; time-domain dielectric response; matrix pencil algorithm; extended Debye model

1. Introduction

With the leap in power infrastructure and the increase of social power consumption, the long-distance, large capacity and low loss UHV transmission project has gradually become an important part of the transmission network [1–4]. As the key equipment of reactive power compensation and filtering, power capacitor is widely used in power system. However, during their decades long operation cycles, frequent switching will expose capacitors to high operating overvoltages and impulse voltages. The cumulative effects of these impulse voltages will lead to deterioration and insulation failure of the oil-film dielectric of the capacitor, which seriously threatens the stable operation of the power system. Therefore, it is of great practical significance to study the failure laws of oil-film dielectrics and to be able to judge the degree of aging of capacitors under multiple impulse voltage exposure.

In 1956, Standring firstly proposed the phenomenon whereby the insulation state of a dielectric will degenerate under the action of impulse voltages, which he named "accumulative effect" [5]. However, his study concept was limited to solid and liquid dielectrics in combination. Since then, more and more researchers have devoted their efforts to investigating this phenomenon in composite dielectrics, such as oil-impregnated paper insulation and oil-film insulation. In order to reflect the cumulative effect of repeated impulses accurately, the most effective method is the U-N characteristic one. In [6], the cumulative effect of oil-film dielectrics was studied and their U-N characteristic was plotted as in Figure 1. The U-N curve shows that the repeated effect of impulse voltages made the breakdown voltage of the oil-film dielectric decrease. Figure 1 is divided into two parts, Area I and Area II. When the oil-film dielectric is subjected to less impulse voltages, its breakdown voltage decreases swiftly, which is represented by Area I. On the contrary, in Area II, with the accumulation of multiple instances of impulse voltage, the breakdown voltage of oil-film dielectrics drops slowly. Atomic force microscopy (AFM) was used in [7] to observe the surface morphology of oil-impregnated paper insulation. With the continual accumulation of impulse voltages, the roughness of the surface morphology increased. Moreover, more and more wart-like protuberances appeared and the height difference between different points became very obvious.

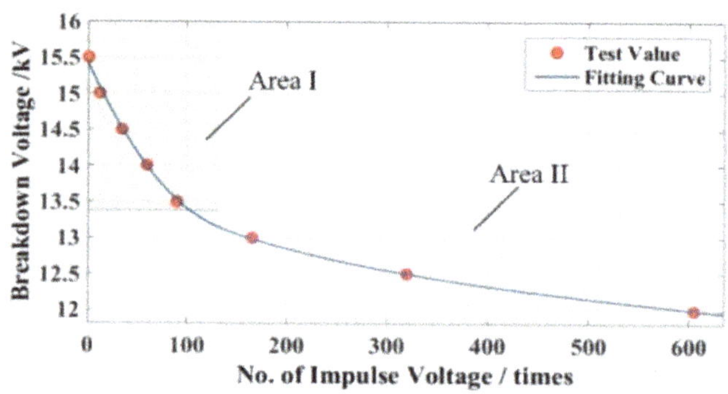

Figure 1. The U-N characteristics of an oil-film dielectric [6].

With regard to the research of the accumulative effect on dielectric breakdown, this field had been dug into deeply, and many breakthrough results were summarized, but when the dielectric was not broken down, a simple and efficient method was needed to reflect the accumulative effect and comprehend the aging mechanism. The traditional methods to detect the insulation state of power equipment are mainly divided into electrical, chemical and physical characteristic parameter measurements, including dissolved gas analysis (DGA), average degree of polymerization (DP), partial discharge (PD), etc. [8–10]. However, sampling for these methods is more or less difficult, the parameters are hard to detect, and measurements are easily affected by external factors. In recent years, the polarization-depolarization current method based on the time-domain dielectric response has been paid much attention by many scholars [11–13]. Under the action of an external electric field, the dipoles in the dielectric will change direction under the action of the external force, which is called a polarization process, mainly including turning polarization, ion displacement polarization and interface polarization. When the external electric field is removed, the dipole will return to the discrete state due to its relaxation property, which requires a certain time and energy loss. The dielectric response refers to the lossy relaxation polarization response [14,15]. When the insulation performance of dielectric materials decreases, the relaxation polarization response will change significantly, the degree of change can be used as a standard to judge the degree of aging of dielectric materials. The PDC method has the advantages of being non-destructive, with fast test speed and so on. The degree of aging of oil-film dielectrics can be effectively evaluated by using the PDC curve to extract the corresponding index and fitting calculations [16].

Based on the phenomenon of dielectric polarization, the dielectric response function of the oil-film insulation system is quantified in the extended time domain, and an equivalent circuit model (such as the extended Debye model) can also be established. Therefore, through reasonable parameter identification of the equivalent circuit model, the oil-film dielectric in different aging states can be judged. Zhang et al. [17] use the characteristic quantity of the initial slope of the recovery voltage as the basis to judge the state of aging of transformer oil paper insulation, but that paper fixed the number of branches of the Debye equivalent circuit, which is not universal. Conclusions can be drawn from [18] that the resistance R and capacitor C corresponding to the maximum and minimum time constants in the equivalent Debye model of oil-paper insulation can be used as the aging characteristic quantities. However, in [19], the depolarization current is only fitted and calculated by three exponential polynomials. In fact, there should be more types of interface polarization. The least squares method that identifies the parameters in the equivalent circuit as discussed in [20] contains integral operations, which is not only complicated to calculate, but also requires a high sampling frequency and has a great impact on the precision of parameters. An artificial intelligence algorithm is adopted by the authors in [21] to apply the hybrid algorithm of information entropy and particle swarm optimization to the identification of equivalent parameters, but the solution process is complex and prone to local optimization.

In order to solve the deficiencies mentioned above, the matrix pencil algorithm is adopted in this work to accurately identify the parameters of the extended Debye model of an oil film medium. The algorithm firstly discretized the depolarization current, constructed the Hankel matrix from the discretized values and decomposed the singular values. The number of large singular values was used as the RC branch number of the Debye equivalent circuit of oil-film dielectric, and on this basis, the characteristic parameters of each branch and the maximum depolarization current amplitude were obtained. Finally, the feasibility of the method was verified by experiments and simulations.

2. Basic Principle and Selection of Characteristic Quantity

2.1. The Dielectric Response Theory

When a voltage is applied to a measured dielectric, the composite dielectric can be equivalent to the capacitance, which can be understood as the applied electric field E(t) acting on the sample capacitance. According to the full current equation, the current inside the dielectric material can be expressed as [22,23]:

$$i(t) = C_0 \left[\frac{\sigma_0}{\varepsilon_0} U(t) + \varepsilon_\infty \frac{dU(t)}{dt} + \frac{d}{dt} \int_0^t f(t-\tau) U(\tau) d\tau \right] \quad (1)$$

where, C_0 represent the equivalent capacitance of the sample. σ_0 and ε_0 represent the DC conductivity and the vacuum dielectric constant of the dielectric, respectively. ε_∞ is the relative permittivity. *f(t)* is an attenuation function, which is used to express the response ability of dielectric polarization process, and the attenuation depends on the dielectric material and external factors.

When the applied voltage is replaced by the DC voltage, the dielectric is in the polarization state, that is, a constant dc source *U*(*t*) is used to charge the sample capacitor. Since it is a DC voltage, there is no differential term in Equation (1), and the polarization current i_p can be expressed as:

$$i_p(t) = C_0 U(t) \left[\frac{\sigma_0}{\varepsilon_0} + f(t) \right] \quad (2)$$

When the DC power supply is removed, the effect of the external electric field also disappears and it enters the depolarization process. The charged particle in the dielectric generates a depolarized current i_d contrary to the polarization current due to its relaxation property, then i_d can be expressed as:

$$i_d(t) = -C_0 U[f(t+t_d) - f(t)] \quad (3)$$

In Equation (3), t_d represents the polarization time. The PDC original curve is shown in Figure 2.

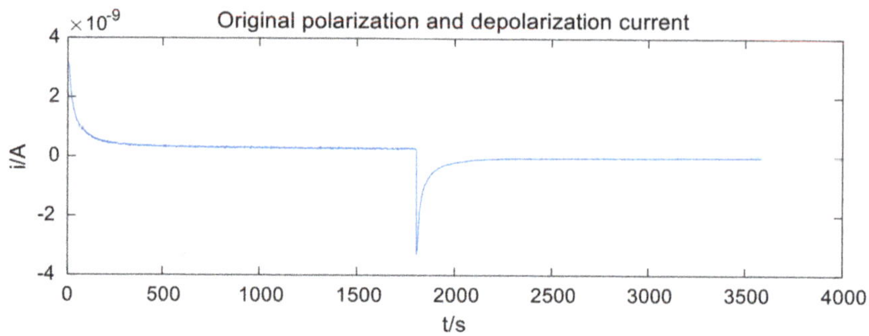

Figure 2. The PDC curve.

Therefore the insulation resistance R_0, one of the important parameters characterizing the insulation of power capacitors, can be derived from Equations (2) and (3):

$$R_0 = \frac{(i_p(t) - i_d(t))}{U(t)} \tag{4}$$

This parameter is sensitive to the state of the material, and is generally applicable to the aging detection of the insulation material. So it can be used as the basis to judge the aging degree of the oil-film dielectric.

The law of action of oil-film dielectric under an applied electric field can be equivalent with the extended Debye model [24], as shown in Figure 3. In the figure, the geometric capacitance of the dielectric is represented as C_0. R_0 is the insulation resistance of the composite medium. The equivalent resistance capacitance of each relaxation branch is replaced by R_i and C_i (i = 1, 2, . . . , N).

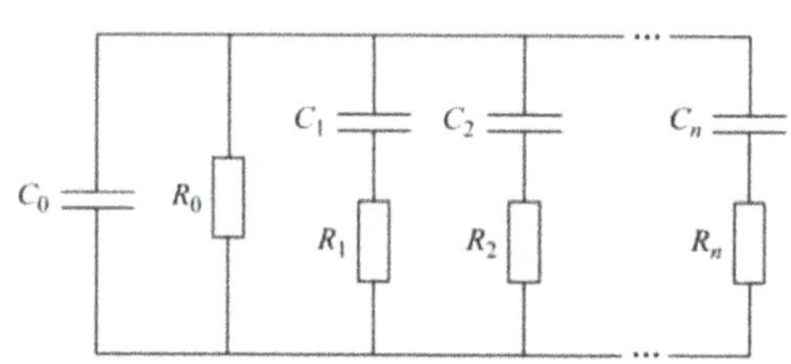

Figure 3. Extended Debye model.

Thus, it can be deduced that the equivalent admittance of the model is:

$$\frac{1}{Z} = j\omega C_0 + \frac{1}{R_0} + \sum_{i=1}^{n} \frac{1}{R_i + 1/(j\omega C_i)}, \tag{5}$$

Therefore, the dielectric complex capacitance can be deduced as:

$$C^* = \frac{1}{j\omega Z} = C_0 - j\frac{1}{\omega R_0} + \sum_{i=1}^{n} \frac{C_i(1 - j\omega R_i C_i)}{1 + (\omega R_i C_i)^2}, \tag{6}$$

The real and imaginary parts of the complex capacitor are:

$$C'' = \mathrm{Re}C^* = C_0 + \sum_{i=1}^{n} \frac{C_i}{1 + (\omega R_i C_i)^2}, \tag{7}$$

$$C''' = -\text{Im}C^* = \frac{1}{\omega R_0} + \sum_{i=1}^{n} \frac{\omega R_i C_i{}^2}{1 + (\omega R_i C_i)^2}, \tag{8}$$

so the tangent value of the dielectric loss angle tan δ, another characteristic quantity to judge the aging degree of the dielectric in frequency domain, can be expressed as the ratio between the imaginary part and the real part of the complex capacitance [25]:

$$\tan\delta = \frac{C'''}{C''} = \frac{\frac{1}{R_0} + \sum\limits_{i=1}^{n} \frac{R_i \omega^2 C_i{}^2}{1+(\omega R_i C_i)^2}}{\omega C_0 + \sum\limits_{i=1}^{n} \frac{\omega C_i}{1+(\omega R_i C_i)^2}}, \tag{9}$$

That is to say, as long as the polarization-depolarization current curve is collected and the parameters of each branch in the extended Debye model are identified more accurately, the above two physical quantities in time domain and frequency domain can be calculated, so as to judge the aging degree of insulating medium in different aging states.

2.2. Matrix Pencil Algorithm

In the PDC curve of oil-film dielectric, the polarization current is composed of the conductance current and the depolarization current. The depolarization current contains more information to reflect the dielectric insulation state than the conductance current. In order to avoid the influence of conductance current, depolarization current is used as the research object of the algorithm. According to the extended Debye model, as shown in Figure 3, the depolarization current $y(t)$ can be expressed as the discharge of n RC branches. That is, the linear superposition of n attenuation exponential functions:

$$\begin{aligned} y(t) &= x(t) + n_s(t) = \sum_{i=1}^{n} \frac{U_C}{R_i}\left(1 - e^{-\frac{t_p}{\tau}}\right) e^{-\frac{t}{R_i C_i}} + n_s(t) \\ &= \sum_{i=1}^{n} A_i e^{-\frac{t}{\tau_i}} + n_s(t) \end{aligned} \tag{10}$$

In the above equation, $x(t)$ is the noiseless signal under ideal conditions. $n_s(t)$ is the interference signal caused by environmental noise. Uc is the magnitude of the dc voltage. A_i is the amplitude coefficient of depolarization current. τ_i is the decay time constant. By discretization of Equation (10), the following expressions can be obtained:

$$\begin{aligned} y(kT_s) &= x(kT_s) + n(kT_s) \\ &= \sum_{i=1}^{n} A_i e^{-\frac{kT_s}{\tau_i}} + n(kT_s) \end{aligned} \tag{11}$$

where T_s stands for the sampling time interval, $k = 0, 1, 2, \ldots, N-1$, N is the maximum sampling number. Therefore, Hankel matrix constructed by sampling sequence y_k ($k = 0, 1, 2, \ldots, N-1$) can be expressed as:

$$Y = \begin{pmatrix} y_0 & y_1 & \cdots & y_L \\ y_1 & y_2 & \cdots & y_{L+1} \\ \vdots & \vdots & \vdots & \vdots \\ y_{N-L-1} & y_{N-L} & \cdots & y_{N-1} \end{pmatrix}, \tag{12}$$

L is the matrix pencil parameter, which is usually evaluated between $N/4$ and $N/3$. Singular value decomposition of Hankel matrix:

$$Y = SVD^T, \tag{13}$$

S is an orthogonal matrix of $(N - L) \times (N - L)$. V is a diagonal matrix of $(N - L) \times (L + 1)$. Its diagonal element σ_i, the singular value of Hankel matrix Y, is arranged in descending order. D is an orthogonal matrix of $(L + 1) \times (L + 1)$.

Noise interference is unavoidable in actual measurements, but if the noise signal is weak compared with the dominant signal, a threshold value can be set to intercept the singular value appropriately, and the negative impact of noise on parameter identification accuracy can be reduced by retaining the previous M large data [26–28]. Therefore, M is usually used as the equivalent RC branch of the extended Debye model. When the number of singular values is M, the matrix V is intercepted and the former M column is reserved to form a new matrix V'. Take the former M major right singular vectors of the matrix D to form the matrix D', remove the last row of D' and call it D_1, remove the first row of D' and call it D_2. Hence, two new matrices of $(N - L) \times L$ can be obtained:

$$Y_1 = SV'D_1^T, \tag{14}$$

$$Y_2 = SV'D_2^T, \tag{15}$$

According to the calculation of Equations (14) and (15), it can be considered that Y_1 and Y_2 no longer contain noise signal. They just contain $x(k)$ in the Equation (10):

$$Y_1 = \begin{bmatrix} x(1) & x(2) & \cdots & x(L) \\ x(2) & x(3) & \cdots & x(L+1) \\ \vdots & \vdots & \vdots & \vdots \\ x(N-L) & x(N-L+1) & \cdots & x(N-1) \end{bmatrix}_{(N-L)\times L}, \tag{16}$$

$$Y_2 = \begin{bmatrix} x(2) & x(3) & \cdots & x(L+1) \\ x(3) & x(4) & \cdots & x(L+2) \\ \vdots & \vdots & \vdots & \vdots \\ x(N-L+1) & x(N-L+2) & \cdots & x(N) \end{bmatrix}_{(N-L)\times L}, \tag{17}$$

The matrix pencil $Y_2 - \lambda Y_1$ consists of Y_1 and Y_2, and its generalized eigenvalue G is:

$$G = Y_1^+ Y_2, \tag{18}$$

Y_1^+ is the pseudo-inverse matrix of Y_1. The M eigenvalues of G can be denoted as $\lambda_1 (I = 1, 2, \ldots, M)$. When M and λ_i are known, the complex amplitude of the signal P_i can be obtained by the least squares method:

$$\begin{bmatrix} y(1) \\ y(2) \\ \vdots \\ y(N) \end{bmatrix} = \begin{bmatrix} 1 & 1 & \cdots & 1 \\ \lambda_1 & \lambda_2 & \cdots & \lambda_M \\ \vdots & \vdots & & \vdots \\ \lambda_1^{N-1} & \lambda_2^{N-1} & \cdots & \lambda_M^{N-1} \end{bmatrix} \begin{bmatrix} P_1 \\ P_2 \\ \vdots \\ P_M \end{bmatrix}, \tag{19}$$

After the P_i are calculated, the depolarization current amplitude A_i and time constant τ_i in the extended Debye model can be obtained:

$$A_i = |P_i|, \tag{20}$$

$$\tau_i = -\frac{T_S}{\mathrm{Re}(\ln \lambda_i)}, \tag{21}$$

where, T_S is the sampling interval.

After obtaining the above parameters in the extended Debye model, the tangent value of the dielectric loss angle can be calculated.

3. The Experimental Setup

3.1. The Oil-Film Dielectric Samples

Unlike the insulation system in a transformer whose pressboard can absorb the insulating oil easily, in order to simulate the internal insulation structure of a capacitor and ensure full contact between the insulating oil and polypropylene film, the oil-film dielectric is made by using the test oilcan as shown in Figure 4. The film can be compressed tightly by the column-column electrodes, which help the insulating oil soak it better. The electrodes and oil-film dielectric are sealed in this container.

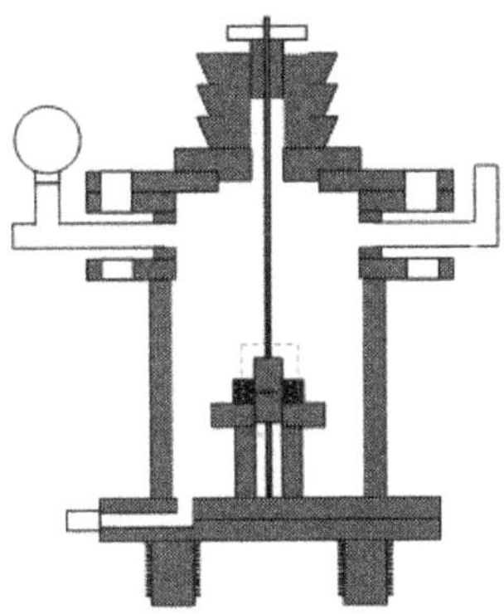

Figure 4. The oil-film dielectric sample.

The oil used in this experiment is C101 manufactured and provided by Mianzhu Xinan Electrotechnical Equipment Co., Ltd. (Mianzhu, China). Its main component is benzyltoluene. The thickness of the polypropylene film is just 12 μm. The film need to be cut into squares with a side length of 60 mm. Considering that the single-layer polypropylene film may be broken down when the impulse voltage was applied a four-layer film was used and immersed in benzyltoluene. Some oil-film dielectric sample pretreatment steps were carried out before subjecting it to an impulse voltage:

(1) Putting the samples into the vacuum drying oven and setting the temperature to 50 °C, which aimed at removing the moisture from the oil and eliminating the bubles in the oil-film interface;
(2) The samples were dried for 48 h with an air pressure of 100 Pa;
(3) Taking out the samples and sealing them with plastic wrap. It can prevent the moisture and impurities in the air;
(4) All the pretreated samples should be sealed in a dry, isolated and well-ventilated area. Lastly, waiting the temperature of the samples dropped to room temperature.

3.2. The Aging Platform

A continuous impulse voltage generator, as shown in Figure 5, was used as an aging platform to apply impulse voltages of different durations to the oil-film dielectric, which aimed to simulate the aging conditions of insulation in the capacitor. In the Figure 5, the schematic diagram of impulse voltage generator, it can be observed that this aging platform is composed of a test transformer T, main capacitor C, high voltage silicon reactor D, spherical gap g, wave head resistance R_f and wave tail resistance R_t. The impulse generator was a traditional Marx type impulse generator, manufactured by HUAGAO Electric Co., Ltd. (Hongan, Hubei, China) This impulses accumulation test platform can generate standard negative switching impulse (250/2500 μs), which rated parameter is

30 kJ/300 kV. The distance between the gaps is changed by the control system of the computer terminal, which ensures that the impulse voltage generator can be fully triggered. The breakdown voltage of the samples was 20 kV. In order to continuously apply impulse voltage and prevent accidental breakdown of oil-film dielectric during the experiment, the amplitude of impulse voltage was set to 17 kV and an impulse was applied every 60 s.

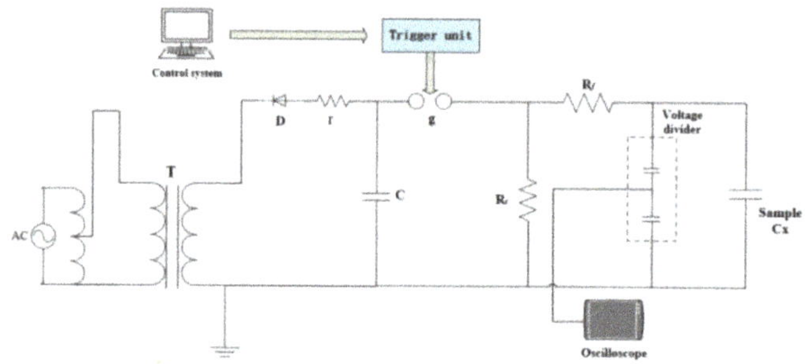

Figure 5. Experimental schematic diagram of impulse voltage generator.

3.3. The PDC Test Platform

The PDC curve of oil-film dielectric was measured and sampled by the measuring instrument as shown in Figure 6. This instrument was designed and manufactured to conduct the PDC measurements. It consisted of a high voltage DC source, a picoammeter and a high voltage double throw switch. The corresponding schematic diagram is shown in the Figure 7. The instrument includes a PDC high-voltage detection device, with a high-voltage output range of 0~10 kV and a current range of 0~3 mA, which needs to be supplied by a power supply of 220 V. In order to make the polarization of oil-film dielectric more sufficient and prevent interference from other factors, the polarization voltage and polarization time were set to 1500 V and 1800 s, respectively.

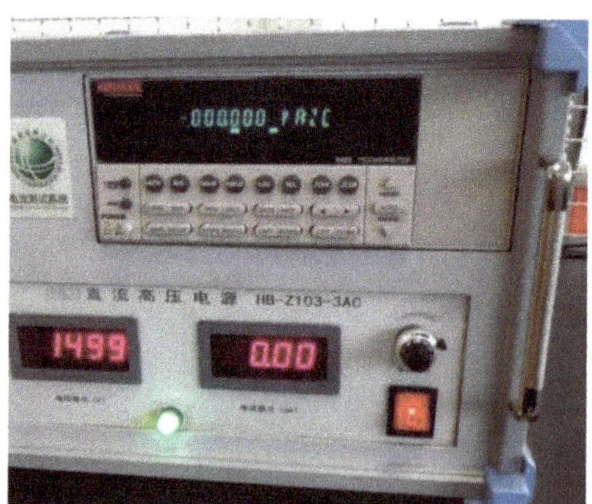

Figure 6. PDC measuring instrument.

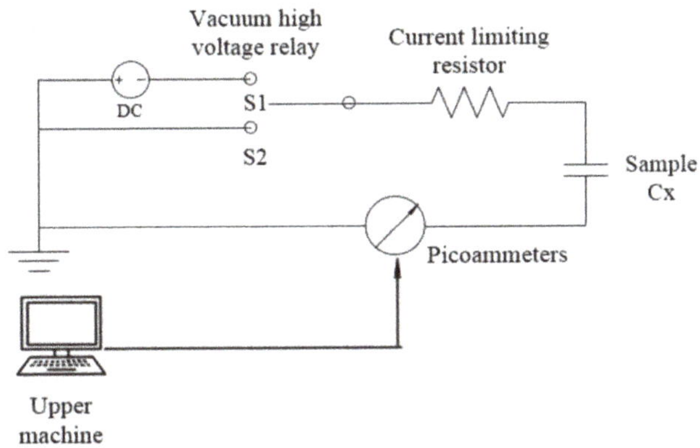

Figure 7. The schematic diagram of PDC measuring instrument.

When the measurement started, according to Figure 7, the vacuum high voltage relay was set to S1, the sample Cx was charged by DC voltage, the dielectric began to polarize. After 1800 s, the switch turned to S2 and the Cx was discharged through insulation resistance, which meant the beginning of depolarization. The sampling procedure applicable to the instrument is installed in the supporting upper computer. When the measurement is finished, a text document is generated to record the sampling data. Finally, the original polarization-depolarization current curve can be plotted as shown in Figure 2. In that figure, the curve is not level and smooth, so the matrix pencil algorithm was needed to process the curves.

4. The Experiment Results

During this experiment, 100, 200, 300 and 400 impulse voltages were applied to the samples in turn. Although all polarization types are included in the whole process, such as electronic polarization, ionic polarization, dipole polarization and interface polarization, however, the time required for the first two polarization types is short extremely, about 10^{-15} s~10^{-13} s. However the dipole and interface polarization need to take a few seconds or even an hour. As the PDC measurement takes a long time (several hundred seconds) and the measurement started 1 or more seconds after the switch action, so the electronic polarization and ionic polarization can't be recorded, and the polarization-depolarization current of oil-film dielectric is mainly caused by dipole and interface polarization.

Every time a new substance is generated, it will correspond to a new interface polarization. Due to the different dielectric properties of each material, the corresponding time constant of interface polarization is not the same, which requires different equivalent RC branches to represent it. Based on this conclusion, the matrix pencil algorithm was used to calculate the number of larger singular values of the matrix constructed by the depolarization current of the samples under different impact times, that is, the RC branch number reflecting the equivalent model of oil-film dielectric, as shown in Figure 8. Figure 8 is the calculation result of matrix pencil algorithm for one sample, which is not damaged. It has three larger singular values in the figure, so we set its number of RC branches to 3. This is different from [17] in that the number of RC branches is calculated by the matrix pencil algorithm. According to [29], when the polarization process of a dielectric lasts a long time, such as in EPR insulated cable and capacitors, a relatively low number (3–4) of RC branches with acceptable error can model the dielectric response better. Then, according to the same method, the number of larger singular values under different aging degrees was counted, so all the curves can be denoised. The fitting situation of original curve and denoised curve is as shown in Figure 9. Table 1 lists the parameter R_{new} between the denoised curves and the original curves. This parameter is used to judge the fitting degree of nonlinear regression equation and its unit is 1. It can be obtained that the feasibility of the algorithm was verified. Finally, the parameters of each branch in the Debye model can be simulated by the matrix pencil algorithm, and the characteristic parameters mentioned above will also be obtained.

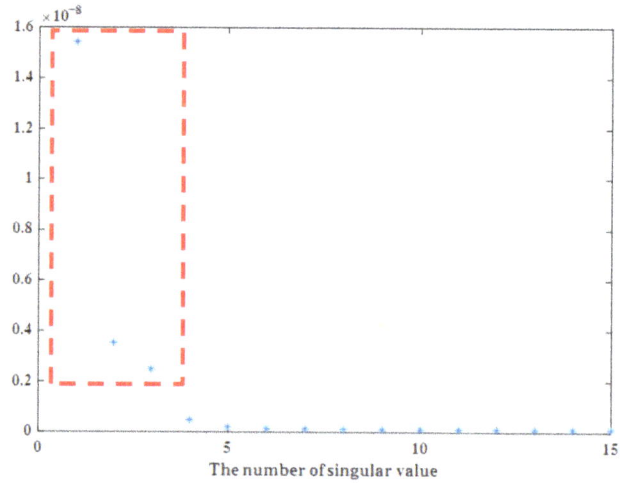

Figure 8. The results of the number of singular values when oil−film dielectric is not damaged.

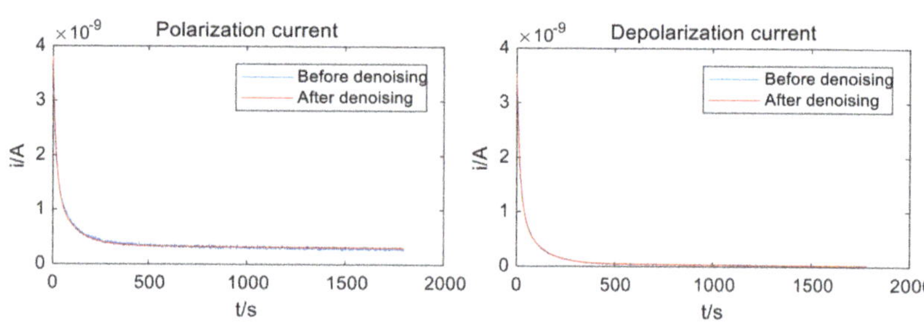

Figure 9. The fitting situation of original curve and denoised curve.

Table 1. Calculation results of R_{new} of an oil-film dielectric under different aging states.

Times of Impulse Voltage/Times	0	100	200	300	400
$R_{new}/1$	0.97491	0.95491	0.95828	0.98086	0.96411

4.1. The Polarization Current of an Oil-Film Dielectric with Different Degree of Aging

The polarization current curves of an oil-film dielectric under different aging degree, denoised by the matrix pencil algorithm were drawn in the same coordinate axis, which is shown in Figure 10. From this coordinate axis, we can see that when the polarization process was begining, the polarization current of oil-film dielectric had a large initial amplitude, which can be explained by the Debye model. As time goes by, the polarization current decreases rapidly. When it falls to a certain level, it will maintain a stable value, which also means that the polarization process is coming to an end. When the oil-film sample is not damaged, its polarization curve is in a lower position. Under the continuous action of impulse voltages, the polarization curve gradually moves up. At the same time, the value of the initial amplitude also increases. Table 2 lists the value of initial amplitude of oil-film dielectrics under different aging states. Due to the polarization current consisted of depolarization current and conductance current, the steady-state value of depolarization current was almost zero. Thus, the conductance current varied with the number of impulse voltages applied, which reflectes the change of insulation resistance.

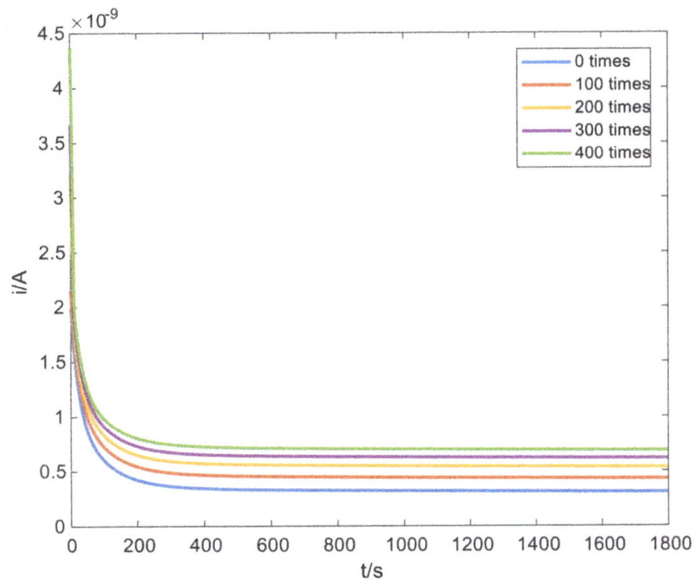

Figure 10. The polarization current curves of oil-film dielectrics under different degrees of aging.

Table 2. The value of initial amplitude of oil-film dielectrics under different aging states.

Number of Impulse Voltages	0	100	200	300	400
The Value of Initial Amplitude/A	1.986×10^{-9}	2.151×10^{-9}	2.536×10^{-9}	3.147×10^{-9}	3.493×10^{-9}

4.2. The Change of Insulation Resistance

According to Equation (4) and Figure 11, the value of the conductance current was equal to the steady-state value of the depolarization current, so the insulation resistance R_0 can be calculated. The variation of insulation resistance R_0 under different numbers of impulse voltages is shown in Figure 10. It can be seen from the figure that the insulation resistance of the oil-film dielectric generally shows a downward trend with different numbers of impulse voltages. When the oil-film dielectric isn't subjected to impulse voltages, its insulation resistance was $3.37 \times 10^{+12}$ Ω. With the accumulation of the number of impulse voltage instances, the damage of the oil-film sample was also increasing. Finally, the insulation resistance fell to $1.45 \times 10^{+12}$ Ω when the number of impulse voltages reached 400. That's down by 57.0%.

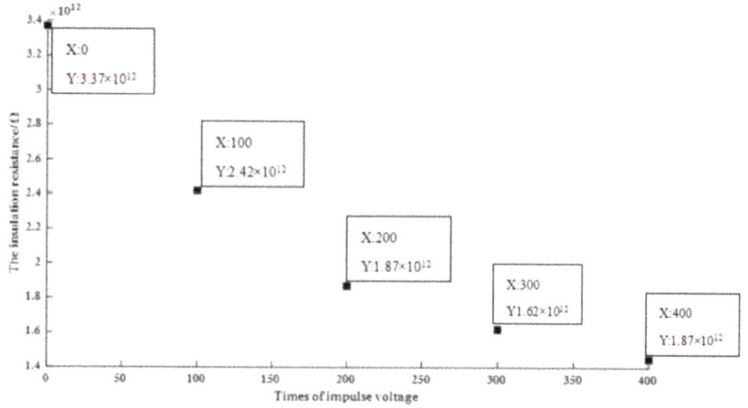

Figure 11. Variation trend of insulation resistance of oil-film dielectric with different numbers of impulse voltages.

4.3. The Results of the Tangent Value of the Dielectric Loss Angle at Low Frequency

For the convenience of this study, the frequency band was set between 0.01 Hz and 100 Hz and we took 50 points where each point had an equal interval in this frequency band. Because the tangent value of the dielectric loss angle of oil-film dielectric with different degrees of aging changes more obviously at low frequency, the tangent value of the dielectric loss angle at low frequency is used as the basis for judging the state of aging.

According to Equation (9), the frequency spectrum curves of tan δ of oil-film dielectric with different aging degrees are shown in Figure 12. Obviously, the frequency spectrum of the oil-film dielectric also shows a downward trend. When the frequency was low, the tan δ of oil-film dielectric changed along with the impulse accumulation aging. With the increase of frequency, the frequency spectrum curves bent to coincide gradually. Lastly, they almost reduced to zero when the frequency reached 100 Hz. With the continuous accumulation of impulse voltage, the tangent value of dielectric loss angle increased at lower frequency. Table 3 shows the tanδ of oil-film dielectric with different degrees of aging at 0.01 Hz.

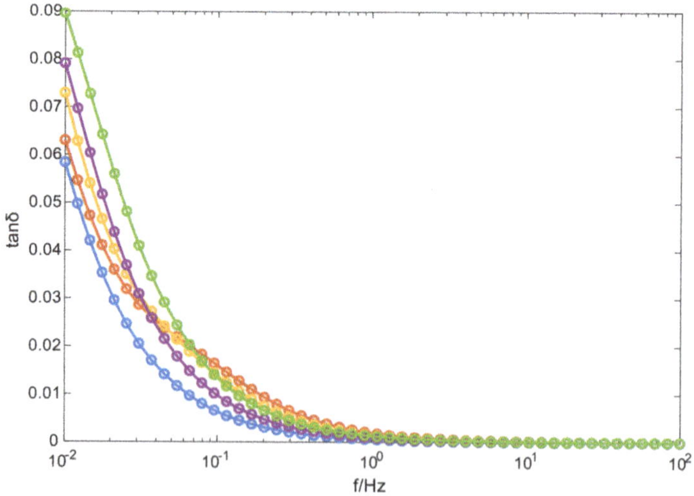

Figure 12. The frequency spectrum curves of tanδ of oil-film dielectric with different aging degrees.

Table 3. The tan δ of oil-film dielectric under different numbers of impulse voltages at 0.01 Hz.

Number of Impulse Voltages	0	100	200	300	400
tan δ at 0.01 Hz	0.0583	0.0631	0.0729	0.0787	0.0896

5. Discussion

We can acquire the change of polarization current curve and aging mechanism from an oil-film dielectric from the experimental results described in Section 3. Firstly, combined with the extend Debye model which consisted of RC branched in parallel, when an extra electric field was applied, the characteristics of RC branches made the initial value of current larger and then decrease rapidly. At the end of the polarization process, the capacitor of the branch C_i was charged, the polarization current finally decreased to a constant due to the existence of a resistance R_i.

From a macro point of view, the continuous effect of impulse voltages makes the oil-film dielectric age over time. The deepening of the degree of aging means that the damage to the oil-film dielectric increases, so the conductive ability of the composite medium composed of polypropylene film and insulating oil is enhanced, which makes the conductance current in the oil-film sample increase. That is, as the conductivity increases, the polarization current curve will move up and the insulation resistance of the entire composite medium will decrease.

From the micro point of view, with the increase of the number of impulse voltages, the polypropylene film and insulating oil changed to varying degrees. The molecular chains of the polypropylene film are broken by the action of impulse voltages, whereby more charged particles and small molecules are generated. The insulating oil will also decompose into water molecules, acids and other substances, so the number of conductive particles in the oil-film dielectric will grow in number. Therefore, under an exterior electric field, more charged particles will move directionally, which makes the initial value of polarization current increase and the polarization current curve move up. At the same time, the polar particles and ions in the new material can enhance the conductivity and polarizability of the oil-film dielectric, which leads to an increase of the conductivity and polarizability loss of the oil-film sample. The sum of the two just represents the tangent value of the dielectric loss angle, thus the tanδ value will increase at lower frequency.

6. Conclusions

Different numbers of impulse voltages were applied to an oil-film dielectric sample by simulating the oil-film insulation in the capacitor, and the polarization-depolarization current (PDC) method was used to test the PDC curves of samples with different degrees of aging. Finally, the matrix pencil algorithm was used to identify and fit the parameters of the measured curves, and the following conclusions were drawn: According to the matrix pencil algorithm, the number of larger singular value represents the RC branches in the Debye model, which will make the calculation results more reasonable. With the increase of the number of impulse voltages applied, the degree of aging of the oil-film dielectric is deepening. The conductivity of the oil-film dielectric is enhanced, so the initial value of polarization current will increase and the polarization current curve will move up. From a microscopic perspective, more moisture and small molecules are generated, and the conductivity of the complex medium is improved. Accordingly, the insulation resistance of oil-film dielectric will decrease. Similarly, the effect of multiple impulse voltages leads to the increase of the conductivity loss and polarizability loss, so the tangent value of dielectric loss angle will increase with the deepening of aging degree at low frequency.

Author Contributions: Investigation, C.Z., K.Z., Y.Z. and S.X.; Methodology, C.Z.; Supervision N.J.; project administration, C.H. All authors have read and agreed to the published version of the manuscript.

Funding: This work is supported by Sichuan Electric Power Company Research Project 521997190010.

Institutional Review Board Statement: Not applicable.

Informed Consent Statement: Not applicable.

Data Availability Statement: The study did not report any data.

Acknowledgments: Thank Zhou Mu, Yalong Xia and Donghui Luo for their advice and help in the process of paper compilation.

Conflicts of Interest: The authors declare no conflict of interest.

References

1. Saha, T.; Purkait, P. Investigation of an expert system for the condition assessment of transformer insulation based on dielectric response measurements. *IEEE Trans. Power Deliv.* **2004**, *19*, 1127–1134. [CrossRef]
2. Koch, M.; Prevost, T. Analysis of dielectric response measurements for condition assessment of oil-paper transformer insulation. *IEEE Trans. Dielectr. Electr. Insul.* **2012**, *19*, 1908–1915. [CrossRef]
3. Wang, M.; Vandermaar, A.J.; Srivastava, K.D. Review of condition assessment of power transformers in service. *IEEE Electr. Insul. Mag.* **2002**, *18*, 12–15. [CrossRef]
4. Fofana, I.; Hadjadj, Y. Electrical-based diagnostic techniques for assessing insulation condition in aged transformers. *Energies* **2016**, *6*, 697. [CrossRef]
5. Standring, W.G.; Hughes, R.C. Breakdown under impulse voltages of solid and liquid dielectrics in combination. *IEEE Proc. Power Eng.* **1956**, *103*, 583–597. [CrossRef]
6. Zhang, C.; Xie, S.; Zhang, Y. Repeated impulse aging of oil-polypropylene insulation. *IEEE Trans. Dielectr. Electr. Insul.* **2019**, *26*, 1581–1587. [CrossRef]
7. Sima, W.; Sun, P.; Yang, Q.; Yuan, T.; Lu, C.; Yang, M. Study on the accumulative effect of repeated lightning impulses on insulation characteristics of transformer oil impregnated paper. *IEEE Trans. Dielectr. Electr. Insul.* **2014**, *21*, 1933–1941. [CrossRef]
8. Linhjell, D.; Lundgaard, L.; Gafvert, U. Dielectric response of mineral oil impregnated cellulose and the impact of aging. *IEEE Trans. Dielectr. Electr. Insul.* **2007**, *14*, 156–169. [CrossRef]
9. Zdanowski, M. Streaming Electrification of Mineral Insulating Oil and Synthetic Ester MIDEL 7131. *IEEE Trans. Dielectr. Electr. Insul.* **2014**, *21*, 1127–1132. [CrossRef]
10. Saha, T.K. Review of Modern Diagnostic Techniques for Assessing Insulation Condition in aged transformers. *IEEE Trans. Dielectr. Electr. Insul.* **2003**, *10*, 903–917. [CrossRef]
11. Saha, T.K.; Yao, Z.T. Experience with return voltage measurements for assessing insulation conditions in service-aged transformers. *IEEE Trans. Power Deliv.* **2003**, *18*, 128–135. [CrossRef]
12. N'cho, J.S.; Fofana, I.; Hadjadj, Y.; Beroual, A. Review of physicochemical-based diagnostic techniques for assessing insulation condition in aged transformers. *Energies* **2016**, *9*, 367. [CrossRef]
13. Liao, R.; Liang, S.; Sun, C.; Yang, L.; Sun, H. A comparative study of thermal aging of transformer insulation paper impregnated in natural ester and in mineral oil. *Eur. Trans. Electr. Power* **2010**, *20*, 518–533. [CrossRef]
14. Hao, J.; Liao, R.; Chen, G.; Ma, Z.; Yang, L. Quantitative analysis ageing status of natural ester-paper insulation and mineral oil-paper insulation by polarization/depolarization current. *IEEE Trans. Dielectr. Electr. Insul.* **2012**, *19*, 188–199.
15. Darveniza, M.; Saha, T.; Hill, D.; Le, T. Investigations into effective methods for assessing the condition of insulation in aged power transformers. *IEEE Trans. Power Deliv.* **1998**, *13*, 1214–1223. [CrossRef]
16. Zhao, K.; Zhang, C.; Deng, Y.; Li, L.; Xie, S.; Zhang, Y. Research on Time-domain Dielectric Response of Multiple Impulse Voltage Aging Oil-film Dielectric. In Proceedings of the 2020 IEEE International Conference on High-Voltage Engineering (ICHVE 2020), Beijing, China, 6–10 September 2020.
17. Zhang, T.; Cai, J. Research on parameters identification for dielectric response equivalent circuit of transformers with oil-paper insulation. *Adv. Technol. Electr. Eng. Energy* **2010**, *29*, 35–39.
18. Tang, P.; Yin, Y.; Wu, J.; Yan, X. Study of oil-paper insulated transformers'aging state based on depolarization current measurements. *Adv. Technol. Electr. Eng. Energy* **2012**, *31*, 39–42.
19. Xu, S.; Middleton, R.; Fetherston, F.; Pantalone, D. A comparison of return voltage measurement and frequency domain spectroscopy test on high voltage insulation. In Proceedings of the 2003 IEEE International Conference on Properties and Applications of Ei Electric, Materials, Nagoya, Japan, 1–5 June 2003; pp. 351–355.
20. Chen, J.; Jiang, X.; Cai, J.; Li, A. Application Circuit Analysis Method to Assess the State of Transformers Oil-paper Insulation. *Proc. CSU-EPSA* **2016**, *28*, 30–35.
21. Yang, F.; Du, L.; Yang, L.; Wei, C.; Wang, Y.; Ran, L.; He, P. A Parameterization Approach for the Dielectric Response Model of Oil Paper Insulation Using FDS Measurements. *Energies* **2018**, *11*, 622. [CrossRef]
22. Andrzej, K. Dielectric Relaxation in Solids: Chelsea Dielectric Press. *J. Phys. Appl. Phys.* **1983**. [CrossRef]
23. Walter, S. Dielectric spectroscopy in time and frequency domain for HV power equipment. I. Theoretical considerations. *IEEE Electr. Insul. Mag.* **2003**, *19*, 5–19.
24. Subocz, J.; Mrozik, A.; Bohatyrewicz, P.; Zenker, M. Condition Assessment of HV Bushings with Solid Insulation based on the SVM and the FDS Methods. *Energies* **2020**, *13*, 853. [CrossRef]

25. Hua, Y.; Sarkar, T. On SVD for estimating generalized eigen values of singular matrix pencil in noise. *IEEE Trans. Signal Process.* **1991**, *39*, 892–900. [CrossRef]
26. Peter, B.; Vasile, S.; Matthias, V. L-Norm Computation for Continuous-Time Descriptor Systems Using Structured Matrix Pencils. *IEEE Trans. Automat. Contr.* **2012**, *57*, 233–238.
27. Liu, Y.; Liu, H.; Nie, Z. Reducing the Number of Elements in the Synthesis of Shaped-Beam Patterns by the Forward-Backward Matrix Pencil Method. *IEEE Trans. Antennas Propag.* **2010**, *58*, 604–608.
28. Yilmazer, N.; Koh, J.; Sarkar, T. Utilization of a unitary transform for efficient computation in the matrix pencil method to find the direction of arrival. *IEEE Trans. Antennas Propag.* **2006**, *54*, 175–181. [CrossRef]
29. Tan, S.; Zhang, C. Aging Detection of Oil Film Insulation for Surge Capacitor Based on Polarization-depolarization Current Method. *Sichuan Electr. Power Technol.* **2019**, *24*, 42.

energies

MDPI

Article

Three-Dimensional Electro-Thermal Analysis of a New Type Current Transformer Design for Power Distribution Networks †

Bingbing Dong, Yu Gu, Changsheng Gao, Zhu Zhang *, Tao Wen and Kejie Li

School of Electrical Engineering and Automation, Hefei University of Technology, Hefei 230009, China; bndong@126.com (B.D.); gyu.wxy@foxmail.com (Y.G.); gcs1681@163.com (C.G.); tao-wen@hfut.edu.cn (T.W.); kejie.li@hotmail.com (K.L.)
* Correspondence: zhuzhang@hfut.edu.cn

† This paper is an extended version of our paper published in 2020 IEEE International Conference on High Voltage Engineering and Application (ICHVE), Beijing, China, 6–10 September 2020; pp.1–4.

Abstract: In recent years, the new type design of current transformer with bushing structure has been widely used in the distribution network system due to its advantages of miniaturization, high mechanical strength, maintenance-free, safety and environmental protection. The internal temperature field distribution is an important characteristic parameter to characterize the thermal insulation and aging performance of the transformer, and the internal temperature field distribution is mainly derived from the joule heat generated by the primary side guide rod after flowing through the current. Since the electric environment is a transient field and the thermal environment changes slowly with time as a steady field under the actual conditions, it is more complex and necessary to study the electrothermal coupling field of current transformer (CT). In this paper, a 3D simulation model of a new type design of current transformer for distribution network based on electric-thermal coupling is established by using finite element method (FEM) software. Considering that the actual thermal conduction process of CT is mainly by conduction, convection and radiation, three different kinds of boundary conditions such as solid heat transfer boundary condition, heat convection boundary condition and surface radiation boundary condition are applied to the CT. Through the model created above, the temperature rise process and the distribution characteristics of temperature gradient of the inner conductor under different current, different ambient temperatures and different core diameters conditions are studied. Meanwhile, the hottest temperature and the maximum temperature gradient difference are calculated. According to this, the position of weak insulation of the transformer is determined. The research results can provide a reference for the factory production of new type design of current transformer.

Keywords: current transformer; finite element analysis; electro-thermal coupling; thermal field

Citation: Dong, B.; Gu, Y.; Gao, C.; Zhang, Z.; Wen, T.; Li, K. Three-Dimensional Electro-Thermal Analysis of a New Type Current Transformer Design for Power Distribution Networks . *Energies* **2021**, *14*, 1792. https://doi.org/10.3390/en14061792

Academic Editor: Issouf Fofana

Received: 5 January 2021
Accepted: 22 February 2021
Published: 23 March 2021

1. Introduction

The new type design of current transformer for distribution network used epoxy resin as the shell of secondary coil, compared with the traditional transformer, has high mechanical strength, miniaturization, maintenance-free advantages [1,2]. In addition to this, the new casting type current transformer internal using air as medium compared with the traditional oil-immersed, SF6 type, it is friendly to the environment.

Insulation performance affects the reliable operation of transformers to a large extent, and it is related to the temperature field distribution under the actual operating condition of the transformer [3,4]. Transformer internal overheating, uneven temperature distribution not only accelerate insulation degradation, reduce its operating life but also makes water molecules in the air inside the transformer to migrate and form condensation, which endangers equipment safety and causes transformer faults [5]. So, this article main studied the new casting type current transformer's electrical and thermal coupling performance, which used in distribution network. Under the actual working conditions, the temperature field calculation of the transformer mainly depends on the ohmic heating of the primary guide rod and the Joule heat generated by each part of the structure, which indicates that the calculation of electrical and thermal environment is a coupled solution process [6–8].

In recent years, scholars have mainly studied the temperature field distribution of electric power equipment through numerical calculation and experimental measurement [9,10]. In some studies, the backward induction method is mostly used to determine the loss. Although it has certain reference value, it is somewhat different from the loss caused by the actual primary heat source due to the deviation of reasonable electrothermal coupling [11–13]. Aiming at SF6 CT, Xiuguang calculated the thermal field distribution characteristics of it by calculating the heat source which generated by the primary guide rod under different external ambient temperatures. It pointed out that after the thermal environment reaches steady-state, the overheating of the CT will be around 35 °C which is independent of the external ambient temperature. However, this model ignores the difference in heat dissipation capacity between different structure positions of the transformer, which is inconsistent with the actual working conditions [14]. Lan established a three-dimensional model of the dry-type transformer by ANSYS commercial software. The model considers three heat dissipation modes, namely conduction, convection and radiation, and sets different convection coefficients for the transformer after taking the on-site installation of the dry-type transformer into consideration. The hottest spot area of the transformer is studied, and the distribution of temperature field of the transformer from bottom to top presents a low-high-low distribution law [15]. Arjona established an axisymmetric model of dry type transformer with load and no load cases by finite element method. The research considers the convection, conduction and radiation effect, accurately predict the steady-state thermal process of dry type transformer and provides the reference for improving equipment heat dissipation performance, the model can also be applied to the transformer under the operation condition of the forced convection heat [16]. Hui took the 110 kV SF6 current transformer as the research object, studied the internal temperature field distribution of the transformer under different external ambient temperatures, and emphatically analyzed the influence of the abrupt decrease of ambient temperature on the transformer. The study shows that, the internal overheating of the transformer remains the same, which has no obvious difference under different external ambient temperatures. Such uneven temperature distribution and the sharp decrease of ambient temperature are very easily to lead to local condensation inside the transformer, thus causing transformer faults [17]. Liu used a quasi-3D electro-thermal field coupling method to study the distribution characteristics of transformer temperature field. In this study, a simplified 3D model was established for the calculation of core loss, a 2D model was established for the calculation of coil loss, and both of them were regarded as heat sources in the form of sequential coupling to study the overall temperature rise of the model [18].

In this paper, a new type design of current transformer for distribution network is studied. According to the actual operation condition, the load is applied to the transformer. Based on the calculation principle of electro-thermal coupled multi-physical field, the finite element calculation software is used to analyzed the electro-thermal coupling field distribution characteristics of transformer under the different current, the different environment temperature and the different guide rod core diameter. The weak links of thermal insulation in the actual operation process of the transformer are obtained, more importantly the research results can provide reference for the design of new casting type current transformer factory production.

2. Electro-Thermal Coupling Model of New Current Transformer

The new current transformer model adopts the pouring casing structure and designed as a coaxial cylinder. It is a complex configuration containing metal guide rod, air, semiconductor materials, ceramics, epoxy resin and other media. The aluminum rod, as the primary coil, is located in the geometric center of the transformer and flows through hundreds or even thousands of amperes. It is the main heat source of the transformer. The structural parameters and physical properties of each part of the CT are shown in Tables 1 and 2, Figure 1 representing the transformer elements.

Table 1. Structural parameter.

Parameters	Guide Bar	The Semiconductor Layer	Epoxy Shell	Bushing	Metal Enclosures
Axial length (mm)	1177	1177	-	410 + 290	-
Radial length (mm)	20	10	110	-	350

Table 2. Physical property parameter.

Parameters	Material	Conductivity (S/M)	Relative Permittivity	Heat Capacity At Constant Pressure (J/Kg·K)	Density (Kg/M^3)	Thermal Conductivity (W/M·K)
Hardware	Copper	5.998×10^7	-	385	8700	400
Guide rod	Aluminum	3.774×10^7	-	900	2700	238
Bushing	Ceramic	1.0×10^{-8}	5.5	426	1750	0.06
Coil shell	Epoxy-resin	2.0×10^{-14}	4.5	1400	980	0.276

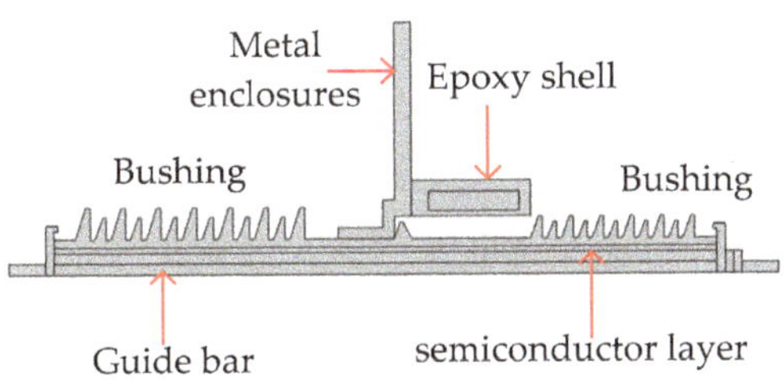

Figure 1. The transformer elements.

Before conducting the electro-thermal coupling simulation, in order to simplify the calculation, the following assumptions were made for the new current transformer electro-thermal coupling model [19,20].

1. Skin effect is not considered for the transformer under the operating environment of power frequency.
2. The new current transformer has an internal quasi-stable field, and the influence of displacement current is not considered in the analysis process.
3. In the model, the relative dielectric constant and conductivity of different materials are constant.
4. The air area inside the transformer is incompressible gas.

2.1. The Coupling Calculation Flow of the Electro-Thermal Coupling Model

In this paper, the Newton iterative calculation method is used in the finite element simulation software to realize the electro-thermal coupling calculation of the transformer. Firstly, the 3D geometry model of the new current transformer is established in the calculation environment, and the physical parameters of the electro-thermal coupling are determined according to the materials of each part of the transformer. The unit volume loss of the current conductor was calculated through the current field module, and then the Joule heat loss caused by the primary guide rod was used as the excitation source in the temperature field calculation environment for simulation solution. After that the overall temperature field distribution of the transformer was calculated and thermal parameters such as hot temperature and temperature gradient in the transformer were extracted. Such calculated iteratively until the two calculated results meet the requirements of control accuracy. Figure 2 shows the coupling calculation process of the new current transformer's electro-thermal coupling model.

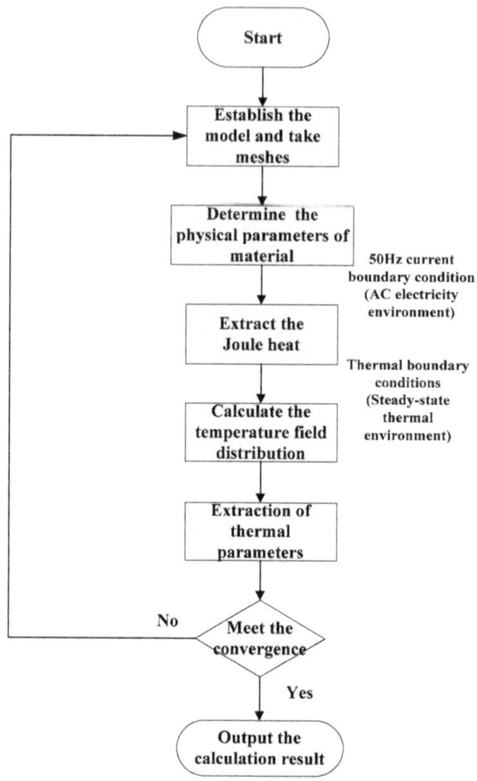

Figure 2. Flow chart of electro-thermal coupling calculation.

2.2. Physical Modeling

Based on the above assumptions, the governing equation of the new current transformer electro-thermal coupling model is as follows:

$$\begin{gathered} \rho C_p u \cdot \nabla T + \nabla \cdot q = Q \\ q = -k \nabla T \end{gathered} \tag{1}$$

where: ρ is the material density, kg/m^3; Cp is the heat capacity at constant pressure of each part of material of the transformer, J/(kg·K); q is the conductive heat flux vector; u is the fluid velocity vector; T is temperature K; k is thermal conductivity, W/(m·k); Q is the volumetric Joule heat (W/m^3) generated by the guide rod.

Considering the new current transformer normal operation in the 35 kV power distribution network system, then the frequency of China's distribution network is 50 Hz. In order to get close to the actual operating conditions and calculate the actual thermal field distribution characteristics of the transformer accurately, the transient AC current field is used to simulate the actual electrical environment. The guide rod as a large current input, its frequency choice for AC 50 Hz, and transformer thermal environment parameters changing with time is relatively slow, which can be considered as steady field. Since the transformer is located in the distribution network switch cabinet, the external environment basically remains unchanged. Therefore, it can be assumed that all boundary temperatures contacted by the transformer and the external environment remain constant.

2.3. Take Meshes

The new current transformer structure is a complex 3D model, in order to guarantee both calculation accuracy and can control the time needed for the simulation calculation, this article adopts the method of differential meshes subdivision, using different size precision tetrahedral meshes to the new current transformer model. The meshes' maximum size is 110 mm, and minimum size is 13.7 mm. The model consists of 115,882 tetrahedral units, 48,562 triangular units, 12,911 edge units and 900 vertex units. The mesh generation of the transformer model in this paper is shown in Figure 3.

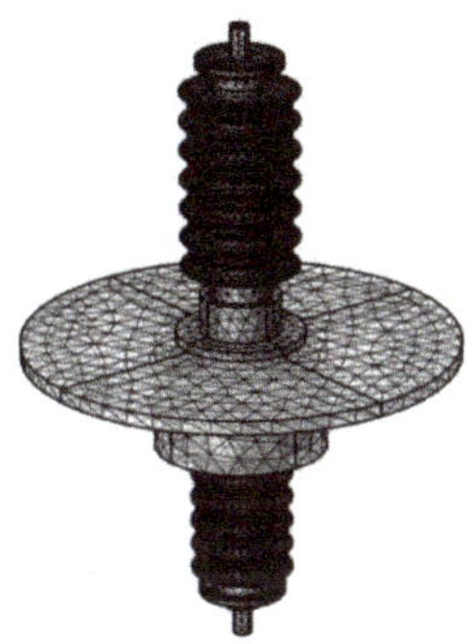

Figure 3. The new current transformer meshes.

3. Heat Source and the Modes of Heat Transfer

3.1. The Analysis of Heat Generation

New current transformer temperature rise is mainly from the loss of the transformer operation, it will be converted into heat energy lifting transformer temperature distribution. Heat energy in the form of conduction, radiation and convection diffusion, when heat production and heat dissipation reach equilibrium, transformer temperature will be stable [21]. The heat sources in the new current transformer are guide rod and secondary coil. The guide rod can pass through hundreds of amps in normal operation, and the current can reach up to thousands of amps. However, the current of the secondary coil is usually within 5 A, so the loss generated by the secondary coil can be ignored, and only the Joule heat generated after the flow of the guide rod can be considered.

Besides, there are some additional losses in the CT, mainly including eddy current loss, circulation loss and stray loss. These additional losses are not considered in this paper because they are relatively small and difficult to calculate.

3.2. Heat Transfer Analysis of CT

The heat transfer of the new current transformer mainly consists of conduction heat transfer, convection heat transfer and radiation heat transfer. Among them, the heat conduction mainly occurs in the guide rod and insulating bushing, and the heat is transferred from the higher temperature to the place with lower temperature. Convective heat transfer occurs mainly in outer insulation umbrella skirt, and hardware, and the surface of the carbon steel shell in contact with air. In order to calculate the actual condition of the transformer temperature field distribution characteristics more accurately, in this paper, considering the actual ventilation in the high voltage cabinet where the transformer is located, the convection heat transfer coefficient h of the upper and lower fittings and the outer insulation umbrella skirt of the transformer are respectively set in a differentiated way. Radiant heat transfer occurs mainly on the wall in contact with air, radiating heat to the surrounding fluid, and the main influencing factor is surface emissivity ε [22].

4. Results and Analysis

According to the actual operation and ventilation conditions of the transformer, the differential convection heat transfer coefficient h was set for the upper and lower ends of the transformer and the outer insulation umbrella skirt. Considering that the transformer was installed in the high voltage cabinet, the ventilation condition was that the external cold air flowed from the bottom of the cabinet, and after the transformer, the hot air flowed out from the top of the cabinet. Based on this, in the case of natural convection, the lower end of the transformer was at the cold air inlet, and the convection heat transfer coefficient was the largest, which was 13 $W/(m^2 \cdot K)$. The upper fixture was located at the hot air outlet, and the convection heat transfer coefficient was the smallest, 10 $W/(m^2 \cdot K)$. The insulated umbrella skirt was in direct contact with the surrounding air, which was the main channel for natural cooling. Its convective heat dissipation intensity was less than that at the entrance of the bottom, but larger than that at the exit of the top. It was set at 12 $W/(m^2 \cdot K)$. The external surface of the transformer was always radiative heat transfer with the air environment, and its surface emissivity ε was set at 0.3.

When the external environment temperature was 293.15 K and the current of the guidance rod connected to the primary side bus was 2000 A, the overall temperature field distribution under the final steady-state condition is shown in Figure 4.

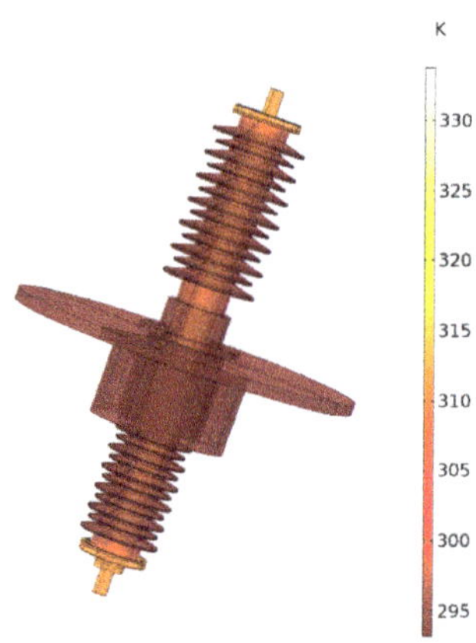

Figure 4. Overall temperature distribution field map.

In order to observe the temperature distribution characteristics of various parts of the structure of the new current transformer more intuitively, the temperature field distribution cloud map of Y-Z section was selected, which is shown in Figure 5, and the temperature gradient distribution cloud map of the transformer is shown in Figure 6. As can be seen from Figure 5, the overall temperature field distribution rule of the new current transformer was high in the middle and low at the end, high inside and low outside. Assuming the total height of the transformer was H, then the hottest point of the transformer was located 0.4 h away from the bottom, and the hottest point temperature reached 334 K, the maximum overheating was 41 K.

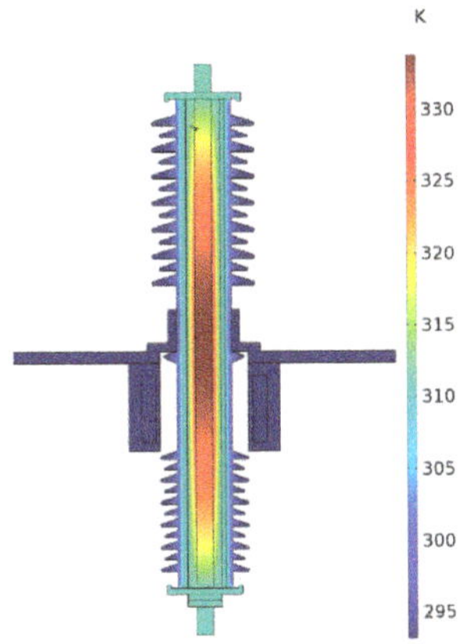

Figure 5. Y-Z temperature distribution field map.

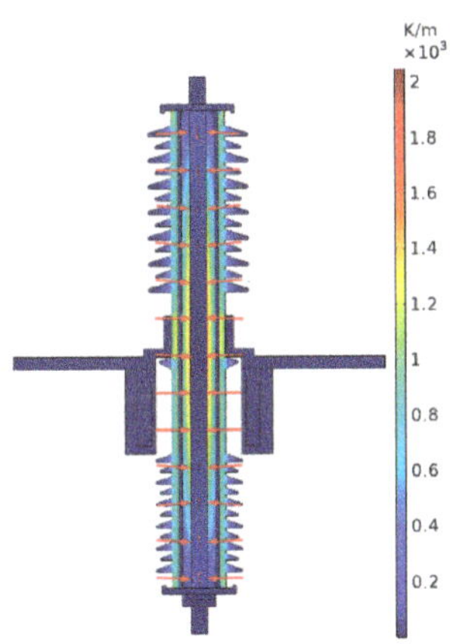

Figure 6. Y-Z temperature gradient field map.

It can be seen from Figure 6 that after the current passed through the transformer, the internal and external temperature gradient difference iwass large, the maximum temperature gradient difference could reach 2111 K/m, and the heat was mostly concentrated in the secondary coil wrapping. The reason is that the heat dissipation mode at the secondary coil of the transformer was mainly convection heat transfer with the external environment, but the heat dissipation performance was poor because of the airway was narrow there. In contrast, the head and tail ends were located in the vent and had sufficient convective heat transfer with the air, so the heat dissipation was obviously better than the middle part of the transformer.

4.1. The Temperature Field Distribution Characteristics with Different Current

The internal guide rod of the new current transformer was connected to the primary side bus through the upper tool. The current usually passed through hundreds of amps or even thousands of amps, which was the main heat source of the transformer. In order to study the overall temperature of the transformer under different current loads, the electromagnetic environment of the transformer was controlled at the power frequency of 50 Hz and the external environment temperature remained unchanged at 293.15 K. Figure 7 shows the temperature rise curve of the guide rod. The external insulation umbrella skirt temperature rise curve of the new current transformer under different current load is shown in Figure 8.

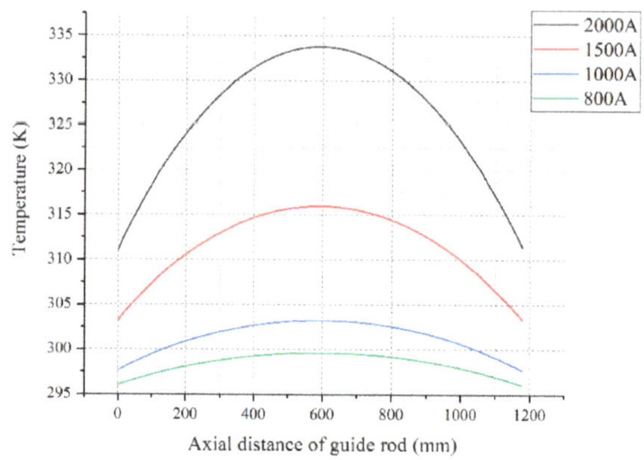

Figure 7. The temperature curves of guide rod at different currents.

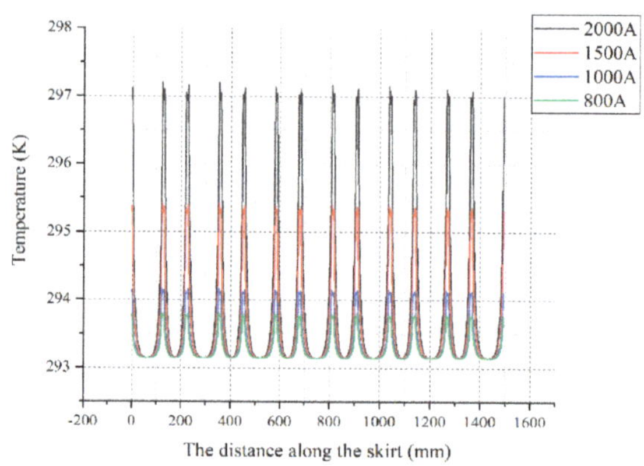

Figure 8. The temperature curves of skirt under different currents.

As shown in Figure 7, the overall temperature rise curve of the guide rod was in the form of "parabola", and the peak temperature occurred at the axial center of the guide rod, that is at 0.5 H. The maximum temperature of the guide rod was up to 334 K, when passing through 2000 A, and the difference was 34 K compared with 800 A. At 2000 A, the maximum overheating could reach 23 K, which was 5.75 times of the maximum overheating at 800 A. It can be seen from Figure 8 that, under different current loads, the minimum temperature of the external insulation umbrella skirt was consistent. This is because the umbrella skirt was in contact with the external environment, and its minimum temperature remained consistent with the external environment temperature after reaching equilibrium between heat generation and heat dissipation.

It can be seen from Figures 9 and 10 that the hot spot temperature and the maximum temperature gradient of the new current transformer increased with the increase of the current load, and increased almost linearly. After data fitting, the growth rate k1 of the hotspot temperature of the transformer was 0.029, and its determination coefficient R12 was 0.984. The growth rate K2 of the maximum temperature gradient of the transformer was a rate of 1.480, and its determination coefficient R22 was 0.987. It can be seen that the trend line of the transformer after fitting of the hot temperature and the maximum temperature gradient was more reliable, and compared with the hot temperature, the maximum temperature gradient changed with the current load more significantly.

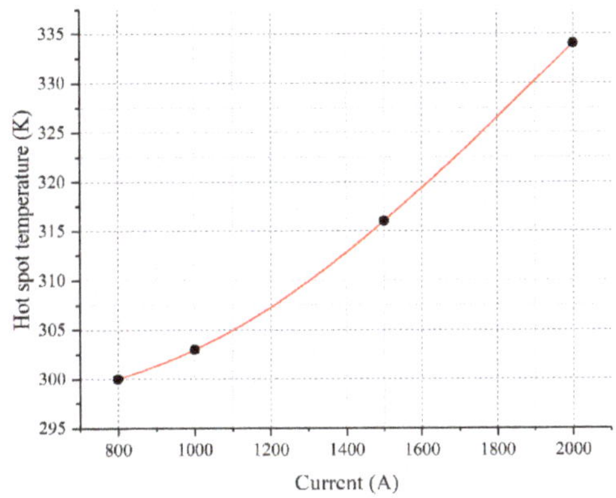

Figure 9. Hot spot temperatures at different currents.

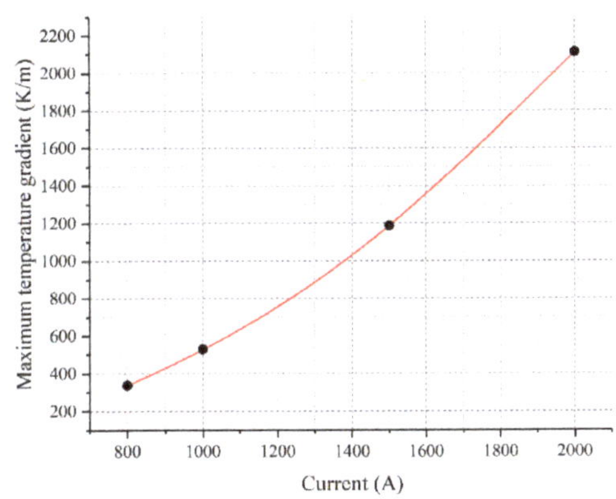

Figure 10. Maximum temperature gradients at different currents.

4.2. The Characteristics of Temperature Distribution Field Under Different External Ambient Temperature

Since the CT was located in the high voltage cabinet and the internal environment of the CT is closed, different external environment boundary temperatures were set to simulate the temperature field distribution characteristics of the CT when it reached steady-state under different external environment temperatures. As shown in Figures 11 and 12 and Tables 3 and 4, the maximum overheating and maximum temperature gradient of the new current transformer occurred when the external environment temperatures were 273 K, 283 K, 293 K and 306 K respectively in this study.

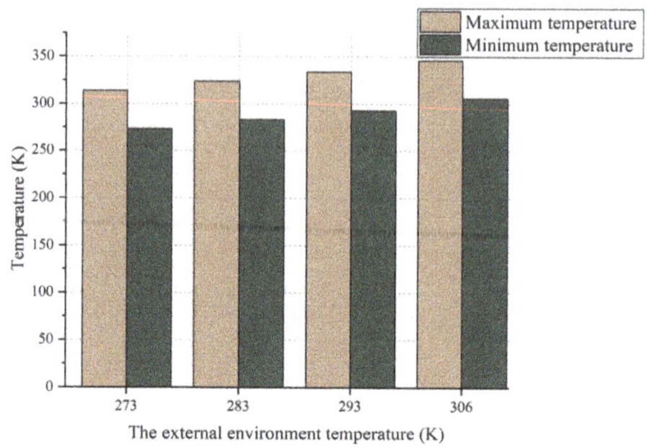

Figure 11. Maximum overheating at different external environment temperatures.

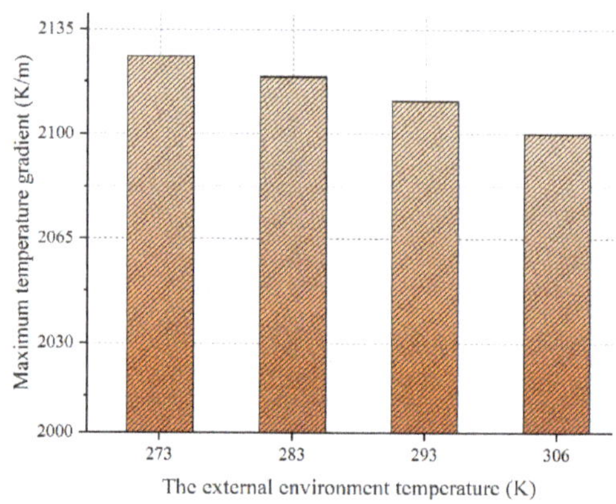

Figure 12. Maximum temperature gradient at different external environment temperatures.

Table 3. Maximum overheating of CT under different external environment.

The External Environment Temperature (K)	273	283	239	306
Hot Spot Temperature (K)	314	324	334	346
Temperature Difference (K)	41	41	41	40

Table 4. Maximum temperature gradient of CT under different external environment.

The External Environment Temperature (K)	273	283	239	306
Maximum Temperature Gradient (K/M)	2126	2119	2111	2100

It can be seen from Figure 11 and Table 3 that regardless of the external ambient temperature, the maximum internal overheating of the new current transformer was between 40 and 41 K. The hot spot temperature and overall temperature rise distribution of the CT were positively correlated with the external ambient temperature. According to the Figure 12 and Table 4, with the gradual increase of the external ambient temperature, the maximum temperature gradient of the transformer gradually decreased, but the decrease was small and both remained above 2100 K/m. When the external ambient temperature was 306 K, the maximum temperature gradient dropped by only 1.2% compared with 273 K. So the maximum overheating and maximum temperature gradient of the new current transformer both had no obvious relation with the external environment temperature. However, due to the uneven distribution of internal temperature of the transformer under various external ambient temperatures, water molecules in the air inside the transformer may have migrated, and the material in the lower temperature area had a strong ability to absorb water molecules, which may form condensation so that it would endanger the safe and reliable operation of the transformer equipment.

4.3. Distribution Characteristics of Temperature Field Under Different Core Diameter Guide Rod

The metal guide rod connected to the primary side bus could pass through hundreds of amps or even thousands of amps during normal operation, which was the most important heat source of the new current transformer. Guide rods with different core diameters had different current carrying and heat conduction capacities. In order to obtain the distribution law of thermal parameters and realize the optimal design of the new type current transformer, it was necessary to study the temperature field distribution characteristics of the guide rods with different core diameters.

It can be seen from Figure 13 that, with the increase of the core diameter of the new current transformer, its hot spot temperature ws negatively correlated with it. When the guide rod core diameter was 15 mm, its hot spot temperature was 407 K, which increased by 26% when compared with the guide rod core diameter was 22 mm. After fitting, the data of hot spot temperature could be approximately regarded as a linear relationship, with its determination coefficient R2 = 0.939. The hot spot temperature decreased with the increase of the core diameter, and the rate of decline was 12.1.

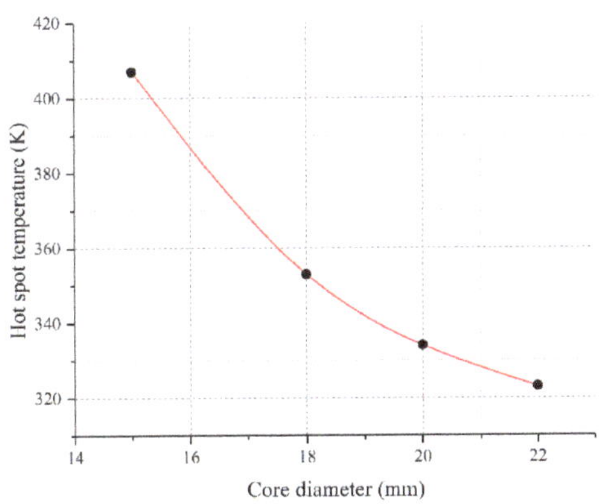

Figure 13. Hot spot temperature at different core diameters.

As shown in Figure 14, the maximum temperature gradient of the new current transformer could be approximately regarded as an exponential decrease with the increase of core diameter. When the diameter of guide rod was 15~18 mm, the maximum temperature gradient decreased rapidly, while the decrease tended to be moderate in the range of 18–22 mm. The maximum temperature gradient at 18 mm core diameter was 2978 K/m, which was 61.5% lower than that at 7726 k /m at 15 mm core diameter. The maximum temperature gradient at 22 mm core diameter was 1759 K/m, which was only 40.9% lower than the maximum temperature gradient at 18 mm core diameter.

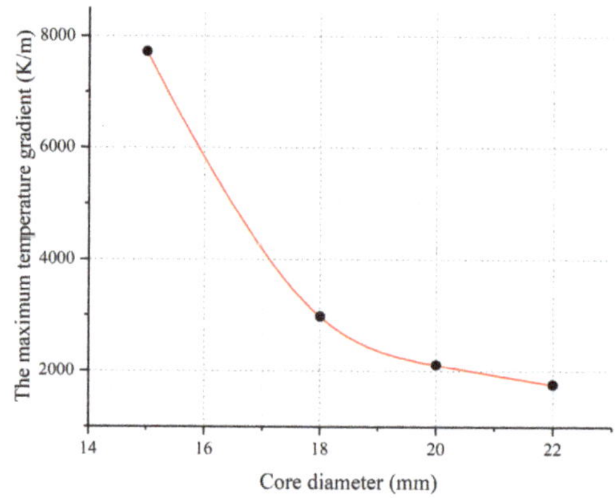

Figure 14. Maximum temperature gradient at different core diameters.

It is not difficult to see that when the core diameter of the new current transformer was 18–20 mm, the change rate of hotspot temperature and maximum temperature gradient of the transformer tended to be saturated and both are in a reasonable range of values.

5. Conclusions

In this paper, a new type design of current transformer for distribution network is taken as the research object. Based on the finite element multi-physical field coupling method, the thermal field distribution law of the current transformer when it reaches the thermal steady state in the actual electromagnetic environment is calculated and analyzed, and the following conclusions are drawn:

1. The temperature distribution of the main heat source guide rod of the new current transformer is in a "parabola" form. The hot temperature and the maximum temperature gradient of the transformer change linearly under different current.
2. Under different external ambient temperatures, the overall maximum overheating of the transformer remains unchanged between 40 and 41 K. Although the maximum temperature gradient decreases gradually with the increase of external ambient temperature, the decline can be ignored. Due to the uneven distribution of temperature, water molecules in the internal may migrate and form condensation, which will harm the safe operation of the current transformer.
3. For the guide rod, which is the main heat source of the transformer, in the case of different core diameters, its hot spot temperature and maximum temperature gradient decrease with the increase of core diameter. Among them, the maximum temperature gradient decreases exponentially with the increase of core diameter, and the rate of change tends to saturation at 18–20 mm. At this point the core diameter size is the most reasonable.

Author Contributions: Conceptualization and Formal analysis, B.D.; Writing—Original draft, Y.G.; Writing—Review & editing, C.G., Z.Z., T.W. and K.L. All authors have read and agreed to the published version of the manuscript.

Funding: This research was funded by State Grid Corporation of China, grant number W2020JSKF0677.

Data Availability Statement: Data sharing not applicable. No new data were created or analyzed in this study. Data sharing is not applicable to this article.

Conflicts of Interest: The authors declare no conflict of interest.

References

1. Guo, L.; Yu, Z.; Bo, L.; Tao, D.; Yongyi, Z. Optimal design of insulation of current transformers cast by epoxy resin at high altitude. *High Volt. Electr. Appl.* **2010**, *46*, 57–60.
2. Yu, G.; Dong, B.; Zhang, Z.; Xiang, N.; Bin, D.; Zhu, Z. Simulation calculation and analysis on three-dimensional electro-thermal coupling of a new type of current transformer for distribution network. In Proceedings of the 2020 IEEE International Conference on High Voltage Engineering and Application (ICHVE), Beijing, China, 6–10 September 2020; pp. 1–4.
3. Shling, Z.; Zongren, P.; Peng, L.; Wei, H.; Haoran, W. The electrothermal coupling model is applied to the calculation of radial temperature and electric field distribution of high voltage dry DC bushing. *Chin. J. Electr. Eng.* **2013**, *22*, 191–200.
4. Shiling, Z.; Yongsheng, H.; Naiyi, L.; Zongren, P. Finite element simulation analysis of SF6 gas insulated composite bushing structure for 1100 kV GIS. *Insul. Mater.* **2019**, *52*, 50–56.
5. Pengyuan, L. Study on the Influence of Ambient Temperature Change on SF6 CT Operation Performance and Countermeasures. Ph.D. Thesis, North China Electric Power University, Beijing, China, 2016.
6. Zhang, S. Evaluation of thermal transient and overload capability of high voltage bushings with ATP. *IEEE Trans. Power Deliv.* **2009**, *24*, 1295–1301. [CrossRef]
7. Ramu, T.S.; Reddy, C.C. On the computation of electric field and temperature distribution in HVDC cable insulation. *IEEE Trans. Dielectr. Electr. Insul.* **2006**, *13*, 1236–1244.
8. Jyothi, N.S.; Ramu, T.S.; Mandlik, M. Temperature distribution in resin impregnated paper insulation for transformer bushing. *IEEE Trans. Dielectr. Electr. Insul.* **2010**, *17*, 931–938. [CrossRef]
9. Rachek, M.; Nait Larbi, S. Magnetic Eddy-Current and Thermal Coupled Models for the Finite-Element Behavior Analysis of Underground Power Cables. *IEEE Trans. Magn.* **2008**, *44*, 4739–4746. [CrossRef]
10. Eslamian, M.; Vahidi, B.; Eslamian, A. Thermal analysis of cast-resin dry-type transformers. *Energy Convers. Manag.* **2011**, *52*, 2479–2488. [CrossRef]
11. Azizian, D.; Bigdeli, M.; Fotuhi-Firuzabad, M. A Dynamic Thermal Based Reliability Model of Cast-Resin Dry-Type Transformers. In Proceedings of the 2010 International Conference on Power System Technology, HangZhou, China, 24–28 October 2010; pp. 1–7.
12. Jang, S.-M.; Park, H.C.; Cho, S.K.; Lee, S.-H.; Cho, H.W. Thermal Analysis of Induction Heating Roll With Heat Pipes. *IEEE Trans. Magn.* **2003**, *39*, 3244–3246. [CrossRef]
13. Lee, M.; Patel, D.; Abdullah, H.A.; Jofriet, J.C.; Fahrioglu, M. Air temperature effect on thermal models for ventilated dry-type transformers. *Electr. Power Syst. Res.* **2011**, *81*, 783–789. [CrossRef]
14. Xiuguang, L. Finite element analysis of temperature field of SF6 CT. *High Volt. Electr. Appl.* **2015**, *51*, 121–124.
15. lan, X.; Yanlong, Z.; Zikang, Y.; Daojuun, S.; Chaohui, X.; Wei, H. Analysis and calculation of temperature rise of resin cast dry transformer. *High Volt. Technol.* **2013**, *39*, 265–271.
16. Arjona, M.A.; Hernandez, C.; Escarela-Perez, R.; Melgoza, E. Thermal analysis of a dry-type distribution power transformer using FEA. In Proceedings of the 2014 International Conference on Electrical Machines (ICEM), Berlin, Germany, 2–5 September 2014; pp. 2270–2274.
17. Hui, J.; Hao, W.; Zhiyong, L.; Xin, W.; Peng, Z. Research on the influence of ambient temperature change on SF6 CT operation performance. *Power Grid Clean Energy* **2017**, *33*, 65–69.
18. Ruan, J.; Liu, C.; Gong, R.; Liao, C. Temperature rise of a dry-type transformer with quasi-3D coupled-field method. *IET Electr. Power Appl.* **2016**, *10*, 598–603.
19. Yang, Z.; Naiqiu, S.; Xiaoqing, L. Numerical calculation and analysis of temperature rise of direct-buried gas-insulated transmission lines based on finite element method. *J. Wuhan Univ.* **2015**, *48*, 820–825, 830.
20. Wen, X.; Zhang, J.; Lu, H. Automatic J–A Model Parameter Tuning Algorithm for High Accuracy Inrush Current Simulation. *Energies* **2017**, *10*, 480. [CrossRef]
21. Guojian, L.; Fenghua, W. Calculation and analysis of temperature field distribution of resin cast dry transformer. *High Volt. Electr. Appl.* **2016**, *52*, 83–89.
22. Yongchun, L.; Qiaoling, W.; Caihong, Y.; Jin, Z.; Yanming, L.; Jinyuan, W. Application of three-dimensional finite element method in the calculation of temperature field and carrying capacity of locally buried cable through pipe. *High Volt. Technol.* **2011**, *37*, 2911–2917.

energies

MDPI

Article

Dielectric and AC Breakdown Properties of SiO_2/MMT/LDPE Micro–Nano Composites

Hongtao Jiang, Xiaohong Zhang *, Junguo Gao and Ning Guo

Key Laboratory of Engineering Dielectrics and Its Application, Ministry of Education, Harbin University of Science and Technology, Harbin 150080, China; jianghongtao012@163.com (H.J.); gaojunguo@hrbust.edu.cn (J.G.); tad@hrbust.edu.cn (N.G.)

* Correspondence: x_hzhang2002@hrbust.edu.cn

Abstract: Low-density polyethylene (LDPE) is an important thermoplastic material which can be made into films, containers, wires, cables, etc. It is highly valued in the fields of packaging, medicine, and health, as well as cables. The method of improving the dielectric property of materials by blending LDPE with inorganic particles as filler has been paid much attention by researchers. In this paper, low-density polyethylene is used as the matrix, and montmorillonite (MMT) particles and silica (SiO_2) particles are selected as micro and nano fillers, respectively. In changing the order of adding two kinds of particles, a total of five composite materials were prepared. The crystallization behavior and crystallinity of five kinds of composites were observed, the ε_r and tanδ changes of each material were investigated with frequency and temperature, and the power frequency (50 Hz) AC breakdown performance of materials were measured. The differential scanning calorimetry (DSC) and X-ray diffraction (XRD) results show that the crystallinity of the composites is higher than that of LDPE. Experimental data of dielectric frequency spectra show that the dielectric constants of micro–nano composites and composites with added MMT particles are lower than LDPE, the dielectric loss of composites can be improved by adding MMT particles. The experimental data of dielectric temperature spectra show that the permittivity of SiO_2-MMT/LDPE is still at a low level under the condition of 20~100 °C. In terms of breakdown field strength, the SiO_2/LDPE composite material increased by about 17% compared with the matrix LDPE, and the breakdown field strength of the materials SiO_2-MMT/LDPE and MMT-SiO_2/LDPE increased by about 6.8% and 4.6%, respectively.

Keywords: micro and nanoparticles; adding order; dielectric properties; AC breakdown strength

Citation: Jiang, H.; Zhang, X.; Gao, J.; Guo, N. Dielectric and AC Breakdown Properties of SiO_2/MMT/LDPE Micro–Nano Composites. *Energies* **2021**, *14*, 1235. https://doi.org/10.3390/en14051235

Academic Editor: Carlos Miguel Costa

Received: 25 January 2021
Accepted: 19 February 2021
Published: 24 February 2021

1. Introduction

Low-density polyethylene (LDPE) is diffusely employed in many fields, such as wire and cable industry and environmental engineering, because of its outstanding insulation, fine mechanical properties, process performance, and adsorption [1,2]. With the rapid economic development of all countries in the world, the cable industry is moving towards high voltage and ultra-high voltage. Therefore, it is very important to enhance the dielectric property of high-voltage insulation materials and d evelop higher-grade high-voltage insulation materials [3].

In the interest of meeting the increasing demand for science and technology progress and the development of society on the insulating electrical properties of polymers, researchers have tried to ameliorate the physicochemical properties of polymers by the method of adding inorganic nano-fillers. A great deal of research work shows that the mixing of nano-fillers and polymer materials could polish up the breakdown field strength [4–6], conductivity [7,8], electrical tree resistance, and partial discharge performances [9,10], as well as inhibit the accumulation of space charges [11,12]. Nevertheless, the result of the aggregation of nanoparticles in polymers has made achieving good dispersion difficult. Thereby, different researchers have different conclusions. Up to now, the improvement mechanism of added inorganic particles to matrix is still not very clear. For the sake of acquisition of maximized dispersion of nanoparticles in polymers, scholars make an attempt to blend micro- and nanoparticles with matrix materials. It is expected that the superior properties of composites will be obtained through the synergistic effect between micro- and nanoparticles. Research shows that the introduction of micron-particles and nanoparticles to matrix materials can not only avoid the agglomeration of nano-particles, but also improve the properties of the matrix materials [13–15]. At a certain proportion, the adding of the micron particle will highlight unique performances of the nanoparticle to a certain degree. This makes certain performances of micro–nano composites resemblant to nano-composites, and even superior to nano-composites. Ma et al. synthesized micro–nano Co_3O_4 ceramic materials by liquid phase and calcination method and studied its microstructure and dielectric properties, and found that micro–nano composite materials have more excellent dielectric properties [16]. Espinoza-Gonzalez et al. doped $CaCu_3Ti_4O_{12}$ (CCTO) with ZrO and ZrO_2 as micro–nano fillers, and found the micro–nano composite exhibited lower dielectric loss and a smaller dielectric constant [17]. Nazir et al. added micro and nano boron nitride (BN) particles to the matrix ethylene propylene diene monomer (EPDM) by melt blending method. It was found that the composite with 25 wt% micron particles and 5 wt% nanoparticles had good dielectric loss performance; the composite with 29 wt% micron particles and 1% nanoparticles showed low permittivity [18].

Adding nano-SiO_2 particles into LDPE can increase the crystal nucleus density, reduce the spherulite size, and enhance the resistance to partial discharge [19]. Montmorillonite (MMT) is a layered silicate mineral with one-dimensional nano-structure. After being organically treated, MMT is introduced into LDPE by melt intercalation method. MMT can scatter and block electrons generated in the discharge process, thus improving the partial discharge resistance of LDPE [20]. In this paper, SiO_2 was selected as inorganic nanoparticles, MMT as micron particles, LDPE as polymer matrix, with 1 wt% of added content, LDPE, MMT/LDPE, SiO_2/LDPE, SiO_2-MMT/LDPE, and MMT-SiO_2/LDPE composite materials were prepared by melt blending method. The adding order of micro–nano particles in SiO_2-MMT/LDPE and MMT-SiO_2/LDPE was changed. SiO_2-MMT/LDPE involved adding the SiO_2 particle first and then the MMT particle, while MMT-SiO_2/LDPE involved adding the MMT particle first and then the SiO_2 particle. The effects of different composite systems and adding order of particles on the crystalline structure, dielectric properties, and AC breakdown properties of the composites were studied.

The research content of the article is a continuation of the research of reference [21]. The research results of the conduction current and space charge characteristics of SiO_2/MMT/LDPE micro–nano composites are published in reference [21], which concluded that MMT-SiO_2/LDPE composite materials can well hinder carrier migration and have a strong inhibitory effect on space charges, but many contents could not be perfected because the previous writing and publications are in the COVID-19 epidemic period. In this paper, the polarizing microscope (PLM) experiment and the differential scanning calorimetry (DSC) experiment were supplemented and analyzed, and X-ray diffraction (XRD) experimental results were combined to verify the crystallization behavior and crystallinity of each material. The characteristics of dielectric spectrum and dielectric temperature spectrum of each experimental material were analyzed, and the breakdown field strength characteristics of each composite material were studied. The research done in this paper is a great supplement to the previous research, which can make readers have a more comprehensive understanding of our research work, which has vital function in understanding the mechanism of improving properties of micro and nano composites.

2. Experimental Method

2.1. Surface Modification of SiO_2

The nano silica (SiO_2) and silane coupling agent (KH570) used to prepare the samples were both from Beijing Deke Daojin Science and Technology (Beijing, China), and the particle size of nano-SiO_2 was 30 nm. SiO_2 particles were treated with coupling agent KH570. Firstly, the SiO_2 powder was dissolved with the mixed solution of anhydrous ethanol and deionized water, and stored in a three-neck boiling flask. Secondly, the ultrasonic wave was used to oscillate for 1 h, silane coupling agent KH570 was added, and heated in a constant temperature water bath. The temperature was set at 80 °C, stirring uniformly at high speed, and applying ultrasonic oscillation for 2 h again. Finally, after filtering, washing, drying, and grinding, the surface-modified SiO_2 particles were obtained.

2.2. *Organic Treatment of MMT*

The montmorillonite (MMT) was purchased from Qinghe Chemical Plant in Zhangjiakou City of China. The particle size of the original MMT was about 40–70 μm, and the cation exchange capacity was 0.9–1.2 mol/kg. Firstly, MMT was mixed with weak acid solution in a certain proportion, and then heated in a water bath at 80 °C and stirred at high speed to obtain a uniform MMT suspension. Then, it was centrifuged and purified, and then intercalated with octadecyl trimethyl ammonium chloride solution. The mixture was placed in the water bath again. The temperature was set at 80 °C, and high-speed stirring and ultrasonic treatment were carried out to obtain the MMT with intercalation suspension. The suspension was allowed to stand, then washed with deionized water until no white precipitate appeared after being tested by $AgNO_3$ solution. After a series of operations such as drying, grinding, and sieving, the intercalated solid MMT particles were obtained.

The MMT particles prepared above were used to prepare the MMT suspension with 95% ethanol solution as solvent in a ratio of 1 g:10 mL, and the solution was stirred at high speed and oscillated by ultrasonic for 3 h with the temperature of 70 °C. After that, the silane coupling agent KH570 was used to modify the surface. The specific operation steps are as follows: In acidic environment with pH value of 3~5, the coupling agent KH570 was pre-hydrolyzed with 95% ethanol aqueous solution. While stirring the MMT suspension, the hydrolyzed KH570 was slowly added into the suspension in three times. After the solution reaction is completed, it should be left to stand for a period of time through the separator funnel. Then, the MMT layer was taken out and dried in an oven under 80 °C for 24 h, and then ground and screened to obtain the organic MMT required in this paper.

2.3. *Preparation of Composite Materials*

Experimental fabrication of materials required for this article is reported in [21].

2.4. *Crystallization Behavior*

The five kinds of materials were soaked in the mixed solution of potassium permanganate and concentrated sulfuric acid with mass fraction of 5% for 5 h to etch the surface of the sample. After that, the samples were cleaned with ultrasonic cleaning apparatus. The microscopic morphology of each sample was detected by LeicaDM2500 polarizing microscope (PLM, Leica Microsystems, Wetzlar, Germany), and the magnification was 500 times.

A DSC-1 differential scanning calorimeter produced by Mettler Toledo company was utilized to measure the crystallization and fusion temperature of each material. The determination condition was nitrogen atmosphere, and the lifting and cooling rates were all 10 °C/min. The temperature selection range was 25~140 °C, and 10~15 mg was weighed for each sample.

The LDPE and composite materials were tested by X-ray diffraction (XRD) using EMPYREAN X-ray diffractometer (Panako, The Netherlands). Cu target was used in the experiment. The tube voltage was 40 kV, the tube current was 30 mA, the experimental scanning speed was 2°/min, the scanning range was $2\theta = 5°\sim50°$, and the X-ray wavelength was 0.154 nm.

2.5. *Dielectric Frequency Spectra and Dielectric Temperature Spectra*

At ambient temperature, the ε_r and $\tan\delta$ of all samples were tested with frequency f by broadband dielectric/impedance spectrometer (Novocontrol Technologies, Montabaur, Germany), and the test frequency range was 10^{-1} Hz~10^6 Hz. For reducing the influence of moisture and residual charge on the test results, it is necessary to short-circuit each experimental material in advance.

The QS30I high-precision high-voltage capacitor bridge (Shanghai Peicheng Electronic Technology Development Co., Ltd., Shanghai, China) and high-low temperature alternating test box (Shanghai Guangpin Test Equipment Manufacturing Co., Ltd., Shanghai, China) were used to test the change of ε_r and tanδ with T by three electrode system. The temperature range was 20~100 °C, and the temperature gradient was 10 °C. The pretreatment method of the samples was consistent with the dielectric spectrum. During the time of test, electrodes and samples should be placed in a high and low temperature alternating test chamber and the temperature should be kept constant.

2.6. Breakdown Strength

According to the Chinese standard GB/T1408.1-2005, the power frequency AC breakdown performance of each sample was tested by plate–plate electrode, as shown in Figure 1. The diameter and height of the upper electrode were 20 mm and 25 mm, respectively. The diameter and height of the lower electrode were 60 mm and 10 mm, respectively. The edges of the electrodes were rounded and the radius was 2 mm. The sample thickness was 50 μm. In order to prevent residual charge from affecting the test results, it is necessary to short-circuit each sample to be tested before testing the breakdown field strength. In order to prevent the edge discharge of metal electrode, the sample with thickness of 50 μm and electrode should be immersed in cable oil. In the test, 50 Hz alternating current was used, and the voltage was slowly increased at a speed of 1 kV/s until composites were broken down, and the voltage was recorded. By the formula E = U/D, the corresponding power frequency AC breakdown field strength was calculated, where U is the voltage at the breakdown of the sample, D is the thickness of the sample, and E is the breakdown field strength of each sample.

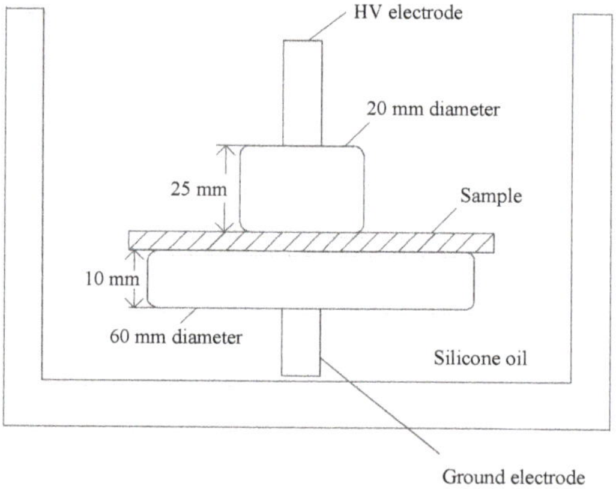

Figure 1. Schematic diagram of breakdown experimental device.

3. Results and Discussion

3.1. Crystalline Morphology

Figure 2 shows the crystalline morphology of each material under PLM after being etched by concentrated sulfuric acid mixed with potassium permanganate. As both the previous writing and the ICHVE 2020 conference were held during the COVID-19 epidemic period, universities were closed, so the crystal cell size of each material was not marked in time [21,22]. Now, the epidemic situation in China is well controlled and universities have been opened. We went back to the laboratory and marked the crystal cell size of each composite material, and the statistical distribution of the crystal cell size of each material was calculated; the results are shown in Figure 3. From Figure 2, the crystal of all materials are spherical structures. It can be seen from Figure 3 the average and standard deviation of cell size of each sample. In Figures 2 and 3, LDPE has the largest crystal cell size, and the crystal cell size varies, the diameter of crystal cell is about 9~19 μm, and the average size is 12.541 μm. After adding MMT into the matrix, the crystal cell size decreases, the average size is 6.849 m, and the crystal cell spacing decreases to some extent. The crystal cell size of composites with SiO_2 particles is further reduced, with an average diameter of 6.345 μm, and the crystal cell size of SiO_2/LDPE is uniform, but its crystal cell spacing is larger than that of micro–nano composites SiO_2-MMT /LDPE and MMT-SiO_2/LDPE. The crystal cell size and spacing of SiO_2-MMT/LDPE material prepared by the method of adding SiO_2 particle first and then MMT particle are further reduced. The diameter of the crystal cell is distributed around 4~8 μm, and the average diameter of the cell is 5.341 μm. The internal structure of the material is tight. The crystal size and spacing of MMT-SiO_2/LDPE material prepared by the method of adding MMT particle first and then SiO_2 particle becomes smaller than the former, and the crystal cell diameter is distributed around 3~7 μm with the average crystal cell diameter as 4.765 μm. The reason why the crystal cell size of MMT-SiO_2/LDPE is smaller than that of SiO_2-MMT/LDPE is that the small-sized particle is added first, which makes it easier for the small-sized particles to enter the machine body and play a heterogeneous nucleation role, thus reducing the unit cell size. However, a small-sized particle is added first, and then a large-sized particle is added, which will damage the structure formed by the small-sized particle added first, and weaken the heterogeneous nucleation effect, so the cell size is reduced.

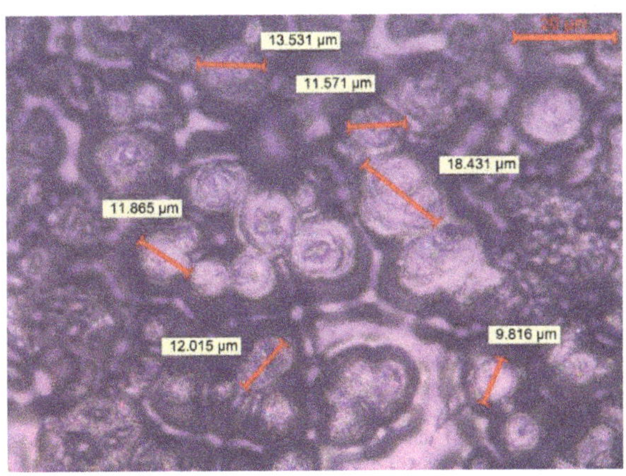

(**a**) LDPE

Figure 2. *Cont.*

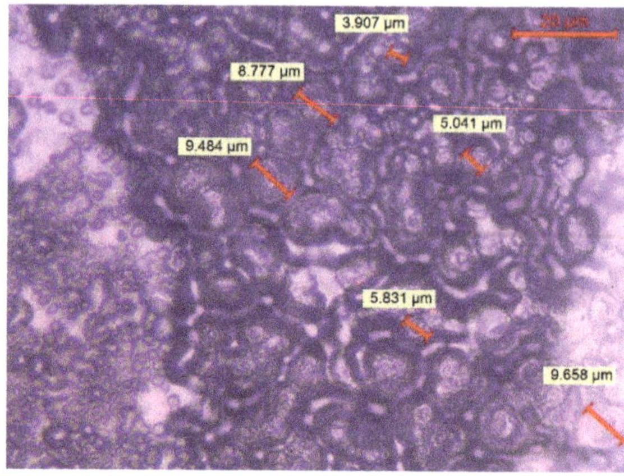

(**b**) MMT/LDPE

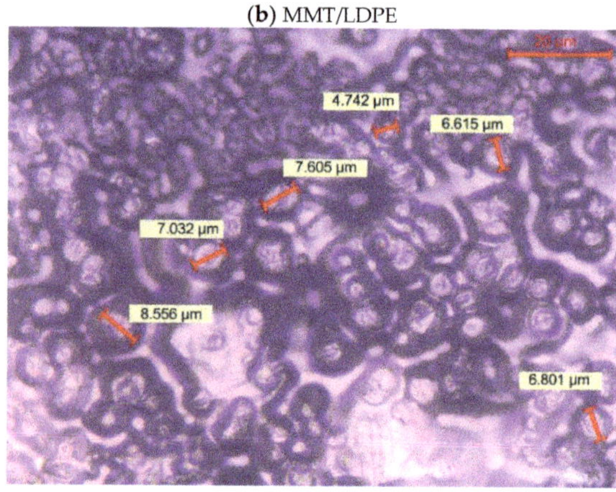

(**c**) SiO_2/LDPE

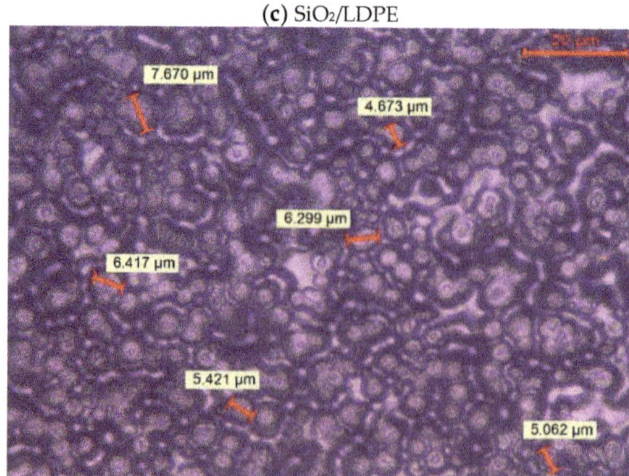

(**d**) SiO_2-MMT/LDPE

Figure 2. *Cont.*

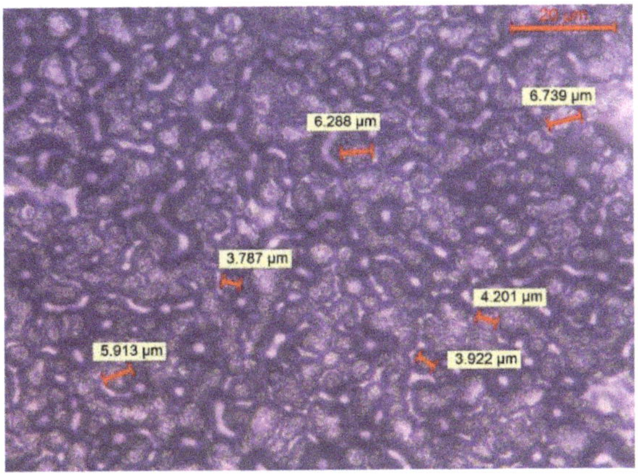

(**e**) MMT-SiO_2/LDPE

Figure 2. The crystalline morphology of various samples.

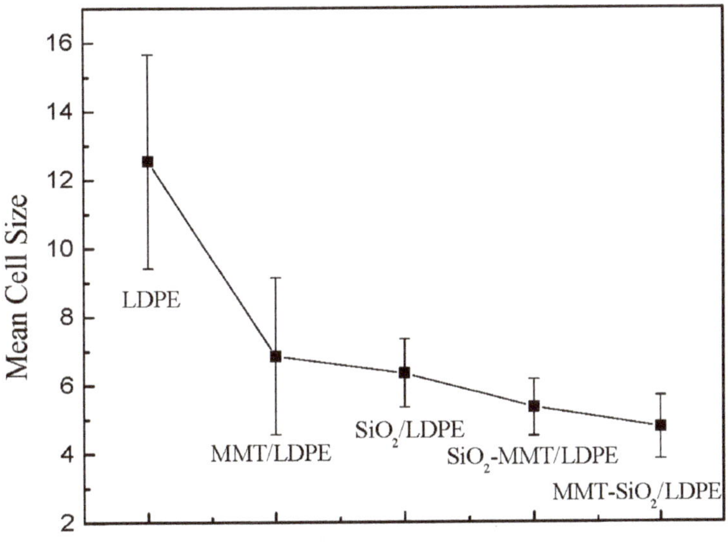

Figure 3. Statistical distribution of crystal cell size of each sample.

3.2. *Crystallinity of Composite Materials*

This part of the experiment has been given some analysis in reference [22]. The experimental results and analysis of XRD, dielectric spectrum, dielectric temperature spectrum, and breakdown in this paper need to be combined with this part of the experiment. In order to make the analysis of this paper more comprehensive, further experimental analysis is given here again. Considering the particles do not contribute to the enthalpy of crystallization, we can obtain the crystallinity of five samples through the calculation of Equation (1) [23].

The fusion temperature and crystallinity results of all samples are listed in Table 1.

$$X_c = \frac{\Delta H_m}{(1-\omega)\Delta H_0} \times 100\% \quad (1)$$

where $H_0 = 293.6\ J/g^{-1}$ and w is the particles content in the composites.

Table 1. Fusion peaks and crystallinity of samples.

Sample	Fusion Peak Temperature T_m/°C	Crystallinity X_c/%	Normalized Heat/J·g^{-1}.
LDPE	108	31.20	90.69
MMT/LDPE	108.18	34.07	99.03
SiO_2/LDPE	108.01	35.69	103.72
SiO_2-MMT/LDPE	107.77	33.05	95.10
MMT-SiO_2/LDPE	107.49	32.23	92.75

According to Table 1, the order of fusion temperature is SiO_2/LDPE > MMT/LDPE > SiO_2-MMT/LDPE > MMT-SiO_2/LDPE > LDPE. The order of crystallinity is SiO_2/LDPE > MMT/LDPE > SiO_2-MMT/LDPE > MMT-SiO_2/LDPE > LDPE. The increase in the crystallinity of the sample is due to heterogeneous nucleation after doped particles. The heterogeneous nucleation of SiO_2 with smaller particle size is obvious, and the crystallinity of SiO_2/LDPE material has a larger increase. During the experiment, the addition order of micro- and nanoparticles was changed to research the crystallinity of micro–nano composite materials. According to the order of adding the SiO_2 particle first and then the MMT particle, the crystallinity of SiO_2-MMT/LDPE composites are weaker than that of the composites with only SiO_2 particles. This is result of the large size of the MMT particles; the addition of MMT particles will squeeze the composite materials crystalline structure of samples, which will change the crystalline structure of the sample. Furthermore, the addition of MMT particles will cause expansion of the original crystal area, and the new structure may increase the hindrance to the movement of molecular chains. According to the order of adding the MMT particle first and then the SiO_2 particle, the crystallinity of MMT-SiO_2/LDPE composites are weaker than the SiO_2-MMT/LDPE composites. This is because the size of the MMT particle is larger than the SiO_2 particle. In the preparation of micro–nano composites, small particles added later are harder to squeeze into the matrix than large ones. Simultaneously, the composites filled with particles may add multiple heat conduction channels, which significantly improves the fusion temperature of the material. Experimental data shows that the smaller the particle size, the more significant the improvement of the fusion temperature. When SiO_2-MMT/LDPE is compared with MMT-SiO_2/LDPE composite materials, both of them are micro–nano composite materials with two kinds of particles, but the fusion temperature of the former is higher than that of the latter. The reason is that the large-sized MMT particle is added later in the former, and more heat conduction channels will be introduced in composite process than the latter, so that the fusion temperature of SiO_2-MMT/LDPE composites is high. However, the particles added later will squeeze the original heat conduction channels, resulting in the fusion temperature of micro–nano composites not being as good as that of adding one particle alone.

3.3. *XRD Analysis of Composite Materials*

Figure 4 shows the X-ray diffraction curves of composite materials. From Figure 4, there are two obvious diffraction peaks in each composite, which are 21.35° and 23.65°. According to reference [24], these two diffraction peaks correspond to (110) and (200) crystal surfaces, respectively.

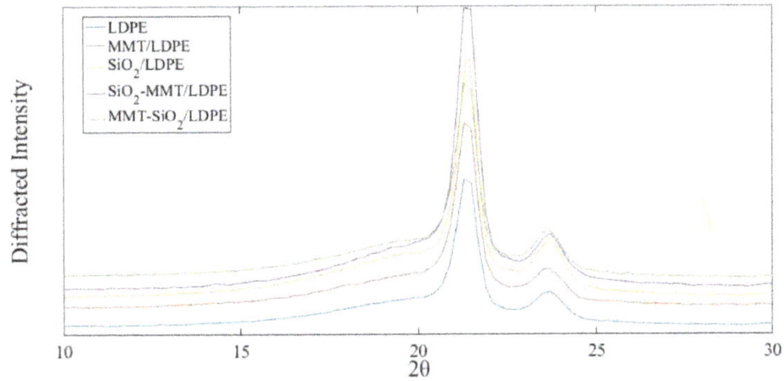

Figure 4. XRD curves of each sample.

According to Equation (2), the average grain thickness of the composite with different crystal planes can be calculated [25].

$$D = \frac{0.89\lambda}{\sqrt{\beta^2 - \beta_i^2 cos\theta}} \tag{2}$$

where λ is the X-ray wavelength and the value is 0.154 nm; β is half height and width of diffraction peak on diffraction crystal surface; β_i is the broadening factor of the instrument; θ is the diffraction angle.

According to Equation (3), the crystallinity of each composite material after XRD test is calculated [26].

$$X_c = \frac{I_{110} + 1.42I_{200}}{I_{110} + 1.42I_{200} + 0.68I_A} \times 100\% \tag{3}$$

where I_{110}, I_{200}, and I_A are diffraction peak integral area of (110) crystal plane, (200) crystal plane, and amorphous scattering diffraction peak integral area, respectively.

The XRD parameter calculation results of each sample are listed in Table 2.

Table 2. XRD parameters of each sample.

Sample	$\beta_{110}/°$	D_{110}/nm	$\beta_{200}/°$	D_{200}/nm	X_c/%
LDPE	0.628	12.48	1.339	7.39	36.99
MMT/LDPE	0.637	12.29	0.906	9.06	40.66
SiO_2/LDPE	0.598	9.29	1.026	8.75	41.10
SiO_2-MMT/LDPE	0.612	12.59	0.714	10.37	38.79
MMT-SiO_2/LDPE	0.634	12.32	1.300	7.51	37.95

In Table 2, the crystallinity order of composites is SiO_2/LDPE > MMT/LDPE > SiO_2-MMT/LDPE > MMT-SiO_2/LDPE > LDPE, which is consistent with the DSC test results. From the average grain thickness of each composite, it can be seen that the average grain thickness both of SiO_2/LDPE and SiO_2-MMT/LDPE composites along the (110) and (200) crystal plane directions have little difference, which indicates that the crystal growth of the two composites is uniform. SiO_2-MMT/LDPE composite is the thickest of the average grain thicknesses in terms of both the direction along the (100) crystal plane and the (200) crystal plane, indicating that the addition of large-size MMT particles broaden the crystallization area to a certain extent, and thus increases the average thickness of the grain, which confirms the analysis results of the DSC experiments above.

3.4. Dielectric Frequency Spectra of Composite Materials

Two major statements for the change of permittivity exist in composites with certain particles. First, the interaction between particles and composites matrix limits the moving of composite macromolecular chains and reduces the polarization of matrix [27,28]. Second, the interfacial polarization would occur in the interface between particles and composites matrix, which will make the value of permittivity rise [29]. Figures 5 and 6 show the variation of ε_r and t tanδ with f of each sample. In Figures 5 and 6, we can see that LDPE is a non-polar material, the polarization of electron displacement occurs primarily in its interior, and therefore ε_r is small. At the same time, it does not vary with the frequency. SiO_2 particles were added into LDPE, and a number of interfaces will be formed. There will be interface polarization in materials, hence ε_r and tanδ of SiO_2/LDPE are higher than LDPE. As the frequency increases, the electrical fields change gradually faster, and the relaxation polarization is too late to keep abreast of the electrical fields. In consequence, polarization establishment is incomplete, and ε_r has a downward trend. As stated in the PLM results, MMT/LDPE and MMT-SiO_2/LDPE have compact structures. Simultaneously, on account of the especially stratified structure of MMT, segment activity and polarization establishment of composite materials are not easy. As a consequence, the value of their dielectric constant is small. The structure of SiO_2-MMT/LDPE restricts severely the movement of macromolecular chains, and as a result, the ε_r and tanδ values of SiO_2-MMT/LDPE are at low levels in five materials.

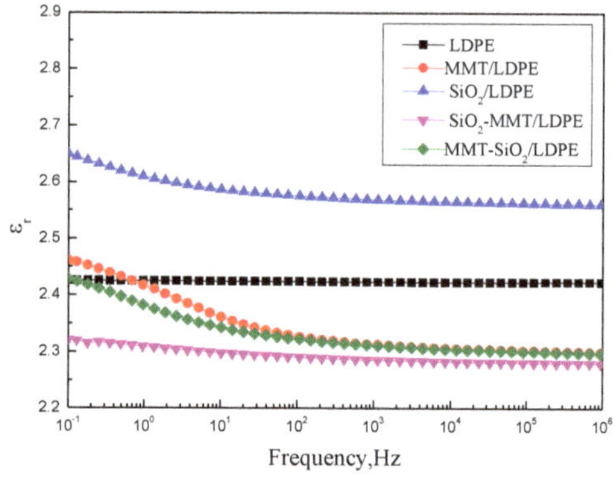

Figure 5. Curves of ε_r with respect to f.

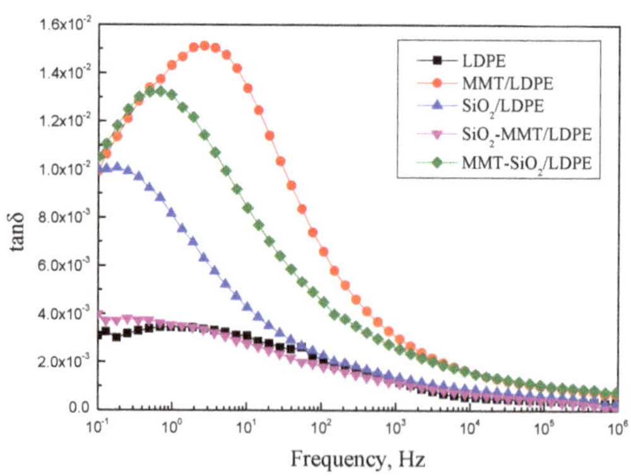

Figure 6. Curves of tanδ with respect to f.

3.5. Dielectric Temperature Spectra of Composite Materials

Figures 7 and 8 show the changes of ε_r and tanδ with T of the five samples, and the test frequency is 50 Hz. It is observed that the dielectric constant of composites reduces with the increment of temperature. This is because of the intensification of molecular thermal motion, which results in directional polarization difficulty. When a micro–nano particle is introduced into the matrix, interface areas will appear, hindering the thermal motion of molecular chains under high temperature, and therefore dielectric constant and loss of composites will raise. Nevertheless, SiO_2-MMT/LDPE could ruin the original interface areas with adding of small-size fillers first, and then large-size fillers. As a result, interfacial regions have less obstruction to molecular chains' thermal movement under high temperature, and dielectric constant of composites decline accordingly. As the increment of temperature, the loss is chiefly decided by the conductance loss, and tanδ increases along with temperature. Since the main components in MMT are SiO_2, Al_2O_3, and H_2O, after introducing MMT particles into composites, lots of interfaces formation. Relaxation polarization has a great influence under low temperature and relaxation loss is large. Therefore, the loss of composites with MMT increases first, and then decreases with the temperature rises.

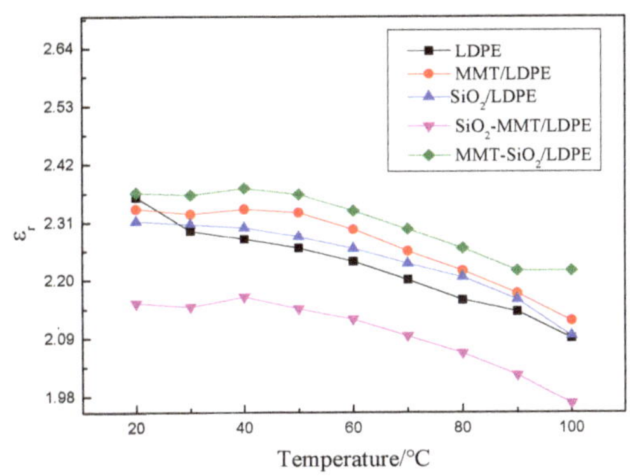

Figure 7. Curves of ε_r with respect to T.

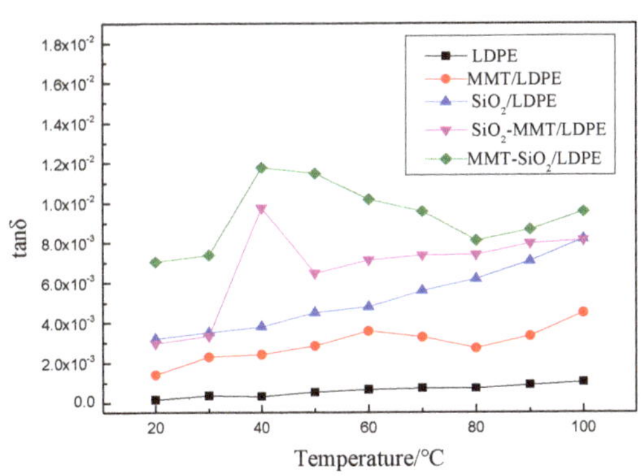

Figure 8. Curves of tanδ with respect to T.

3.6. Breakdown Field Strength

The breakdown test data of each composite material are processed by two-parameter Weibull distribution [30,31]. The relationship between the cumulative breakdown probability $P(E)$ and the breakdown electric field intensity E is:

$$P(E) = 1 - \exp\left(\frac{-E}{E_0}\right)^{\beta} \tag{4}$$

In Equation (4), E_0 is the standard breakdown field strength when the cumulative breakdown probability is 63.2%; β is the shape parameter, which characterizes the dispersion of breakdown data.

The Weibull distribution analysis of breakdown field strength data of each sample measured by the experimental device in Figure 1 was analyzed by MINITAB software; 15 samples were measured for each material, and the results are shown in Figure 9.

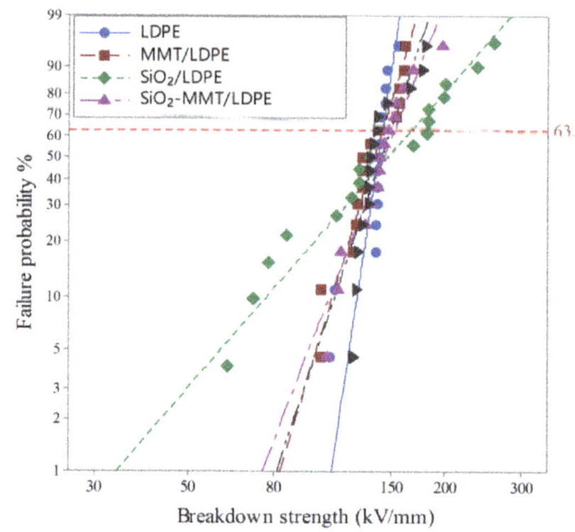

Figure 9. Weibull curve of breakdown field strength of each sample.

The logarithm of both sides of Equation (4) are taken at the same time, and then simplified to get:

$$ln[-ln(1-P)] = \beta(lnE - lnE_0) \tag{5}$$

The breakdown field strength E_0 can be calculated by Equation (5). The characteristic breakdown field strength values of each sample are shown in Table 3.

Table 3. Characteristic breakdown field strength of each sample.

Sample	Characteristic Strength (kV/mm)	β
LDPE	142.2	16.930
MMT/LDPE	142.6	8.387
SiO_2/LDPE	166.5	2.883
SiO_2-MMT/LDPE	151.8	6.468
MMT-SiO_2/LDPE	148.7	7.549

β represents the degree of data dispersion. The smaller the value, the more dispersed the data. From Table 3, it can be seen that the β values of each material are greater than 1. At present, regarding the mechanism by which inorganic particles improve the breakdown strength of polymer insulating dielectrics, scholars at home and abroad have basically reached a certain consensus. Lewis et al. proposed that the interface is the main factor determining the electrical properties of nano-dielectrics, and an interface charged layer will form between particles and polyethylene matrix; the interface effect will improve the dielectric properties of the dielectric [32].

From the PLM experiment and dielectric spectrum analysis, it serves to show that the addition of micron MMT particle will impede the movement of molecular chains and inhibit the migration of carriers, which will weaken the collision of electrons on the macromolecular chains to a certain extent, and lead to the improvement of the breakdown performance of MMT/LDPE composite materials. However, due to the weak interface between micron particles and matrix, the breakdown performance is not significantly improved.

The addition of nano-SiO_2 particles will form an independent interface domain in the polyethylene matrix, which will affect the space charge transport and improve the breakdown field strength. By mixing and ultrasonic treatment of the twin-screw extruder, nano-SiO_2 particles can obtain better uniform dispersion [33,34], which are easy to form more interface domains. In the previous study [21], it was proved that the addition of SiO_2 particles will introduce deep traps into the matrix LDPE. Under the action of the electric field, the charge cannot be trapped and is limited on the surface of dielectric materials. Moreover, the structure formed by SiO_2/LDPE composite materials has a certain blocking effect on carrier migration, thereby reducing the energy accumulation of electrons and weakening its impact on macromolecular chains, thus improving the impact breakdown field strength by about 17% [35].

For SiO_2-MMT/LDPE and MMT-SiO_2/LDPE, due to the introduction of two kinds of particles, a great many interfaces are introduced into the matrix, so the breakdown field strength is increased by about 6.8% and 4.6%, respectively. However, due to the introduction of micron particles that will form part of the weak interface with the matrix, the breakdown field strength of them is lower than SiO_2/LDPE. When MMT-SiO_2/LDPE is compared with SiO_2-MMT/LDPE, according to the PLM experimental results, the MMT-SiO_2/LDPE composite materials formed by adding large-sized MMT particle first and then small-sized SiO_2 particle have a compact crystal structure, and the crystallinity is small from the crystallinity results of DSC and XRD. This will increase the free volume of the materials, increase the chance of electron collisions, and decrease the breakdown field strength. At the same time, according to DSC experiment and analysis, the heat dissipation of MMT-SiO_2/LDPE composite materials is worse than that of SiO_2-MMT/LDPE, so MMT-SiO_2/LDPE easily accumulates heat inside the materials, which improves the probability of thermal breakdown of the materials. Therefore, the breakdown field strength of MMT-SiO_2/LDPE is about 2% lower than that of SiO_2-MMT/LDPE.

4. Conclusions

From PLM experiment, DSC experiment, and XRD experiment results, it can be seen that the introduction of micro- and nanoparticles together will make the internal structure of the materials more compact than the matrix. The adding method of first small and then large can more effectively limit the molecular chains activity than the composite materials prepared by adding large first and then small. The addition of a variety of particles will destroy the original crystal structure, which makes the crystallinity of materials added with varieties of particles slightly lower than materials added with a single particle. The addition method of first small and then large will expand the crystallization region of the composites, and the crystallinity is higher than that of the composite materials prepared by first large and then small addition method.

From the experimental results of dielectric spectrum, it can be seen that the addition of micro- and nanoparticles will make the materials structure more compact, resulting in serious restriction of molecular chain motion, difficulty in establishing partial polarization, and a decrease in the dielectric constant and loss to a certain extent. Among them, the effect of adding small particles first and then large particles is better.

From the experimental results of dielectric temperature spectrum, it can be seen that with the change of temperature, the original interface area may be destroyed by the addition method of first small and then large, and the interface area is not bound enough to the thermal motion of molecules, so the dielectric constant will decrease.

From the experimental results of AC breakdown field strength, it can be seen that the interface effect is the main factor affecting the breakdown performance of composite materials. By adding nano-SiO_2 particles alone in LDPE, the breakdown field strength is greatly increased by about 17%. Adding micro- and nanoparticles into LDPE at the same time will introduce a large number of interfaces into the matrix, which will also improve the breakdown field strength of micro–nano composite materials. The lifting amplitude of SiO_2-MMT/LDPE composite materials is about 6.8%, and that of MMT-SiO_2/LDPE composite materials is about 4.6%.

Author Contributions: H.J.: conceptualization, methodology, software, and writing—original draft. X.Z.: funding acquisition, reviewing, resources, validation, supervision, and editing. J.G.: data curation, visualization, and investigation. N.G.: reviewing, formal analysis, and editing. All authors have read and agreed to the published version of the manuscript.

Funding: This research was funded by the National Natural Science Foundation of China, grant number 51577045.

Acknowledgments: We acknowledge the financial support from the National Natural Science Foundation of China (51577045).

Conflicts of Interest: The authors declare no conflict of interest.

References

1. Du, B.X.; Han, C.L.; Li, J. Research Status of Polyethylene Insulation for High Voltage Direct Current Cables. *Trans. China Electrotech. Soc.* **2019**, *34*, 180–191.
2. Lei, P.; Zhu, J.J.; Pan, K.; Zhang, H. Sorption kinetics of parent and substituted PAHs for low-density polyethylene (LDPE): Determining their partition coefficients between LDPE and water (K_{LDPE}) for passive sampling. *J. Environ. Sci.* **2020**, *87*, 351–362. [CrossRef]
3. Zhen, Z.H.; Li, M.L.; Liu, Z.; Ma, C.X. Discussion on the way to improve the operation reliability of cables and cable accessories. *China Sci. Technol. Expo.* **2011**, *30*, 453.
4. He, L.J.; Zeng, J.J.; Huang, Y.W.; Yang, X.; Li, D.W.; Chen, Y.; Yang, X.Y.; Wang, D.B.; Zhang, Y.X.; Fu, Z.D. Enhanced Thermal Conductivity and Dielectric Properties of h-BN/LDPE Composite. *Materials* **2020**, *21*, 4738. [CrossRef]
5. Tian, F.; Lei, Q.; Wang, X.; Wang, Y. Investigation of electrical properties of LDPE/ZnO nanocomposite dielectrics. *IEEE Trans. Dielectr. Electr. Insul.* **2012**, *19*, 763–769. [CrossRef]
6. Le, H.; Schadler, L.S.; Nelson, J.K. The influence of moisture on the electrical properties of crosslinked polyethylene/silica nanocomposites. *IEEE Trans. Dielectr. Electr. Insul.* **2013**, *20*, 50–53. [CrossRef]
7. Cheng, Y.J.; Yu, G.; Yu, B.Y.; Zhang, X.H. The Research of Conductivity and Dielectric Properties of ZnO/LDPE Composites with Different Particles Size. *Materials* **2020**, *18*, 4136. [CrossRef]
8. Mi, R.; Xing, Z.L.; Hao, J.H.; Hu, X.N.; Min, D.M.; Li, S.T.; Wu, Q.Z. Effect of morphology and traps on DC conductivity and breakdown of polyethylene nanocomposites. *IEEE Trans. Dielectr. Electr. Insul.* **2020**, *27*, 489–497. [CrossRef]
9. Fleming, R.J.; Ammala, A.; Casey, P.S. Conduction and space charge in LDPE/TiO2 nanocomposites. *IEEE Trans. Dielectr. Electr. Insul.* **2010**, *18*, 15–23. [CrossRef]
10. Ai, Y.; Li, C.Y.; Zhao, H.; Yang, J.M.; Zhang, C.C.; Han, B.Z. Effects of nano SiO2 on AC/DC breakdown strength and electrical treeing resistance of cross-linked polyethylene. *J. Compos.* **2019**, *36*, 2031–2041.
11. Fan, Y.H.; Song, W.; Li, L.; Hua, Y.; Yang, J.M.; Sun, Z.; Han, B. Space charge characteristics of micron- and nano-$BiFeO_3$/LDPE composites under a magnetic field. *J. Mater. Sci. Mater. Electron.* **2020**, *32*, 1–13.
12. Castellon, J.; Banet, L.; Preda, I.; Agnel, S.; Toureille, A.; Fréchette, M.; David, E.; Sami, A. Space charge measurements on polymer nanocomposite HDPE/SiO_2 materials. In Proceedings of the 2010 Annual Report Conference on Electrical Insulation and Dielectic Phenomena, West Lafayette, IN, USA, 17–20 October 2010; pp. 1–4.
13. Fabiani, D.; Montanari, G.C.; Krivda, A. Epoxy based materials containing micro and nano sized fillers for improved electrical characteristics. In Proceedings of the 2010 10th IEEE International Conference on Solid Dielectrics, Potsdam, Germany, 4–9 July 2010; pp. 1–4.
14. Li, Z.; Okamoto, K.; Ohki, Y.; Tanaka, T. Effects of nano-filler addition on partial dis-charge resistance and dielectric breakdown strength of micro-Al_2O_3/epoxy composite. *IEEE Trans. Dielectr. Electr. Insul.* **2010**, *17*, 653–661. [CrossRef]
15. Li, Z.; Okamoto, K.; Ohki, Y. The role of nano and micro particles on partial discharge and breakdown strength in epoxy composites. *IEEE Trans. Dielectr. Electr. Insul.* **2011**, *18*, 675–681. [CrossRef]
16. Ma, M.X.; Wei, L.; Zhang, S.L.; Lu, Y.H. Synthesis and Dielectric Properties of Micro-nanostructure Co_3O_4 Ceramics. *J. Ceram.* **2019**, *40*, 101–105.
17. Rodrigo, E.G.; Mosquera, E. Influence of micro- and nanoparticles of zirconium oxides on the dielectric properties of $CaCu3Ti_4O_{12}$. *Ceram. Int.* **2017**, *43*, 14659–14665.
18. Nazir, M.T.; Phung, B.T.; Zhang, Y.Y.; Li, S.T. Dielectric and thermal properties of micro/nano boron nitride co-filled EPDM composites for high-voltage insulation. *Micro Nano Lett.* **2019**, *40*, 150–153. [CrossRef]
19. Sridhar, A.; Thomas, M.J. Electrical treeing in polyethylene: Effect of nano fillers on tree inception and growth. In Proceedings of the 2010 International Conference on High Voltage Engineering and Application, New Orleans, LA, USA,, 11–14 October 2010.
20. Zhang, X.H.; GAO, J.G.; Zhang, J.M.; Liu, J.Y.; Ji, Q.Q. Characteristics of microstructure and electrical treeing in PE/MMT nanocomposites. *Trans. China Electrotech. Soc.* **2009**, *24*, 1–5.
21. Jiang, H.T.; Zhang, X.H.; Gao, J.G.; Guo, N. Conductance Current and Space Charge Characteristics of SiO2/MMT/LDPE Micro-Nano Composites. *Materials* **2020**, *13*, 4119. [CrossRef]

22. Jiang, H.T.; Zhang, X.H.; Gao, J.G.; Guo, N. Dielectric Properties of SiO2/MMT/LDPE Micro-nano Composites. In Proceedings of the 2020 IEEE International Conference on High Voltage Engineering, Beijing, China, 6–10 September 2020.
23. Kuang, X.W.; Gao, Q.; Zhu, H. Effect of calcination temperature of TiO2 on the crystallinity and the permittivity of PVDF-TrFE/TiO2 composites. *J. Appl. Polym. Sci.* **2013**, *129*, 296–300. [CrossRef]
24. Panaitescu, D.; Ciuprina, F.; Iorga, M.; Frone, A.; Radovici, C.; Ghiurea, M.; Sever, S.; Plesa, I. Effects of SiO_2 and Al_2O_3 Nanofillers on Polyethylene Properties. *J. Appl. Polym. Sci.* **2011**, *122*, 1921–1935. [CrossRef]
25. Guo, J.L.; Shen, Y.N. Several problems that should be paid attention to in calculating grain size by Scherrer formula. *J. Inner Mong. Normal Univ. (Nat. Sci. Ed.)* **2009**, *38*, 357–358.
26. Mo, Z.S.; Zhang, H.F. measurement of crystallinity OF Polymers by X-ray diffraction method. *Poltmeric Mater. Sci. Eng.* **1993**, *3*, 9–20.
27. Tanaka, T. Dielectric nanocomposites with insulating properties. *IEEE Trans. Dielectr. Electr. Insul.* **2005**, *12*, 914–928. [CrossRef]
28. Singha, S.; Thomas, M.J. Influence of filler loading on dielectric properties of epoxy-ZnO nanocomposites. *IEEE Trans. Dielectr. Electr. Insul.* **2009**, *16*, 531–542. [CrossRef]
29. Zhang, L.; Zhou, Y.X.; Huang, M.; Sha, Y.C.; Tian, J.H.; Ye, Q. Effect of nanoparticle surface modification on charge transport characteristics in XLPE/SiO2 nanocomposites. *IEEE Trans. Dielectr. Electr. Insul.* **2014**, *21*, 424–433. [CrossRef]
30. Chauvet, C.; Laurent, C. Weibull statistics in short-term dielectric breakdown of thin polyethylene films. *IEEE Trans. Electr. Insul.* **1993**, *28*, 18–29. [CrossRef]
31. Liao, R.J.; Liu, H.B.; Bai, G.; Hao, J.; Li, X.; Lin, Y.D. Space Charge Characteristics and Dielectric Properties of Nano-SiO_2/Aramid Paper Composite. *Trans. China Electrotech. Soc.* **2016**, *31*, 40–48.
32. Lewis, T.J. Interfaces are the dominant feature of dielectrics at the nanometric level. *IEEE Trans. Dielectr. Electr. Insul.* **2004**, *11*, 739–753. [CrossRef]
33. Yang, J.M.; Zhao, H.; Zheng, C.J.; Wang, X.; Yang, C.C. Effects of nano-particles dispersion on the DC dielectric properties of SiO2/LDPE nanocomposite and its effect on direct current conduction properties. *Proc. CSEE* **2015**, *35*, 5087–5094.
34. Chen, Z.Z.; Wang, J.c.; Ma, Q.Y.; Wang, Y.; Wang, M.Z. Influence of Nano-SiO2 Ultrasonic Dispersion Conditions on the Properties of Composite Films. *Packag. Eng.* **2020**, 85–90.
35. Tanaka, T.; Kozako, M.; Fuse, N.; Ohki, Y. Proposal of a multi-core model for polymer nanocomposite dielectrics. *IEEE Trans. Dielectr. Electr. Insul.* **2005**, *12*, 669–681. [CrossRef]

energies

MDPI

Article

A Novel Fault Location Method for Power Cables Based on an Unsupervised Learning Algorithm †

Mingzhen Li [1,*], Jialong Bu [1], Yupeng Song [1], Zhongyi Pu [1], Yuli Wang [2] and Cheng Xie [3]

1 School of Electrical Engineering, Nantong University, No. 9, Seyuan Road, Nantong 226019, China; 1712021069@stmail.edu.ntu.cn (J.B.); 1708031027@stmail.ntu.edu.cn (Y.S.); 1712021010@stmail.edu.ntu.cn (Z.P.)
2 China Electric Power Research Institute, Wuhan 430074, China; wangyuli@epri.sgcc.com.cn
3 Zhejiang Electric Power Research Institute, Hangzhou 310014, China; chengxie@zj.sgcc.com.cn
* Correspondence: mzli@ntu.edu.cn

† This work was extensions of our paper published in Xie, C.; Shao, X.; Jin, Y.; Wang, Y.; Li, M. A Novel Fault Location Method for Power Cables Based on Unsupervised Learning. In Proceedings of 2020 IEEE International Conference on High Voltage Engineering and Application (ICHVE), Beijing, China, 6–10 September 2020.

Abstract: In order to locate the short-circuit fault in power cable systems accurately and in a timely manner, a novel fault location method based on traveling waves is proposed, which has been improved by unsupervised learning algorithms. There are three main steps of the method: (1) build a matrix of the traveling waves associated with the sheath currents of the cables; (2) cluster the data in the matrix according to its density level and the stability, using Hierarchical Density-Based Spatial Clustering of Applications with Noise (HDBSCAN); (3) search for the characteristic cluster point(s) of the two branch clusters with the smallest density level to identify the arrival time of the traveling wave. The main improvement is that high-dimensional data can be directly used for the clustering, making the method more effective and accurate. A Power System Computer Aided Design (PSCAD) simulation has been carried out for typical power cable circuits. The results indicate that the hierarchical structure of the condensed cluster tree corresponds exactly to the location relationship between the fault point and the monitoring point. The proposed method can be used for the identification of the arrival time of the traveling wave.

Keywords: power cable; fault location; sheath current; traveling wave; unsupervised learning

Citation: Li, M.; Bu, J.; Song, Y.; Pu, Z.; Wang, Y.; Xie, C. A Novel Fault Location Method for Power Cables Based on an Unsupervised Learning Algorithm . *Energies* **2021**, *14*, 1164. https://doi.org/10.3390/en14041164

Academic Editor: Issouf Fofana

Received: 25 January 2021
Accepted: 17 February 2021
Published: 22 February 2021

1. Introduction

Due to the limitations of urban land resources and the requirements for power reliability, power cables are widely used in urban power transmission and distribution systems [1–3]. With the rapid increase in the scale of power cable applications, the number of short-circuit faults has also increased [4,5]. Although the overall short-circuit fault rate of the power cables is low, the origin of the short-circuit faults has many possible factors, which are difficult to predict, and there will be wide social impacts and large economic losses caused by the faults. In order to shorten the repair time of the faulty power cables, efforts are needed to develop reliable and accurate fault location methods.

According to the fault location principles, currently, online fault location methods used in power systems are mainly based on the impedance and the traveling wave. Both of the impedance-based and the traveling-wave-based methods originated from the fault location for overhead lines. The impedance-based method [6–8] mainly uses the relay protection device to collect fault voltage/current data, and locates the fault based on the parameter identification of the system impedance. Since the impedance monitored by this method may have a nonlinear relationship with the distance between the fault point, it is difficult to directly locate the fault accurately through the relationship between the impedance and the distance between the fault point. The traveling-wave-based methods [9–11] locate the fault through the identification of the propagation time of the fault traveling wave. When the line distribution parameters are large or non-uniform, it is difficult to accurately extract the head and/or the arrival time of the traveling wave.

Recently, efforts have been made to improve the accuracy and applicability of impedance-based methods. In [12], mutual inductance effects have been taken into account to make the method suitable for the medium voltage distribution network in the double-circuit lines. In [13], the fault location criteria for different fault sections of the distributed network have been proposed, taking into account the electrical characteristics of nodes in the distributed network under fault and non-fault conditions. In [14], a fault line selection method of distributed network has been proposed based on the direction vector characteristics of the fault currents. In [15], a Fibonacci search algorithm has been used to analyze a specific network structure under the premise of low fault resistance, the results indicate that this method improves the robustness of the impedance-based method with respect to the fault resistance. In [16], the influence of the load uncertainty on the accuracy of fault location has been studied, and a method to reduce the influence has been proposed. The research above can be summarized in Table 1.

Table 1. The summary of the improvements of the impedance-based method.

References	[12]	[13]	[14]	[15]	[16]
Line model	π-type	Distributed	π-type	Distributed	Distributed
Double or multiple loops	$\surd$ [1]	$\surd$	$\surd$	- [2]	-
Fault section location	-	$\surd$	-	$\surd$	-
Non-uniform transmission line	$\surd$	-	$\surd$	-	-
Branches	$\surd$	$\surd$	$\surd$	$\surd$	$\surd$
Robustness to fault resistance	-	-	-	$\surd$	$\surd$
Distributed Power	-	-	$\surd$	-	$\surd$
Load uncertainty	-	-	$\surd$	-	$\surd$

([1] $\surd$ means applicable; [2] - means not applicable).

As shown in Table 1, the improvements of the impedance-based method mainly focus on the applicability of different application scenarios. With the different focuses, the fault location criterion cannot be uniform or mutually applicable.

Additionally, efforts have been made to improve the accuracy and the applicability of the traveling-wave-based method in recent years. In [17], the regular pattern of the reflection of voltage and current traveling waves at the end of the line has been analyzed, then a fault searching algorithm has been proposed based on mid-point time, which can be used for the fault location of mixed line with discontinuous wave velocity. In [18], time-frequency analysis of fault traveling waves has been carried out using continuous wavelet transform method, so that the arrival time of the traveling wave could be identified according to the characteristic frequency of the fault traveling wave. In [19], the identification accuracy of the high frequency part of the fault traveling wave has been improved using a multilayer neural network. In [20], the problem of multi-channel signal synchronization has been solved by using the time difference between the first arrival wave and the first reflected wave. In [21], a normalized fault location criterion has been proposed, which does not require external time reference and accurate traveling wave velocity. However, the time difference may be aliased in multiple discharges of the fault, which is common in cable breakdown discharges. The above studies have improved the traveling-wave-based method in terms of traveling wave propagation, time-frequency analysis, and feature recognition, which can be summarized in Table 2.

Table 2. The summary of the improvements of the traveling-wave-based method.

References	[17]	[18]	[19]	[20]	[21]
Frequency range	10~100 kHz	2.5 kHz~1 GHz	- [2]	Within 100 kHz	Within 1 MHz
Mixed lines	√ [1]	√	√	-	-
Non-uniform transmission line	√	-	√	√	-
Branches	√	√	√	-	-
No signal synchronization required	-	-	√	√	√
No traveling wave speed required	-	-	-	√	√
No line parameters required	-	-	-	-	√
Clock skew	-	-	-	-	√

([1] √ means applicable; [2] - means not applicable).

It can be seen from the above-mentioned literature analysis that, as time goes by, there are continuous researches to improve the traveling-wave-based method, which also makes the method supplemented and improved. With the improvement of computing performance and signal synchronization accuracy, many applications of intelligent algorithms have emerged in recent years, but these improvements and applications are limited to a single type of line or specific application scenarios, and it is difficult to form a unified fault location criterion. The improvements have not fundamentally solved the inherent problems of traveling wave head attenuation and noises.

The authors of this work previously proposed an unsupervised learning method for fault location of power cables and to improve on the traditionally applied traveling-wave-based method [22,23]. The present paper further expands on their previously published work, and makes modifications of the fault location process. There are four steps for the fault location in the previous contribution [23], i.e., (1) build a matrix of the traveling waves associated with the sheath currents of the cables; (2) apply t-distributed Stochastic Neighbor Embedding (t-SNE) to reconstruct the matrix into a low dimension; (3) cluster the data in the matrix according to its closeness, using Density-Based Spatial Clustering of Applications with Noise (DBSCAN); (4) search for the maximum slope point of the non-noise cluster with the fewest samples to identify the arrival time of the traveling wave. The improvement mainly lies in the high-dimensional clustering for the identification of the arrival time of the traveling waves of multiple signals. Thus, Steps 2 and 3 can be merged as one, the information of the original signals can be reserved without the dimensionality reduction, and the waveforms of the traveling waves can be identified directly via an unsupervised learning algorithm. In addition, the presented method can output a condensed cluster tree which is believed to correspond with the traveling waves. The specific relation between the condensed cluster tree and the traveling wave will be studied in further research, which could help to make the process of fault occurrence and traveling wave propagation clearer.

2. Analysis of the Traveling Wave Signals and the Fault Location

2.1. Typical Cable Structures and the Monitoring of The Traveling Wave Signals

2.1.1. Typical Cable Structures

A typical power cable for high voltage (HV: 110 kV or more) has a single-core structure, and a typical power cable for medium voltage (MV: 10 to 35 kV) has a three-core structure, as shown in Figure 1a,b, respectively.

Since the metal sheath of the HV cable has been reliably grounded at least one end, the sheath voltage is close to the ground level during normal operation. In addition, when the load current flows through the core conductor, the direction of the electric field on the cable cross section is radial, shown as $\boldsymbol{E}$ in Figure 1a, and the direction of the magnetic field is shown as $\boldsymbol{H}$. Therefore, the direction of the Poynting vector of the energy transmission is the $\boldsymbol{E} \times \boldsymbol{H}$ direction, which indicates that the energy propagates along the line axis between the main insulation between the two layers of metal. When a short-circuit fault occurs at any position of the cable, there will be fault signals on the core conductor and the metal sheath. Therefore, regardless of whether the core conductor or the metal sheath is monitored, the arrival time of the traveling wave is the same. For safety and convenience, the metal sheath has been chosen to monitor for the traveling wave signals in the paper. For the MV cable shown in Figure 1b, both the metal sheath and the metal armor are grounded at the same grounding position. Therefore, the method of obtaining traveling waves by monitoring the sheath current is also suitable for MV cables.

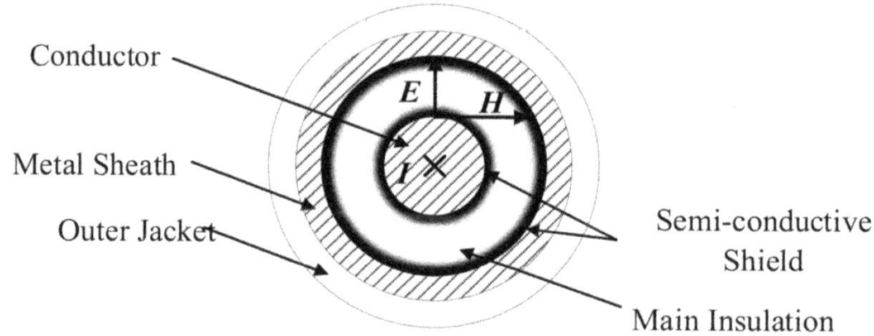

(a) A typical structure for HV cable.

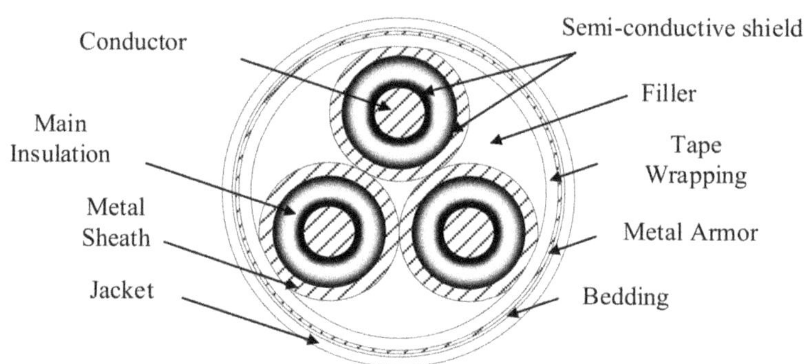

(b) A typical structure for a MV cable.

Figure 1. Cross-sectional structure diagram of typical power cables.

2.1.2. The Monitoring of the Traveling Wave Signals

Figure 2 shows the sheath current monitoring system, where a typical HV cable circuit is taken as an example. The system consists of four parts, namely, the data acquisition module, the communication module, the location analysis software installed in the cloud server and the interface for cable maintenance engineers. Among them, the data acquisition module includes sheath current sensors, and it is installed in the grounding boxes and the cross-bonding boxes. When a short-circuit fault occurs in the cable, the sheath current in the loop will rise to the level of the fault current. At the same time, the data acquisition module will be awakened and will start to upload the data. The communication module of the data acquisition system can transmit the recorded data to a designated cloud server, where the data analysis software will perform data analysis, after that, the location results will be sent to the maintenance engineer. The recoded data can also be downloaded from the server for further analysis.

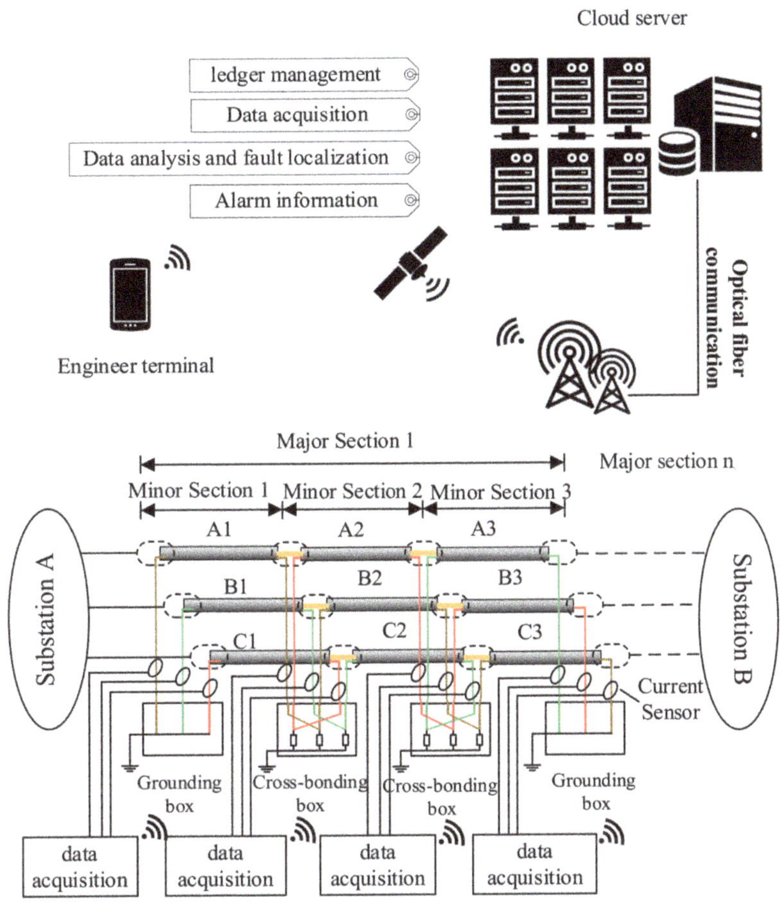

Figure 2. The designed sheath currents monitoring system (The yellow line represents phase A, the green line represents phase B, the red line represents phase C. A1–C3 are the labels of different cable sections).

2.2. The Fault Location

An autonomous learning mechanism based on the two-terminal traveling wave method is proposed. Thus, the arrival time of the fault traveling wave can be identified accurately with multiple monitored data. Since the error of the two-terminal traveling wave method mainly comes from the identification of the arrival times at the two terminals of the cable, moreover, the main source of identification errors is randomness, such as synchronization time delay, noise interference, etc., and more monitoring points can help to reduce random errors, thereby improving accuracy. In order to reduce random errors, the sheath currents of the three-phase cables and adjacent cables in the same cable passage are designed to be monitored. Thus, a multi-dimensional data matrix can be built. Through the analysis of the multi-dimensional matrix, the arrival times can be identified more accurately. The process of the identification of the arrival times at the two terminals of the cable is shown in Figure 3.

There are three main steps of the proposed method, as presented in Figure 3. Firstly, a high-dimensional data matrix needs to be built using the monitored sheath currents. The amplitude of each sheath current is taken as the row vector of the matrix, and the number of sampling points of the sheath current is the number of columns of the matrix.

Secondly, the high-dimensional matrix can be directly clustered using HDBSCAN, which is a density-based hierarchical clustering algorithm. The specific process of HDBSCAN is relatively complicated, which will be demonstrated in Sections 2.3 and 2.4. Finally, the arrival time of the traveling wave can be identified by searching for the characteristic cluster point(s) of the two branch clusters with the smallest density level.

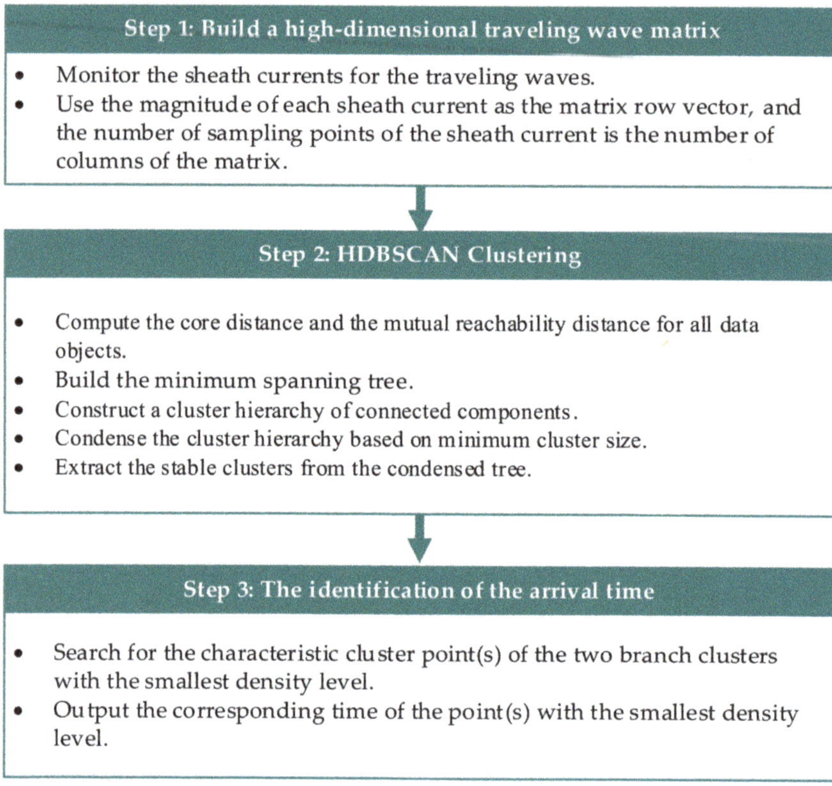

Figure 3. The process of the identification of the arrival time at the two terminals of the cable.

It is to be noticed that, in [23], unsupervised learning (t-SNE + DBSCAN) has been used to improve the identification of traveling wave arrival time. The results show that even if there is noise and/or random time errors in the recorded data, the arrival time of the traveling wave can be identified. However, when the matrix dimension is greater than 3, the dimensionality reduction process will sacrifice the integrity of the data information. After the dimensionality reduction, the data will lose the original information to a certain extent, for example, the physical meaning. The paper omits the step of the dimensionality reduction and uses the recorded data directly for identification. This allows the original data information to be retained to the greatest extent, and in principle makes the error as small as possible.

2.3. The Construction of the Traveling Wave Matrix

As shown in Figure 4, the sheath currents can be monitored at the grounding boxes and the cross-linking boxes in actual HV cross-bonded cable systems. Usually, there are multiple cables sharing the same passage/duct. If a short-circuit fault occurs in a cable, an inductive coupling signal can be detected on the sheath of the adjacent cable. Thus, multiple sheath currents can be analyzed together, which is believed to make the identification accuracy of the arrival time to be improved. Hence, a high-dimensional data matrix can be built using the amplitude of each monitored sheath current as the row vector, and the number of sampling points of the sheath current is the number of columns of the matrix.

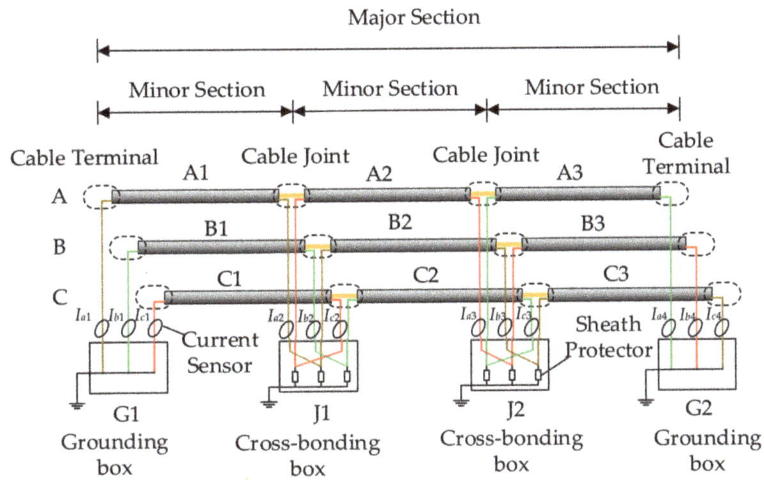

Figure 4. The schematic diagram of a typical high voltage (HV) cross-bonded cable circuit (The yellow line represents phase A, the green line represents phase B, the red line represents phase C. A1–C3 are the labels of different cable sections).

A typical HV cross-bonded cable circuit model, as shown in Figure 4, has been established using PSCAD. The cross-sectional structure parameters of the cable are shown in Table 3. As shown in Figure 4, 12 current sensors have been used in a major section to monitor the sheath current. The sheath currents measured at the 12 sensors are, respectively, I_{a1}, I_{b1}, I_{c1}, the sheath currents recorded by the current sensors of phase A, B, C at G1; I_{a2}, I_{b2}, I_{c2}, the sheath currents recorded in phase A, B, C at J1; I_{a3}, I_{b3}, I_{c3} the sheath currents recorded in phase A, B, C at J2; I_{a4}, I_{b4}, I_{c4}, the sheath currents recorded in phase A, B, C at G2. A short-circuit fault has been set in cable section A1, 300 m away from G1. The 12 sheath currents generated in the simulation are presented in Figure 5. In the case, the sheath current matrix I can be presented in (1).

$$I = \begin{bmatrix} I_{a1} & I_{b1} & I_{c1} & \dots & I_{c3} \end{bmatrix}^T \tag{1}$$

$$v_0 = \frac{1}{\sqrt{\mu\varepsilon}} \tag{2}$$

Table 3. The cross-sectional structure parameters of the HV cable.

	Structure	Outer Radius/mm
1	Core conductor (Copper)	17.0
2	Inner semi-conductor (Nylon belt)	18.4
3	Main insulation (Ultra-clean XLPE)	34.4
4	Outer semi-conductor (Super-smooth semi-conductive shielding material)	35.4
5	Water-blocking layer (Semi-conductor)	39.4
6	Metal sheath (aluminum)	43.9
7	Jacket (PVC)	48.6

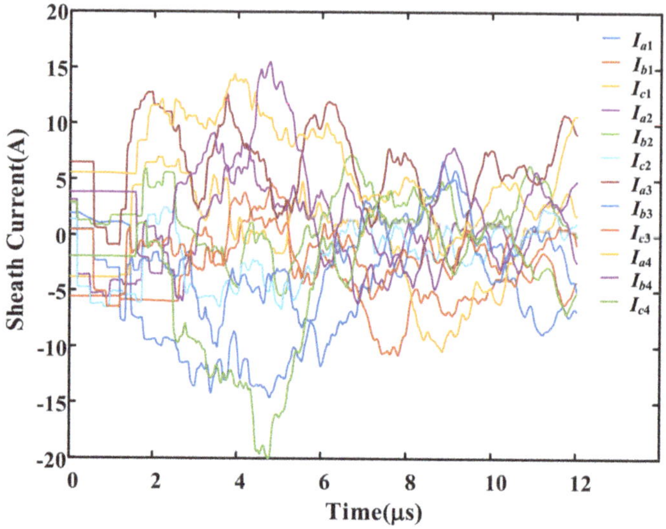

Figure 5. The simulated 12 sheath currents of the HV cable.

Since the velocity of the traveling wave depends on the material parameters of the cable, it can be calculated using (2), where μ is the magnetic permeability and $\mu\varepsilon$ is the insulation permittivity. In the case, $\mu = 4\pi \times 10^{-7}$ and $\varepsilon = 3.63 \times 10^{-11}$, so $v_0 = 1.48 \times 10^8$ m/s. Hence, the arrival time of the traveling wave is expected to be 2.03 μs at G1 and 1.35 μs at J1.

2.4. The Unsupervised Learning Algorithm

DBSCAN [24] is a typical density-based clustering algorithm, which is able to identify arbitrarily shaped clusters and noise (outliers) in data. However, the main problem of such algorithms in application is the ability to process high-dimensional data. HDBSCAN [25] follows Hartigan's classic density contour clustering/tree model, and improves the existing density-based clustering algorithms for different aspects [26]. Moreover, it provides paths of minimum spanning tree for high-dimensional data [27].

Instead of using parameters *Eps* and *MinPts* (*Eps* is the threshold of the neighborhood distance of a certain sample; *MinPts* is the threshold of the number of samples near the sample whose distance is *Eps*.) to describe the closeness of the sample distribution in the neighborhood as in DBSCAN, HDBSCAN uses parameters *MinPts* and *MinClustSize* (*MinClustSize* indicates the minimize cluster size) to limit the possible clusters, and makes the high-dimensional clustering problem an optimization problem.

2.4.1. The Measurement of High-Dimensional Data

The first problem introduced by high-dimensional data is the measurement method. It has been proven that the Euclidean distances of all data objects in the high-dimensional space tend to be equal [28]. Therefore, the high-dimensional space needs to be measured in a more suitable way.

Let $X = \{x_1, \ldots, x_n\}$ be a dataset containing n data objects, each of which is described by an attribute vector x. The Euclidean distance between x_p and x_q denotes as $d(x_p, \mathbf{x}_q)$. An object $\mathbf{x}_p$ is called a *core object* if its ε-neighborhood (ε means the value of the minimum radius) contains at least *MinPts* objects. The *core distance* of an object x_p, $d_{\text{core}}(x_p)$, is the Euclidean distance from x_p to its *MinPts* -nearest neighbor (including x_p). The *mutual reachability distance* between two objects $\mathbf{x}_p$ and $\mathbf{x}_q$ is defined as (3).

$$d_{\text{mreach}}(x_p, x_q) = max\{d_{\text{core}}(x_p), d_{\text{core}}(x_q), d(x_p, x_q)\} \quad (3)$$

Under the measurement of *mutual reachability distance*, dense objects, whose core distance are small, remain the same distance from each other, but sparser objects have been dispersed to at least the core distance from any other object. It is to be noticed that the effectiveness depends on the value of *MinPts*. The larger *MinPts* values the more data objects are affected by the measurement. While the value of *MinPts* cannot be too large, or the number of the final clusters would be very small.

2.4.2. The Hierarchical Clusters of High-Dimensional Data

With the measurement of *mutual reachability distance*, the density discrimination of the data objects is more significant. As the data objects are high-dimensional, the density cannot be directly expressed on a plane, in other words, it is not "flat". Thus, the data objects can be regarded as a weighted graph, where the data object is a vertex, and the weight of the edge between any two data objects is equal to the *mutual reachability distance*.

Now the clustering process can be regarded as the process of merging vertices and deleting edges. Similar to DBSCAN, a threshold parameter *Eps* is used in the process. This time, *Eps* is not a constant value. The threshold starts from a relatively large value, and is steadily being lowered. Simultaneously, the edges whose weights are above the threshold value are deleted. Eventually, the connected graph is turned into a completely disconnected graph which is composed by a hierarchy of connected components. This process can be achieved by building the minimum spanning tree [29].

2.4.3. The Construction of the Clustering Hierarchy

Since the minimum spanning tree of the high-dimensional data objects is still high dimensionally, it needs to be converted into the hierarchy of connected components to make the clustering hierarchy "flat". This process is completed by the following steps:

(1) Sort the edges of the minimum spanning tree by the weights (in ascending order);
(2) Traverse all the branches;
(3) Create a new merged cluster for each edge.

It is to be noticed that Step 3 can be completed via a union-find data structure [30]. Thus, each data object has a clear corresponding start and end in the hierarchy (with the ordinate of *mutual reachability distance*), which can be used for arbitrary clustering through cutting branches with a single horizontal line. The next step is to find the reasonable horizontal line(s) for the clustering, not just by setting a fixed parameter of the *mutual reachability distance*. Essentially, the clustering process here of DBSCAN is achieved by setting such a fixed parameter, which is very unintuitive. Therefore, a better solution is needed to select the clusters.

2.4.4. Condense the Cluster Hierarchy

The first step in the cluster extraction is to condense the huge and complex cluster hierarchy into a smaller tree, and attach a little more data to each node. Rather than setting a fixed parameter of the *mutual reachability distance*, HDBSCAN uses *MinClustSize* (minimum cluster size) limit the possible clusters. Thus, when the hierarchy are traversed, each split is judged that whether the number of points in one of the new clusters created by each split is less than *MinClustSize*. If the number of the data objects is less than *MinClustSize*, the data object falls outside the cluster, and the larger cluster retains the cluster identity of the parent node. Then, the data objects with its *mutual reachability distance* which fall outside the cluster are marked. On the other hand, if the split is into two clusters each at least as large as *MinClustSize*, the split persists in the tree. Thus, the clusters are easier to view and process, especially for clustering problems with simple data structures. However, the clusters are still needed to be selected as 'flat' clusters.

2.4.5. Extract the Clusters

In fact, the 'flat' clusters have the highest stability. To illustrate the notion of the stability, several important mathematical measurement parameters are introduced. First, a different measurement is needed to denote the persistence of clusters, as shown in (4), representing the density level.

$$\lambda = \frac{1}{\text{distance}} \tag{4}$$

For a cluster C_i, the *excess of mass* of C_i can be defined by (5), where $f(\mathbf{x})$ is a density function, representing the density of each cluster.

$$E(C_i) = \int_{\mathbf{x}\in C_i} (f(\mathbf{x}) - \lambda_{\min}(C_i))d\mathbf{x} \tag{5}$$

The *relative excess of mass* of a cluster C_i, which appears at level $\lambda_{\min}(C_i)$ can be defined by (6), where $\lambda_{\max}(\mathbf{x}, C_i) = \min\{f(\mathbf{x}), \lambda_{\max}(C_i)\}$ and $\lambda_{\max}(C_i)$ is the density level at which C_i is split or disappears.

$$E_R(C_i) = \int_{\mathbf{x}\in C_i} (\lambda_{\max}(\mathbf{x}, C_i) - \lambda_{\min}(C_i))d\mathbf{x} \tag{6}$$

The *stability* of a cluster C_i can be defined by (7), where $\lambda_{\min}(C_i)$ is the minimum density level at which C_i exists, $\lambda_{\max}(\mathbf{x}_j, C_i)$ is the density level beyond which object $\mathbf{x}_j$ no longer belongs to cluster C_i.

$$S(C_i) = \sum_{\mathbf{x}_j\in C_i} (\lambda_{\max}(\mathbf{x}_j, C_i) - \lambda_{\min}(C_i)) \tag{7}$$

Therefore, the step can be regarded as an optimization problem, and the objective is to maximize the overall stability of the extracted clusters. After optimization, the most prominent non-overlapping cluster will be the output.

3. Results

3.1. The Arrival Time Identification

There are big differences between the 12 sheath current amplitudes recorded at different locations, as shown in Figure 5, which also constitute the elements of the high-dimensional matrix $\boldsymbol{I}$, the input. The next step is the clustering. In order to make the distinction of the clustering more remarkable, a relatively large *MinClustSize* and a relatively small *MinPts* are required. In this case, *MinClustSize* and *MinPts* are set to 12 and 7, respectively. By applying HDBSCAN to the matrix $\boldsymbol{I}$ (the original data to be input into the algorithm), the condensed cluster tree can be obtained, as plotted in Figure 6.

The red nodes plotted in Figure 6 represent the final stable clusters, and the black nodes represent the clusters which are not extracted after stability optimization. The serial numbers of the red nodes are same as those of the clusters of the tree, and they can be renumbered as 1–9 for convenience. In Figure 6, the abscissa has no physical meaning; the ordinate represents the relative magnitude of the λ value of each cluster, which increases from top to bottom.

The cluster scatter plots of sheath currents I_{a2}, I_{b2} and I_{c2} are presented in Figure 7, where Cluster 1–9 represent the final stable clusters, and Cluster 0 represents "noise". It should be noted that Cluster 0 does not have to be the real noise in practice. As mentioned above, whether the object $\mathbf{x}_p$ is labeled as "noise" only depends on the hierarchical structure of the data, which is related to the density level and stability of the high-dimensional matrix. In the case, the arrival time point of the traveling wave that needs to be extracted has similar characteristics to the "noise". Therefore, Cluster 0 should be analyzed together with the ordinary clusters. For further analysis, Figure 8 shows the number of samples in each of the 10 clusters.

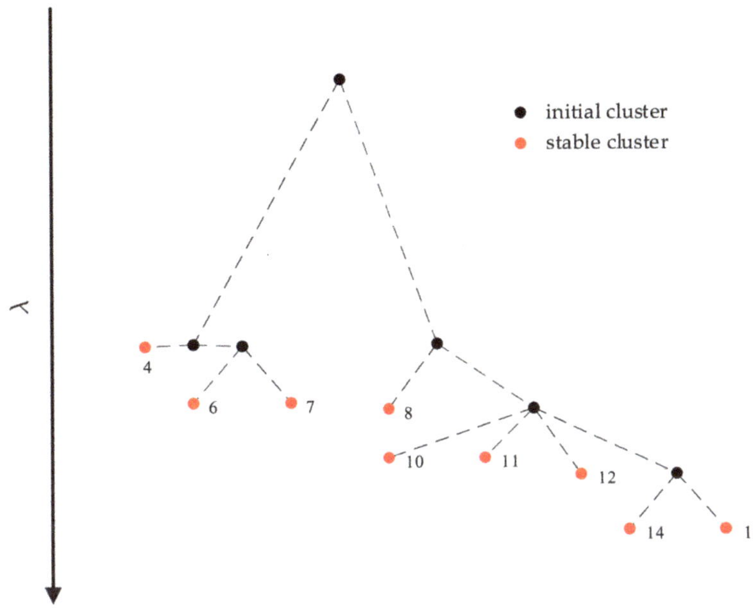

Figure 6. The schematic diagram of the condensed cluster tree for the matrix ***I***.

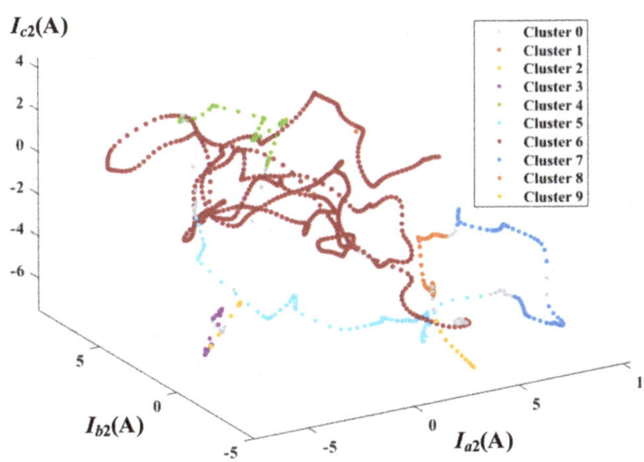

Figure 7. The cluster scatter scheme of sheath currents I_{a2}, I_{b2} and I_{c2}.

Since the traveling waves propagate to the two ends of the cable in two opposite directions, and they are roughly mirrored on the two sides of the fault, the monitoring position closest to the fault will record traveling waves earliest. In addition, there are signal steps in the sheath current when a fault occurs, and they correspond exactly to the arrival time of the traveling wave. For the density-based hierarchical clustering method, they are the two branch nodes with the smallest λ value. In the case, they correspond to node 4 and node 8 in Figure 6. It is noted that the fault is closer to J1, corresponding to node 8 in Figure 6 or Cluster 4 in Figure 7. Therefore, the samples of Cluster 4 and Cluster 0 can be analyzed together. As shown in Table 4, the time stamp of the only sample labeled as

Cluster 0 in Cluster 4 samples is 1.35 μs. It is exactly the expected time of the first arriving traveling wave recorded at J1, corresponding to the sheath current I_{a2}.

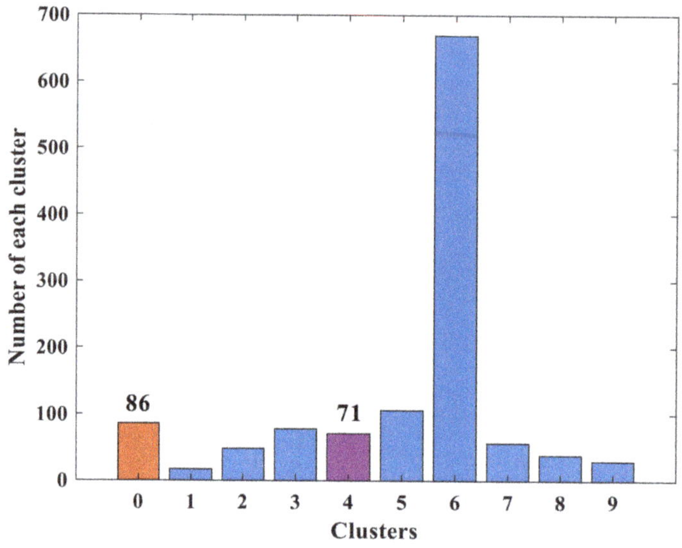

Figure 8. The number of the sheath current samples in each cluster of the HV cable circuit.

Table 4. The clustering results of some data labels.

Serial Number *	Cluster Label	Time Stamp/μs
131	4	1.31
132	4	1.32
133	4	1.33
134	4	1.34
135	0	**1.35**
136	4	1.36
137	4	1.37

(* The sampling rate is 100 MHz).

3.2. The Application for a MV Cable Circuit

The method also can be applied for MV cable circuits, whose radial structure was shown in Figure 1b. The metal sheath and the metal armor are connected and grounded together at the same position of the two terminals, as presented in Figure 9. In this case, the MV cable circuit is 400 m and the fault was set to 200 m from terminal 1. Table 5 shows the parameters of cross-sectional structure of the MV cable. It is to be noticed that the material parameters of the main insulation (XLPE, crosslinked polyethylene) are the same as those in the HV cables. The sheath currents are monitored by the two current sensors, as shown in Figure 9, where I_{m1} is the sheath current recorded by current sensor 1 and I_{m2} is the sheath current recorded by current sensor 2.

In this case, the initial matrix dimension is greatly reduced, and the process is basically the same. As shown in Figure 10, the initial data is relatively monotonous. In the case, *MinClustSize* and *MinPts* are set to 7 and 7, respectively. By applying HDBSCAN to the sheath current matrix of the MV cable circuit, the condensed cluster tree can be presented in Figure 11.

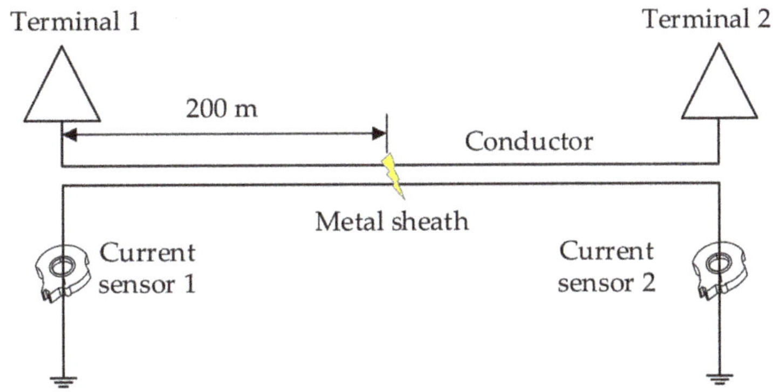

Figure 9. The schematic diagram of a double-end bonded medium voltage (MV) cable system. Note: 'conductor' has three cores, one for each phase.

Table 5. The parameters of cross-sectional structure of the MV cable.

	Structure	Outer Radius/mm
1	Core conductor (Copper)	10.30
2	Main insulation (including inner and outer semi-conductor)	16.60
3	Metal sheath (Copper tape)	16.75
4	Filler and tape wrapping (semi-conductor)	36.50
5	Metal armor (Steel)	38.50
6	Bedding (Water-blocking PVC)	40.10
7	Jacket (Fire-resistant PVC)	44.90

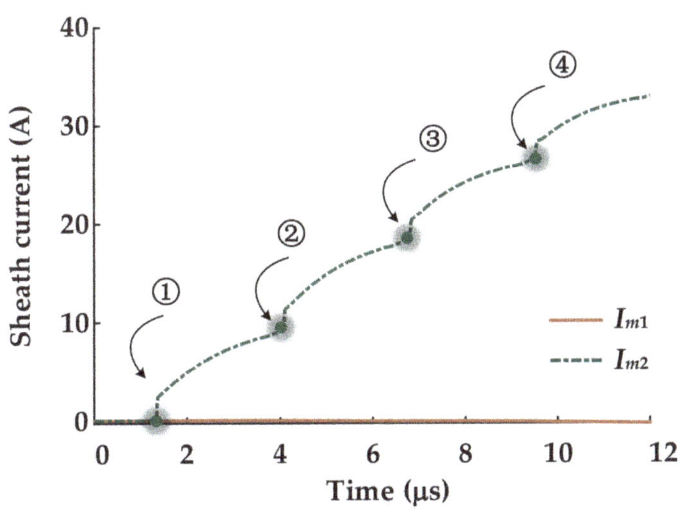

Figure 10. The sheath currents monitored by the current sensors at the two ends. (①~④ represent the traveling wave steps no.1 to no.4).

The meaning of the icons in Figure 11 are the same as Figure 6. There are more clusters in this case, and there are two clusters, namely Cluster 22 and Cluster 23, with much bigger λ value than other clusters. They are the two 'flat' clusters with the highest density. Therefore, they are not the clusters containing the point of the arrival time, as a result of the cluster should be the one with the fewest samples. For further analysis, the number of samples in each of the clusters is shown in Figure 12.

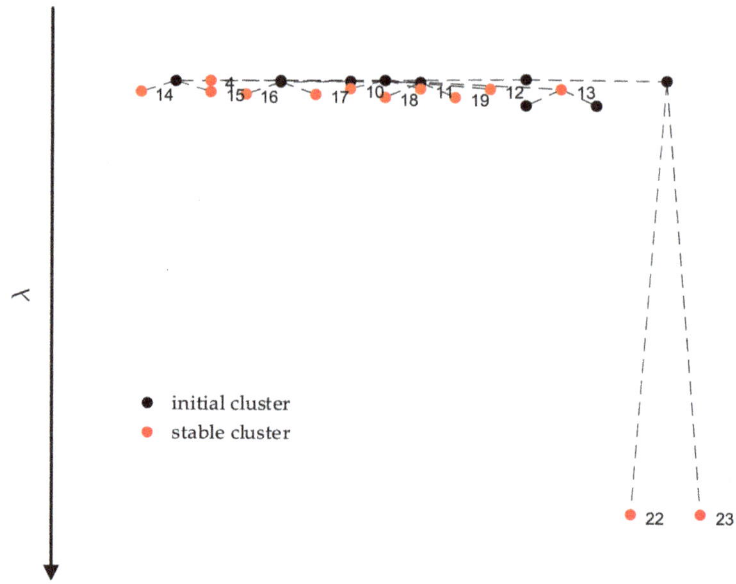

Figure 11. The schematic diagram of the condensed cluster tree for sheath current matrix of the MV cable circuit.

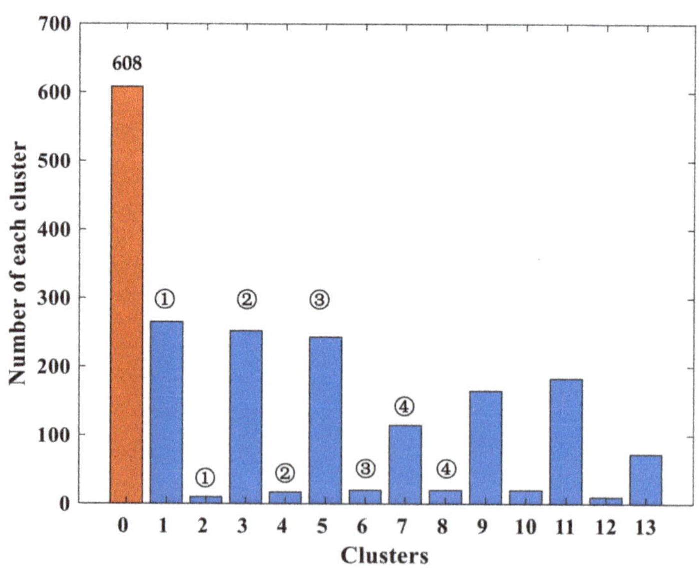

Figure 12. The number of the sheath current samples in each cluster of the MV cable circuit (①~④ represent the traveling wave steps no.1 to no.4).

Similar to the results of the HV cable circuit shown in Figure 6 and Table 4, Cluster 0 still has a large number of the sheath current samples, even with the largest number in this case. As stated, Cluster 0 does not have to be the real noise in practice. Instead, the point of the arrival time can be just in Cluster 0. The two branches of the traveling wave just correspond to the branch tree of Figure 11 and Cluster 1–13 in Figure 12. The traveling wave steps ①~④ in Figure 10 just correspond to Cluster 1–8 in Figure 12.

4. The Superiority of the Presented Method

4.1. The Comparisons for the Arrival Time Identification

In [23], five popular algorithms for the arrival time identification have been compared. They are, respectively, (A) the threshold method; (B) the wavelet-based multi-scale time-frequency analysis method; (C) the cross-correlation algorithm; (D) the cumulative energy method; (E) t-SNE + DBSCAN. Now add the presented method to the comparison, numbered as method F.

Simulations have been carried out for the MV and HV cable circuits, and the arrival time identification has been performed using the 6 algorithms.

Firstly, simulations have been carried out for the MV cable circuit shown in Figure 9, where two signals have been assessed. The fault position has been set from current sensor 1 to 50 m, 100 m, 150 m, 200 m, 250 m, 300 m and 350 m in sequence. Thus, the corresponding time differences are 2.03 μs, 1.35 μs, 0.68 μs, 0 μs, −0.68 μs, −1.35 μs and −2.03 μs, respectively. The time difference results obtained by using the six algorithms under ideal conditions are presented in Table 6.

Table 6. The time difference results obtained by using the six algorithms.

Method	TD * (μs)						
	2.03	**1.35**	**0.68**	**0**	**−0.68**	**−1.35**	**−2.03**
A	2.03	1.35	0.68	0	−0.68	−1.35	−2.03
B	2.03	1.35	0.68	0	−0.68	−1.35	−2.03
C	2.03	1.35	0.68	0	−0.68	−1.35	−2.03
D	2.03	1.35	0.68	0	−0.68	−1.35	−2.03
E	2.03	1.35	0.68	0	−0.68	−1.35	−2.03
F	2.03	1.35	0.68	0	−0.68	−1.35	−2.03

(*: TD is short for the time difference).

As shown in Table 6, under ideal conditions, all the six algorithms can identify the time difference accurately for MV cables.

Secondly, simulations have been carried out for the HV cable circuit shown in Figure 4, where 12 signals have been assessed. The fault position has been set from the left monitoring position (G1) to 300 m. In order to make the simulations more realistic, white noise, between 0 and 50 dB, has been added to the waveforms. The time difference between the traveling waves arriving at the detection points G1 and J1 should be −0.68 μs. The time difference results from the six algorithms are presented in Table 7.

Table 7. The time difference results from the six algorithms with white noise.

Method	SNR *					
	50 dB	**40 dB**	**30 dB**	**20 dB**	**10 dB**	**0 dB**
A	−0.68 μs	−0.59 μs	−0.76 μs	−0.54 μs	−0.47 μs	−0.25 μs
B	−0.68 μs	−0.68 μs	−0.63 μs	−0.59 μs	−0.73 μs	−0.37 μs
C	−0.68 μs	−0.68 μs	−0.66 μs	−0.75 μs	−0.75 μs	−0.35 μs
D	−0.68 μs	−0.68 μs	−0.68 μs	−0.67 μs	−0.55 μs	−0.42 μs
E	−0.68 μs	−0.68 μs	−0.68 μs	−0.68 μs	−0.67 μs	−0.53 μs
F	−0.68 μs	−0.68 μs	−0.68 μs	−0.68 μs	−0.69μs	−0.49 μs

(*: SNR is short for signal-to-noise ratio).

As shown in Table 7, the six algorithms have different sensitivities to noise. Methods A to E/F sequentially have better noise suppression capabilities; while there is no essential difference between method E and F. When the noise level becomes too high, e.g., 0 dB, all six algorithms lose the accuracy to varying degrees. In practice, there are many cables which equip with a minimal number of sensors. The noise recorded by these sensors is too large and the synchronization error is too high for the accurate fault location. In order to reduce the influence of randomly generated errors, multiple recorded data from the power cables are recommended to be analyzed together using method(s) E and/or F.

Still with the simulations for the HV cable circuit, all simulation conditions and parameters are the same. This time, a random error has been added to the recording time of the current sensor I_{a1}, which is, simulating the loss of synchronization time stamp(s). The tolerance range of the random error applied is [−1 μs, +1 μs]. Under this condition, the performances of the six algorithms are shown in Figure 13.

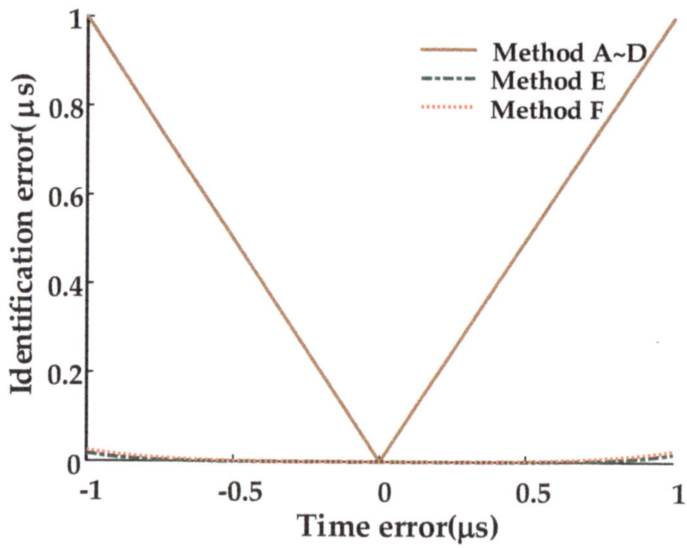

Figure 13. The performances of methods A–F with a recording time error added to I_{a1}. (The ordinate axis represents the absolute value of the identification error).

As shown in Figure 13, when there is a time deviation within the recorded data from I_{a1}, methods A–D have almost no ability to reduce errors, while the errors of method E and F are almost negligible. Therefore, the performances of method E and F are much better than those of methods A–D. Although the issue has a low probability of occurring in practice, when it does occur, it is catastrophic for the fault location using other algorithms. Even if the error is only microseconds, the fault location error is hundreds of meters. The results indicate that method E and F are suitable for use in fault location with a few random time errors in the recorded data.

4.2. The Improvement from t-SNE + DBSCAN

The superiority of the fault location method of the paper mainly lies in the processing of multiple signals, even faster than the former method (t-SNE + DBSCAN) [23]. Simulations have been carried out for the comparison. The typical HV cross-bonded cable circuit shown in Figure 2 has been used for the simulation, the fault has been set at cable section A1–C3, respectively, and the fault position has been set, in turn, at 50 m, 100 m, 150 m, 200 m, 250 m, 300 m, 350 m, 400 m and 450 m from the left end of the cable. All simulations have been carried out on an AMD Ryzen Threadripper 3970X 32-Core, 64 GB RAM computer. The running time of each simulation can be presented in Figure 14.

In Figure 14, the running time of R is 2.23~20.94 times than L. Although the time saved is just dozens of seconds for the typical HV cross-bonded cable circuit, it can be hours for a much more complicated circuit or days for practical cable circuits.

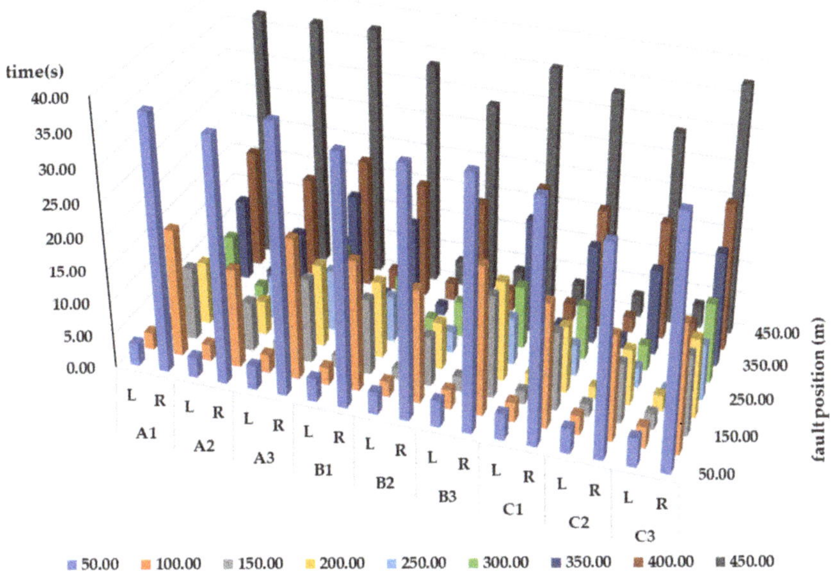

Figure 14. The running time of each simulation. (L represent the method presented in this paper; R represent the algorithm of t-SNE + DBSCAN; A1~C3 represent the cable sections where the fault occurs).

Another improvement mainly lies in the tree structure. The condensed cluster tree is basically the two-branch-tree structure, which is believed to be corresponded to the traveling waves, generated at the fault position, headed to the two opposite directions to the ends of the cable. Thus, the condensed cluster tree makes the method a better visualization, which is more convenient for the analysis of the state at the beginning of the fault.

5. Discussions and Conclusions

5.1. Discussions

HDBSCAN is not only facilitates high-dimensional clustering, but also outputs a condensed cluster tree which is believed to be corresponded to the traveling waves. As the condensed cluster tree is basically the two-branch-tree structure, and the hierarchical structure is related to the density level (or the distance), on the other hand, the fault traveling waves generated at the fault position and headed to the two opposite directions to the ends of the cable, therefore, the reason for the appearance of the condensed cluster tree could be related to the propagation process (catadioptric reflection) of the traveling waves. Although the propagation process of the traveling waves is not the direct or the only influencing factor for the appearance of the condensed cluster tree, it could be one of the root factors. The specific relation between the condensed cluster tree and the traveling wave needs further study in the future work. The specific structure of the condensed cluster tree will be the key research object. The distribution of the λ value of Cluster 0 node needs to be figured out. In turn, the proposed method will be more usable in practice. In addition, the true-type test based on high-dimensional data will be carried out in future research.

5.2. Conclusions

This paper proposed a novel traveling-wave-based method for fault location of power cables. It improved the former method (t-SNE + DBSCAN) by reducing the data processing steps to reduce data loss. The proposed approach consisted of three main steps: (1) build a matrix of the traveling waves associated with the sheath currents of the cables; (2) cluster the data in the matrix according to its density level and the stability, using HDBSCAN; (3) search for the characteristic cluster point(s) of the two branch clusters with the smallest density level to identify the arrival time of the traveling wave. The simulation comparison of the six methods validated the superiority of the proposed method. The following conclusions can be drawn from the results:

(1) HDBSCAN runs faster than the former method (t-SNE + DBSCAN). The running time of the former method is 2.23–20.94 times than the proposed method by the simulation test.
(2) The hierarchical structure of the condensed cluster tree corresponds exactly to the location relationship between the fault point and the monitoring point.
(3) The proposed method can be used for the identification of the arrival time of the traveling wave.

Author Contributions: Conceptualization, M.L.; methodology, M.L. and J.B.; software, Y.S. and Z.P.; validation, Y.W. and C.X. All authors have read and agreed to the published version of the manuscript.

Funding: This research was funded by Key Scientific and Technical Funds of Zhejiang Electric Power Corporation, grant number 5211DS17000Y.

Institutional Review Board Statement: Not applicable.

Informed Consent Statement: Not applicable.

Data Availability Statement: Not applicable.

Conflicts of Interest: The authors declare no conflict of interest.

References

1. Dong, X.; Yang, Y.; Zhou, C.; Hepburn, D.M. Online monitoring and diagnosis of HV cable faults by sheath system currents. *IEEE Trans. Power Del.* **2017**, *32*, 2281–2290. [CrossRef]
2. Li, M.; Zhou, C.; Zhou, W.; Zhang, J.; Zhang, L.; Yao, L. Feasibility Study on Lengthening the High Voltage Cable Section and Reducing the Number of Cable Joints via Alternative Bonding Methods. *High Volt.* **2019**, *4*, 292–299. [CrossRef]
3. Razzaghi, R.; Lugrin, G.; Rachidi, F.; Paolone, M. Assessment of the Influence of Losses on the Performance of the Electromagnetic Time Reversal Fault Location Method. *IEEE Trans. Power Del.* **2017**, *32*, 2303–2312. [CrossRef]
4. Zhou, C.; Yang, Y.; Li, M.; Zhou, W. An integrated cable condition diagnosis and fault localization system via sheath current monitoring. In Proceedings of the2016 International Conference on Condition Monitoring and Diagnosis (CMD), Xi'an, China, 25–28 September 2016. [CrossRef]
5. Wang, K.; Sun, D.; Zhang, J.; Xu, Y.; Luo, K.; Zhang, N.; Zou, J.; Qiu, L. An acoustically matched traveling-wave thermoacoustic generator achieving 750 W electric power. *Energy* **2016**, *103*, 313–321. [CrossRef]
6. Xu, Z.Y.; Jiang, S.J.; Yang, Q.X.; Bi, T.S. Ground distance relaying algorithm for high resistance fault. *IET Gener. Transm. Distrib.* **2009**, *4*, 27–35. [CrossRef]
7. Eissa, M.M. Ground distance relay compensation based on fault resistance calculation. *IEEE Trans. Power Del.* **2006**, *21*, 1830–1835. [CrossRef]
8. Li, M.; Zhou, C.; Zhou, W. A Revised Model for Calculating HV Cable Sheath Current Under Short-Circuit Fault Condition and Its Application for Fault Location—Part 1: The Revised Model. *IEEE Trans. Power Del.* **2019**, *34*, 1674–1683. [CrossRef]
9. Manesh, H.M.; Lugrin, G.; Razzaghi, R.; Romero, C.; Paolone, M.; Rachidi, F. A new method to locate faults in power networks based on Electromagnetic Time Reversal. In Proceedings of the IEEE, International Workshop on Signal Processing Advances in Wireless Communications, Cesme, Turkey, 17–20 June 2012. [CrossRef]
10. Lugrin, G.; Razzaghi, R.; Rachidi, F.; Paolone, M. Electromagnetic time reversal applied to fault detection: The issue of losses. In Proceedings of the IEEE International Symposium on Electromagnetic Compatibility, Dresden, Germany, 16–22 August 2015. [CrossRef]
11. Bawart, M.; Marzinotto, M.; Mazzanti, G. Diagnosis and location of faults in submarine power cables. *IEEE Elect. Insul. Mag.* **2016**, *32*, 24–37. [CrossRef]

12. Dashti, R.; Salehizadeh, S.; Shaker, H.; Tahavori, M. Fault Location in Double Circuit Medium Power Distribution Networks Using an Impedance-Based Method. *Appl. Sci.* **2018**, *8*, 1034. [CrossRef]
13. Bahmanyar, A.; Jamali, S. Fault location in active distribution networks using non-synchronized measurements. *Int. J. Electr. Power Energy Syst.* **2017**, *93*, 451–458. [CrossRef]
14. Dashti, R.; Ghasemi, M.; Daisy, M. Fault location in power distribution network with presence of distributed generation resources using impedance based method and applying π line model. *Energy* **2018**, *159*, 344–360. [CrossRef]
15. Orozco-Henao, C.; Bretas, A.S.; Chouhy-Leborgne, R.; Herrera-Orozco, A.R.; Marín-Quintero, J. Active distribution network fault location methodology: A minimum fault reactance and Fibonacci search approach. *Int. J. Electr. Power Energy Syst.* **2017**, *84*, 232–241. [CrossRef]
16. Dashti, R.; Sadeh, J. Applying Dynamic Load Estimation and Distributed-parameter Line Model to Enhance the Accuracy of Impedance-based Fault-location Methods for Power Distribution Networks. *Electr. Power Compon. Syst.* **2013**, *41*, 1334–1362. [CrossRef]
17. Ji, T. Study on Fault Location of Distribution Feeders Based on Transient Traveling Waves. Ph.D. Thesis, Shandong University, Jinan, China, March 2006. [CrossRef]
18. Borghetti, A.; Bosetti, M.; Nucci, C.A.; Paolone, M.; Abur, A. Integrated Use of Time-Frequency Wavelet Decompositions for Fault Location in Distribution Networks: Theory and Experimental Validation. *IEEE Trans. Power Deliv.* **2010**, *25*, 3139–3146. [CrossRef]
19. Da Silva, A.P.A.; Lima, A.C.S.; Souza, S.M. Fault location on transmission lines using complex-domain neural networks. *Electr. Power Energy Syst.* **2012**, *43*, 720–727. [CrossRef]
20. Lopes, F.V. Settings-free traveling-wave-based earth fault location using unsynchronized two-terminal data. *IEEE Trans. Power Del.* **2016**, *31*, 2296–2298. [CrossRef]
21. Lopes, F.V.; Dantas, K.M.; Silva, K.M.; Flávio, B.C. Accurate two-terminal transmission line fault location using traveling waves. *IEEE Trans. Power Deliv.* **2018**, *33*, 873–880. [CrossRef]
22. Xie, C.; Shao, X.; Jin, Y.; Wang, Y.; Li, M. A Novel Fault Location Method for Power Cables Based on Unsupervised Learning. In Proceedings of the 2020 IEEE International Conference on High Voltage Engineering and Application (ICHVE), Beijing, China, 6–10 September 2020. [CrossRef]
23. Li, M.; Liu, J.; Zhu, T.; Zhou, W.; Zhou, C. A Novel Traveling-Wave-Based Method Improved by Unsupervised Learning for Fault Location of Power Cables via Sheath Current Monitoring. *Sensors* **2019**, *19*, 2083. [CrossRef] [PubMed]
24. Ester, M.; Kriegel, H.; Sander, J.; Xu, X. A Density-Based Algorithm for Discovering Clusters in Large Spatial Databases with Noise. In Proceedings of the 2nd ACM SIGKDD, Portland, OR, USA, 4–8 August 1996; pp. 226–231.
25. Campello, R.J.G.B.; Moulavi, D.; Sander, J. Density-Based Clustering Based on Hierarchical Density Estimates. In *Advances in Knowledge Discovery and Data Mining*; PAKDD 2013. Lecture Notes in Computer Science; Springer: Berlin/Heidelberg, Germany, 2013; Volume 7819. [CrossRef]
26. Hartigan, J.A. Estimation of a convex density contour in two dimensions. *J. Am. Statist. Assoc.* **1987**, *82*, 267–270. [CrossRef]
27. Campello, R.J.G.B.; Moulavi, D.; Zimek, A.; Sander, J. Hierarchical density estimates for data clustering, visualization, and outlier detection. *ACM Trans. Knowl. Discov. Data* **2015**, *10*, 1–51. [CrossRef]
28. Deng, J.; Dong, W.; Socher, R.; Li, L.; Li, K.; Li, F. ImageNet: A large-scale hierarchical image database. In Proceedings of the IEEE Conference on Computer Vision & Pattern Recognition, Miami, FL, USA, 20–25 June 2009. [CrossRef]
29. Carrabs, F.; Gaudioso, M. A Lagrangian approach for the minimum spanning tree problem with conflicting edge pairs. *Networks* **2020**, 1–14. [CrossRef]
30. Todo, S.; Matsuo, H.; Shitara, H. Parallel loop cluster quantum Monte Carlo simulation of quantum magnets based on global union-find graph algorithm. *Comput. Phys. Commun.* **2019**, *239*, 84–93. [CrossRef]

energies

MDPI

Article

Study on Structural Parameters and Analysis Method of Soil Successive Impulse Discharge Channel †

Donghui Luo [1,2,*], Yongxing Cao [1], Yu Zhang [1], Shijun Xie [1], Chenmeng Zhang [1] and Shuping Cao [1]

1 State Grid Sichuan Electric Power Research Institute, Chengdu 610041, China; 87085306@163.com (Y.C.); zy863129@163.com (Y.Z.); sj-xie@163.com (S.X.); zhangchenmeng@126.com (C.Z.); caoshuping2021@163.com (S.C.)

2 Department of Electrical Engineering, Tsinghua University, Beijing 100084, China

* Correspondence: dhluo@foxmail.com; Tel.: +86-028-6999-5621

† This paper is an extended version of our paper published in International Conference on High Voltage Engineering and Application 2020, Beijing, China, 6–10 September 2020.

Abstract: The transient analysis model of grounding systems is an important tool to analyze the lightning characteristics of grounding devices. When lightning enters the soil through the grounding device, there is a centralized discharge channel. The spatial structure of the discharge channel in the soil has a great effect on the accuracy of the transient analysis model of the soil impulse discharge. In this paper, based on the gray information analysis method, the volume of successive impulse discharge channels in the ground under various currents and soil conditions is calculated, and the changing law with time interval is analyzed. According to the experimental results, an analysis method of the model considering the discharge channel structure is proposed, and an example is analyzed. The results show that the time interval has an effect on the volume of the discharge channel. In a certain range, the volume of the channel increases with the time interval of successive impulse discharges. Taking the critical breakdown electric field strength as the judgment condition and the residual resistivity as the variable, the development process of the discharge channel can be simulated. The calculated results of the model are close to the experimental results.

Keywords: grounding; discharge channel; X-ray; successive impulse; gray information; transient analysis model

Citation: Luo, D.; Cao, Y.; Zhang, Y.; Xie, S.; Zhang, C.; Cao, S. Study on Structural Parameters and Analysis Method of Soil Successive Impulse Discharge Channel. *Energies* **2021**, *14*, 877. https://doi.org/10.3390/en14040877

Academic Editor: Issouf Fofana
Received: 6 January 2021
Accepted: 3 February 2021
Published: 8 February 2021

1. Introduction

Soil is the most important medium in the grounding system [1–5]. Its discharge characteristics directly affect the lightning strike performance of grounding devices [6–8]. Strong ionization occurs when lightning injects the soil [9–11], and a centralized discharge channel is generated near the grounding electrode instead of uniform dispersion in all directions [6,7]. The development of the soil discharge directly determines the ground potential rise and the influence on the pipelines in the ground. The morphology and structure of the discharge channel is one of the most important factors to characterize the discharge, and it is also one of the most important factors that affect the accuracy of the grounding device transient analysis model. It is very important to obtain the actual structure of the soil discharge channel for building the lightning transient model of grounding systems and accurately calculate discharge characteristic parameters.

Due to the isolation of light and heat by soil, it is always a difficult point to observe the discharge in the ground accurately and comprehensively. In early research, many scholars proposed observation methods for soil discharge, which can be divided into three general methods. The first method is to embed observation media, such as conductive paper, film or X-ray film, in the soil around the grounding electrode [12–15]. The area where the discharge takes place is observed by means of ablation or exposure. The second method is to use a high-speed camera to capture the surface discharge process, or to use transparent glass sand instead of soil for the photography [16,17]. The third method is to excavate the area of lightning into the ground to observe the morphology of the discharge channel [18,19].

In order to solve the problems of traditional methods in changing the discharge environment as well as incomplete observation and high cost, the literature [1] proposes the use of X-ray transmission imaging technology to carry out non-intrusive imaging observation on the single impulse discharge channel in soil and obtain a gray-scale image that can reflect the structural outline of the discharge channel. The discharge channel is imaged from three perspectives (front view, side view and top view), and its 3D spatial structure is synthesized by using the boundary in the three views. The characteristic parameters of soil discharge are calculated based on the physical structure.

A lightning discharge process contains several subsequent strokes. Under the successive impulse currents, a more complex multi-branch discharge channel is formed in the soil [20,21]. This results in coverage and intersection among multiple channels. It is difficult to achieve the same accuracy as for the case of a single channel be relying only on the boundary of three views for 3D reconstruction. It is necessary to add constraints in the process of 3D reconstruction for correction, so as to ensure the model accuracy. Moreover, it is difficult to get three views at the same time, when taking instantaneous pictures of the discharge between two impulses. We propose an analysis method based on a single view image.

One of the purposes of obtaining the discharge channel structural parameters is to establish a more accurate transient analysis model for the grounding system. A variety of soil discharge models have been proposed, such as the hemispherical structure model [22], cylindrical structure model [23,24], segmented cylinder model [25] and multi-zone model [26]. The characteristic parameters of lightning discharge in soil are calculated based on the discharge channel structure in literature [1,2,7], and the transient analysis model of the grounding system is established. However, the existing models are generally based on the assumption of discharge current being uniform in each direction, without considering the underground discharge channel, which is inconsistent with the observation results of soil discharge with a centralized discharge channel.

It is very difficult to accurately observe the discharge channel structure in the soil and to consider it in the calculation of characteristic parameters. To solve these challenging problems, X-ray transmission imaging technology is used to obtain multi-view images of the discharge channel structure. The gray information analysis method is used to extract the structural parameters of the discharge channel and analyze its changing law with the time interval of successive impulses. At the same time, according to the image results, the structural characteristics of the discharge channel and the characterization methods of the successive impulse discharge process in the parameter calculation process are analyzed to make up for the deficiency of ignoring the discharge channel structure in the existing models. A simple modeling method of soil successive impulse discharge considering discharge channel structure is proposed. Hopefully, it can provide the basis and ideas for follow-up research. This paper is an extended version of our paper published in ICHVE 2020 [7]. Compared with the conference papers, the new research content of this paper is the analysis method and modeling method of the soil successive impulse discharge process considering the structural characteristics of discharge channels.

2. Experimental Platform and Analysis Method

2.1. Experimental Platform

A successive impulse current with several thousand amperes was applied to the soil by using the successive impulse discharge experimental platform, and the time interval was continuously adjustable from 0 to 500 ms. The sand was taken as the test sample and placed in a cube box with a side length of 20 cm. A vertical electrode with a length of 10 cm was placed in the center of the sandbox to discharge evenly on all sides. Previous literature [20,21] gives a detailed description of the structure and parameters of the experimental platform.

A standard gray sandbox (SGS) was placed next to the discharge sandbox (DS), with a size of 20 × 20 × 5 cm, and an oblique baffle was arranged inside, as shown in Figure 1. Its function was to provide a standard correspondence between the soil thickness and the gray value.

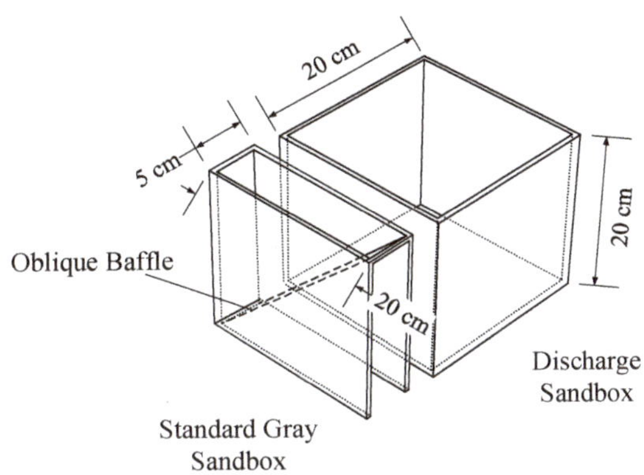

Figure 1. Schematic diagram of main discharge sandbox (DS) and standard gray sandbox (SGS). The SGS was placed next to DS and imaged at the same time. The oblique baffle set in the SGS could produce a soil layer with gradual thickness, which was used to analyze the corresponding relationship between soil thickness and image gray value.

2.2. Extraction Method of Discharge Channel Volume

The gray images of DS and SGS are obtained by an X-ray imaging system. The error caused by X-ray intensity and distance can be reduced by imaging them synchronously. The gray value matrix of the SGS image was extracted and the average value gi of each row of matrix was calculated. The results show that the gray value of the image increased linearly with the soil thickness under the experimental conditions [27]. The slope of the linear function could be obtained by fitting and is the estimated growth rate of gray difference (GRGD), which means that the gray value of the image increases a for every 1 cm increase in the thickness of soil penetrated by X-ray.

The prism, shown in Figure 2, was buried in the soil and then slowly taken out. A standard channel, with known structural information and exactly consistent with the model, could be formed inside the soil to simulate the discharge channel. The gray images with and without discharge channel were cut off with the same size, and then the gray value matrices with the same order were extracted from the gray images to obtain $\mathbf{G}_{\mathrm{n}}$ (no discharge channel) and $\mathbf{G}_{\mathrm{d}}$ (with discharge channel). By calculating the difference value of gray matrix and correcting the background error, we could obtain the change of gray value caused only by the discharge channel, that is, the gray difference matrix $\mathbf{G}_{\mathrm{c}}$ [27].

$$\mathbf{G}_{\mathrm{c}}' = \mathbf{G}_{\mathrm{n}} - \mathbf{G}_{\mathrm{d}} \tag{1}$$

$$\mathbf{G}_{\mathrm{c}} = \mathbf{G}_{\mathrm{c}}' - \mathbf{G}_{0} \tag{2}$$

$$\mathbf{G}_{\mathrm{c}}' = \begin{bmatrix} g_c'(11) & \cdots & g_c'(1n) \\ & \ddots & \\ \vdots & g_c'(ij) & \vdots \\ & \ddots & \\ g_c'(m1) & \cdots & g_c'(mn) \end{bmatrix} \tag{3}$$

$$\mathbf{G}_0 = \begin{bmatrix} g_0 & \cdots & g_0 \\ \vdots & \ddots & \vdots \\ g_0 & \cdots & g_0 \end{bmatrix} \tag{4}$$

$$g_0 = \overline{g_c'(ij)}, (ij \notin channel) \tag{5}$$

where $\mathbf{G}_n$, $\mathbf{G}_d$, $\mathbf{G}_c$, $\mathbf{G}_c{}'$ and $\mathbf{G}_0$ are all matrices of the same size. $\mathbf{G}_n$ is the matrix of gray image before the discharge channel is generated; $\mathbf{G}_d$ is the matrix of gray image after the discharge channel is generated. Because the thicker the soil is, the larger the gray value is, the gray value of the discharge channel position is smaller than that of the image before discharge. Therefore, $\mathbf{G}_c{}'$ represents the change of gray value caused only by the discharge channel and includes the background error caused by the dispersion of X-ray metrology.

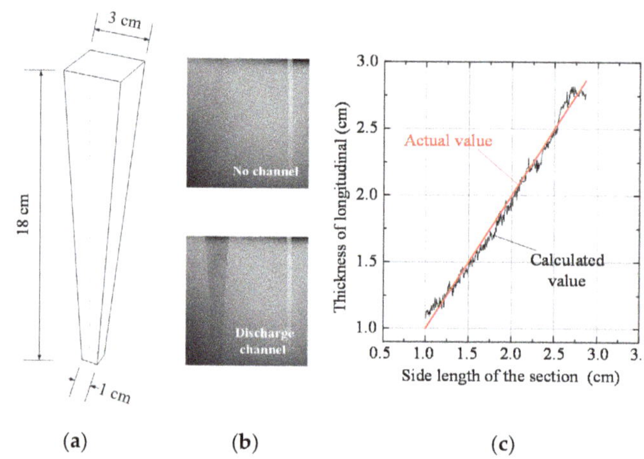

Figure 2. Actual thickness of discharge channel and calculated thickness. (**a**) Structural parameters of standard prism; (**b**) gray images with and without discharge channel; (**c**) the relationship between the calculated value of discharge channel thickness and its actual value.

In order to eliminate the background error, the background error constant was calculated. The average gray value of the non-discharge channel position in the $\mathbf{G}_c{}'$ matrix was taken as constant g_0, which is the background error constant caused by irradiation dose. The constant matrix $\mathbf{G}_0$ is composed of g_0. Therefore, after subtracting $\mathbf{G}_0$ from $\mathbf{G}_c{}'$, the resulting $\mathbf{G}_c$ is the gray difference matrix caused by the discharge channel, and the gray value of each pixel is $g_{c(ij)}$.

According to the GRGD, the calculation between gray difference and discharge channel thickness is as follows:

$$h_{c(ij)} = g_{c(ij)}/a \tag{6}$$

The average value of each row of thickness matrix was calculated and compared with the actual value, as shown in Figure 2. It can be seen that the calculated values were basically the same as the actual values. $h_{c(ij)}$ represents the thickness of any point in the discharge channel. For the area without discharge, $h_{c(ij)}$ is zero after error correction. So the discharge channel volume is

$$V_c = \sum (s_{c(ij)} \times h_{c(ij)}), (ij \in channel) \tag{7}$$

Here, a is the growth rate of gray difference which means that the gray value of the image increases a for every 1 cm increase in the thickness of soil penetrated by X-ray. $h_{c(ij)}$ is the thickness of the discharge channel corresponding to each pixel; $s_{c(ij)}$ is the area of each pixel; V_c is the volume of the discharge channel, which is the sum of the product of the area and thickness of each pixel.

The calculated average volume of the prism is 77.02 cm^3, the actual value is 78 cm^3. It shows that the method is feasible. The implementation and verification of this method are described in detail in [27]. The main content of this paper is to obtain the volume and changing law of successive impulse discharge channel based on this method, which provides data basis and theoretical support for modeling.

3. The Structure Changing Rule of Successive Impulse Discharge Channel

3.1. Influence of Current

It is important to obtain the structural parameters and changing laws of the discharge channel for the accurate establishment of the transient analysis model of the grounding system. Based on the experiment, the images of the discharge channel in the soil under different time intervals can be obtained, and its 3D structural parameters can be calculated. From the point of view of structural parameters, the change rule of successive impulse discharge channel in time domains is verified.

Figure 3 shows the images of the discharge channel structure changing over time under different current peaks. Figure 4 shows the corresponding discharge channel volume, where the red arrow indicates an increasing trend on the whole. Literature [20] indicates that, the larger the time interval is, the greater the probability of a branch discharge channel. It can be found that the experimental results under different current peaks conform to this rule; the probability of a branch discharge channel increases with the time interval. Moreover, within the experimental conditions, the total volume of the successive impulse discharge channel in the soil increases with the time interval.

At the same time, the larger the current peak value, the larger the discharge channel volume. Therefore, it can be concluded that under the successive impulse current, the discharge region in the soil will increase with the energy injected into the ground and the time interval between the two impulses.

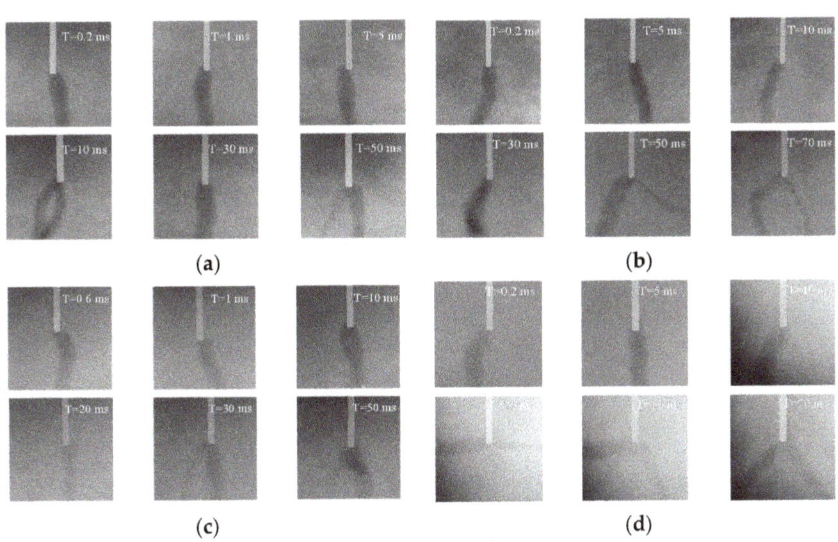

Figure 3. Discharge channels of soil under successive impulse currents with various current peaks. (**a**) Current peak of 1.5 kA; (**b**) current peak of 1.8 kA; (**c**) current peak of 2.0 kA; (**d**) current peak of 2.2 kA.

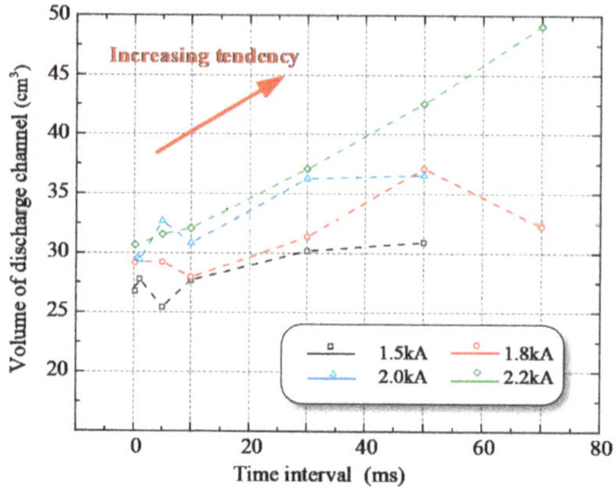

Figure 4. Volume of soil discharge channels under successive impulse currents with various peaks.

3.2. *Influence of Soil Properties*

Soil parameters are also important factors affecting the volume of the discharge channel. The images of discharge channels under different soil conditions were obtained and the 3D structure parameters were calculated. Figure 5 shows the images of the discharge channel structure changing with time intervals under different salt content; for the discharge channel structure of soil with various water content, refer to previous literature [20]. Figure 6 shows the volume change of the successive impulse discharge channel under different soil water content and salt content. With the increase of soil water content, soil resistivity decreased and the discharge channel volume increased. When the soil water content was constant, the volume of the discharge channel increased with the time interval. The effect of soil salt content on the volume of the discharge channel was basically the same as found for varying water content. The larger the soil salt content was, the smaller the soil resistivity and the larger the discharge channel. When the salt content was constant, the volume of the discharge channel increased with the time interval. It can be concluded that the volume of the successive impulse discharge channel increased with the soil conductivity and time interval.

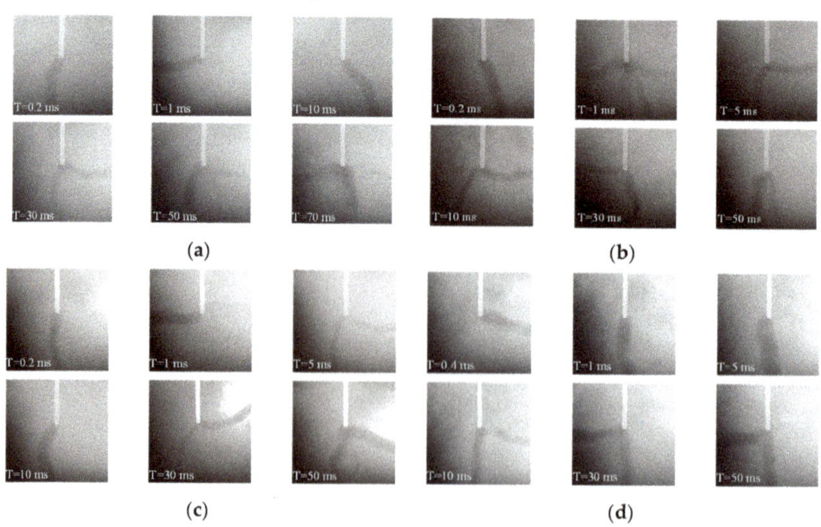

Figure 5. Discharge channels of soil under successive impulse currents with various salt content. (**a**) Salt content of 0%; (**b**) salt content of 0.1%; (**c**) salt content of 0.2%; (**d**) salt content of 0.5%.

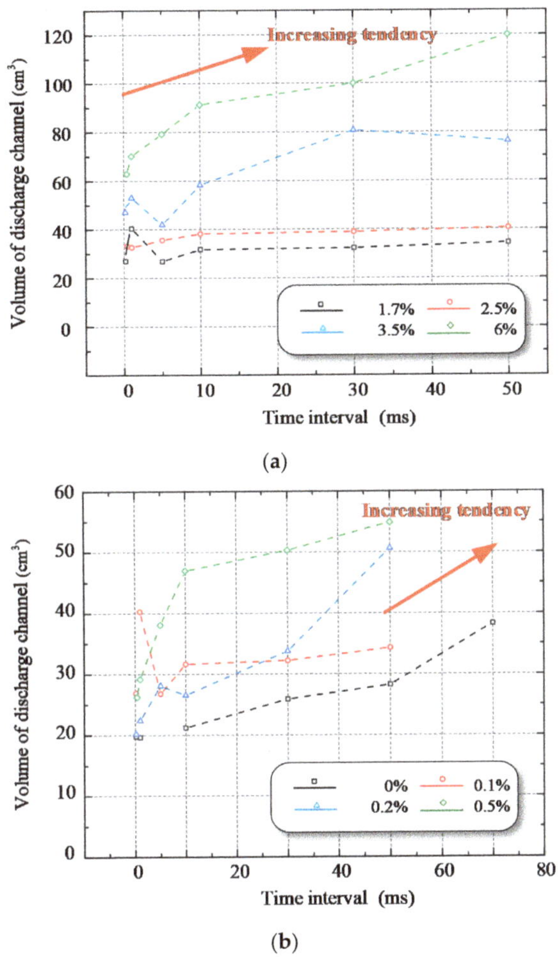

Figure 6. Volume of soil discharge channels under successive impulse currents. (**a**) Influence of soil water content; (**b**) influence of soil salt content.

In addition, compared with the water content, the influence of soil salt content on the volume of the successive impulse discharge channel was smaller, which may be due to the smaller influence of salt content on the recovery characteristics [21].

Based on the above experimental results, it can be found that although there was randomness in the structure of the discharge channel, from the perspective of overall probability, the volume of the discharge channel increased with the time interval. According to the recovery mechanism of soil discharge [20], when the time interval is small, the discharge channel generated by the first impulse is in a highly ionized state, and the conductivity is large, so it is easy to discharge the subsequent shock current. Therefore, a new discharge area is not likely to occur.

With an increase in the time interval, the highly ionized state of the first impulse discharge channel disappears and the resistivity increases. Therefore, the probability of subsequent impulse currents ionizing the soil around the channel or breaking through other areas to produce new discharge channels increases, and the discharge zone also increases.

The above experimental results also verify the discharge recovery mechanism proposed in literature [20] from the perspective of the discharge channel volume and provide the theoretical basis for the establishment of a calculation method.

4. Analysis Method Considering Discharge Channel Structure

4.1. Basic of Model

The accuracy of the lightning transient model of grounding device can be improved by constructing a soil impulse discharge model considering the structure of the discharge channel, which is more suitable for the actual situation. The analysis of the influence range of the soil discharge area can improve the assessment of the impact of lightning on the pipeline near the grounding device. Therefore, this section explores the modeling method for the soil discharge model considering the structure of the discharge channel.

Literature [20] shows that the process of soil discharge under successive impulse currents generally includes three stages: first impulse discharge, recovery process and subsequent impulse discharge. The physical process and initial conditions of each stage are different. The recovery process of soil electrical performance and subsequent impulse discharge should not be ignored, nor can they be regarded as the simple superposition of the single impulse discharge process in the time domain. The subsequent impulse discharge process is greatly affected by the first impulse and recovery process.

The main physical model of the soil successive impulse discharge process is the soil discharge recovery model. When the successive lightning impulse current is injected into the soil through the grounding device, the first impulse forms a centralized discharge channel in the soil [20,21]. Because of the large instantaneous energy of the impact current and the uneven distribution of soil particles in all directions, the impulse current preferentially breaks through the soil at the location where discharge occurs most easily when the electric field intensity near the electrodes reaches the breakdown field strength of the soil, forming a partial discharge channel. Then, the current continues to be injected into the arc channel, breaking through the soil where discharge occurs most easily near the end of the arc channel. The discharge channel continues to grow until the energy injected decreases to such an extent that it cannot break through the soil. When the first impulse current drops to zero, the discharge channel remains highly ionized for some time and recovers gradually.

If the subsequent impulse current is injected before the highly ionized state disappears, there is a great probability that the impulse current will be directly and rapidly discharged along the existing discharge channel with high conductivity. Moreover, when energy reaches the end of the channel, it continues to break through the soil at the most prone location of discharge and continues to grow until it cannot break through the soil.

Experimental images show that there is a centralized discharge channel near the grounding electrode after lightning enters the ground. In the case of a single discharge channel, the residual resistivity of subsequent impulses is a function of the residual resistivity of the first impulse and the recovery coefficient [20].

Therefore, based on the existing soil discharge theory, an improved four-zone structural model of soil impulse discharge is proposed, as shown in Figure 7. The ionization region in the soil can be simply divided into arc channel, ionization area, non-linear conductivity area and constant conductivity area. The characteristics of different regions are as follows:

- The resistivity of soil in the arc discharge channel is the product of initial soil resistivity and the attenuation coefficient.
- When the electric field strength is greater than the critical breakdown electric field strength ($E > E_C$), the resistivity of the soil is much lower than the initial resistivity, but the reduction degree is much lower than that of the soil inside the arc channel.
- When the electric field strength is less than the critical breakdown electric field strength ($E < E_C$), the soil resistivity changes nonlinearly, and decreases exponentially with the increase of electric field intensity, and the attenuation degree is smaller than that of the ionization zone.

- When the electric field intensity is small and can be ignored, the soil resistivity is the initial value.

In this paper, the values of soil discharge parameters in each region can be determined according to the experimental results:

- The soil resistivity in the arc discharge channel can be determined according to the experimental results in literature [2,20]. The residual resistivity can be obtained by iterative calculation when it is first broken down by impulse current [2]. When the second breakdown occurs, the residual resistivity is the product of the calculated value of the first breakdown and the recovery coefficient [20].
- The soil residual resistivity in the ionization zone has been studied in the literature [28–30], and the results show that the residual resistivity is distributed between 1.7% and 50% of the initial resistivity. For example, when the initial soil resistivity is about 50 Ωm, the residual resistivity of the ionization zone is about 10–20% of the initial value [28–30]. However, because the arc channel and ionization zone are not distinguished in the above test analysis, the larger value in the range is selected in this paper.
- The experimental results for the relationship between soil resistivity and electric field in the nonlinear zone are given in literature [31]. Although the results of different organizations are varied, their change trend and value range are basically consistent. Before reaching the critical breakdown field strength, the soil resistivity decreases exponentially with the increase of electric field strength, and the value range of the reduction coefficient is basically decreased from 1 to about 0.5 [31].
- The soil resistivity in the constant conductivity zone can be considered to maintain the initial value unchanged. In this paper, the space coordinate transformation method is used to represent the infinite earth environment.

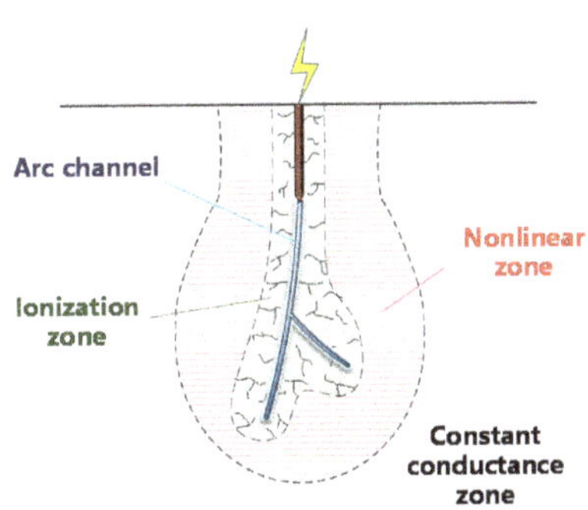

Figure 7. Improved four-zone model of soil under impulse current, including arc channel, ionization area, non-linear conductivity area and constant conductivity area.

4.2. *Numerical Analysis Method*

At the initial time of the effective action of impulse current ($0 + \Delta t$, where Δt is the progressive step of time element in calculation), based on the electric field intensity generated by the instantaneous current value in the soil, the direction where the maximum electric field intensity was located near the grounding device was selected as the initial development direction of the discharge channel, which could simulate the randomness in the actual development process.

The position of the critical breakdown electric field strength value E_C in the development direction of the discharge channel was taken as the growth limit of discharge channel in the time step, Δt. According to the experimental images and the research results in literature [1,2,18,20,21], we analyzed the structural characteristics of the discharge channel. Generally, the diameter of lightning discharge channel in the soil is several centimeters. Under the normal size of a single columnar grounding electrode structure, most extend from the end of the electrode, and the section radius is slightly larger than the equivalent radius of the electrode. At the same time, the section radius of each position of the discharge channel is basically the same, that is, the growth of the discharge channel is basically along the depth direction, and the channel radius can be regarded as constant during the growth process.

We set the growth rule of the discharge channel in the time step Δt according to the above rules. The growth results of the discharge channel structure at the last moment were taken as the initial state of the physical structure in the next time step $(0 + 2\Delta t)$, and the distribution results of the electric field intensity in the ground caused by the instantaneous current were recalculated. The dynamic growth of the discharge channel and the analysis of the electric field intensity were realized by repeated iterative calculation. The modeling process of the above analysis model is shown in Figure 8.

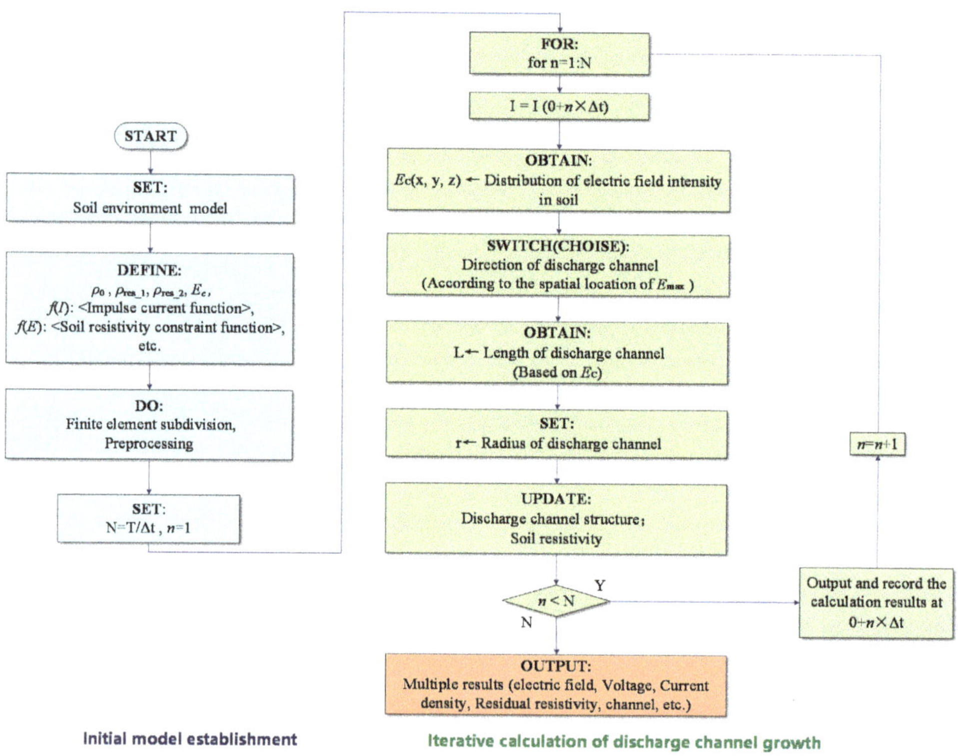

Figure 8. Finite element flowchart of the modeling process of the dynamic growth model of soil impulse discharge channels.

4.3. Basic Case Analysis

In order to verify the rationality of the proposed model and its parameter range, a finite element analysis model was established based on the actual grounding device impulse discharge experiment, and the calculation results were compared with the experimental results.

The equivalent model was established based on the actual impulse discharge experimental parameters in literature [32]. The grounding device was a vertical electrode with a length of 1 m and a section radius of 0.025 m, and the top of the electrode was flat with the ground surface, as shown in Figure 9. The distribution of the initial resistivity of soil was uniform, and its value was between 40.5 Ωm and 43.5 Ωm. In the analysis model, the initial resistivity of soil was set as 43.5 Ωm.

Since the single impulse current is used as the source in literature [28], and factors such as soil water content and compactness are not discussed, according to the calculation method and results of discharge parameters proposed in the latest research of literature [1,2], the critical breakdown field strength of soil under this condition was set as 210 kV/m, and the residual resistivity in the discharge channel at the first breakdown was ρ_{res_1} = 0.05% × ρ_0. Because the discharge area in the ground was not observed in the actual experiment [32], in order to reduce the influence of the randomness of the discharge channel growth direction on the calculation results and to facilitate the demonstration of the case results, the direction of the discharge channel growth direction was conducted in the same direction and the vertical downward growth was set.

In this paper, the analysis model was constructed in the finite element environment. The hemispherical module was used to represent the earth. The infinite element equivalent module was set around the hemispherical module, and the space coordinate transformation method was used to realize the infinite earth equivalent. The enclosure potential of the infinite equivalent module was set to 0, i.e., φ = 0.

Discharging of the impulse current through the electrode into the soil can be described by the quasi-static Maxwell equation [20]:

$$\nabla \times \boldsymbol{E} = 0 \tag{8}$$

$$\nabla \cdot \boldsymbol{J} = 0 \tag{9}$$

$$\boldsymbol{J} = \boldsymbol{J}_c + \boldsymbol{J}_d = \sigma \boldsymbol{E} + \frac{\partial \boldsymbol{D}}{\partial t} \tag{10}$$

$$\boldsymbol{E} = -\nabla \varphi \tag{11}$$

where $\boldsymbol{D}$ is the electric flux density, $\boldsymbol{E}$ is the electric field intensity, φ is the potential, and $\boldsymbol{J}$ is the total current density, which comprises $\boldsymbol{J}_c$ as the conduction current density and $\boldsymbol{J}_d$ as the displacement current density.

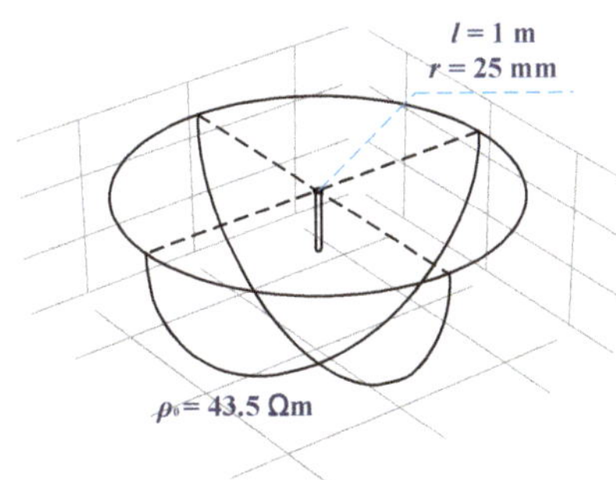

(a)

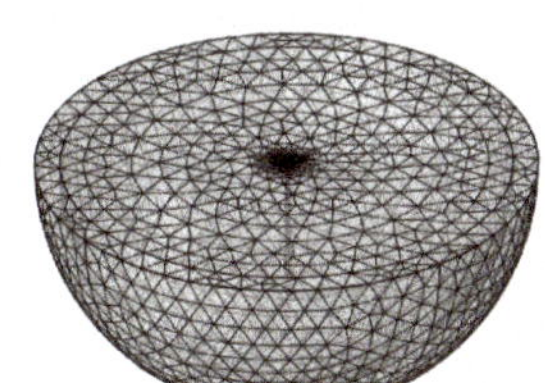

(b)

Figure 9. Model diagram of structure of grounding device in impulse discharge test. (**a**) Structural diagram; (**b**) diagram of subdivision model.

The impulse current experiment results with a peak current of 30.8 kA were selected as the modeling object. The model calculation results of 1 μs, 2 μs, 5 μs and 8 μs were extracted from the starting time of the impulse current to the peak time of the current. Figure 10 shows the evolution process of the spatial structure of the discharge channel, spatial distribution of current density in the ground, spatial distribution of ground potential and surface potential in the time dimension.

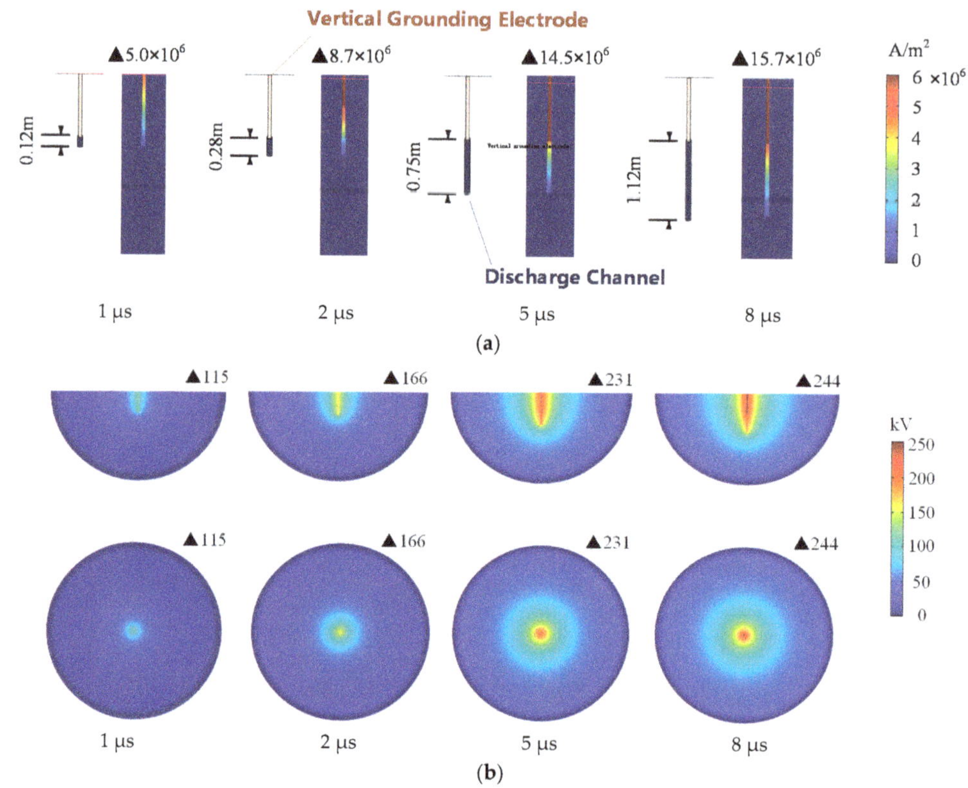

Figure 10. The spatial-temporal evolution of the dynamic growth model of discharge channels. (**a**) Evolution of discharge channel structure and current density. (**b**) Evolution process of ground potential and surface potential.

The results show that in the rising stage of the impulse current, the length of the discharge channel increased gradually with the current value. The discharge channel grew to 1.12 m at the time of the current peak, and it continued to increase until the current was zero. According to the spatial distribution of current density in the ground shown in Figure 10, most of the energy of the impulse current was discharged along the discharge channel longitudinally, and only a small part of the current was discharged from the soil around the discharge channel. The simulation results were consistent with the discharge channel images and theoretical analysis.

At the same time, it was found that the growth of discharge channels also had a significant impact on the potential distribution in the ground. The variation of potential in the soil near the discharge channel was stronger than that in the area far away from the discharge channel, and the main voltage distortion direction was consistent with the growth direction of the discharge channel.

The calculated impulse grounding resistance and maximum grounding potential rise (GPR) were compared with the test results. The voltage at the top of the grounding electrode (current injection point) was selected as the value of GPR. When the impulse current peak value was 30.8 kA, the maximum ground potential rise at the peak time was 244 kV. The calculated impulse grounding resistance of the grounding device was 7.9 Ω, which is similar to the experimental results in the literature, as shown in Table 1.

Table 1. The impulse grounding resistance and ground potential rise of the grounding device calculated by the model compared with the experimental results in literature [32].

I_{max} (kA)	R_i (Ω)		GPR (kV)	
	Experiment	Model	Experiment	Model
30.8	8.1	7.9	248	244

The existence of a centralized discharge channel has significant influence on the calculation results of grounding potential distribution. Figure 11 shows the calculation results based on the above model with the same parameters, but ignoring the existence of the discharge channel and its growth process. The GPR of the model without considering the discharge channel was 305 kV, which increased by 61 kV compared with 244 kV calculated by the model considering the discharge channel, showing significant differences.

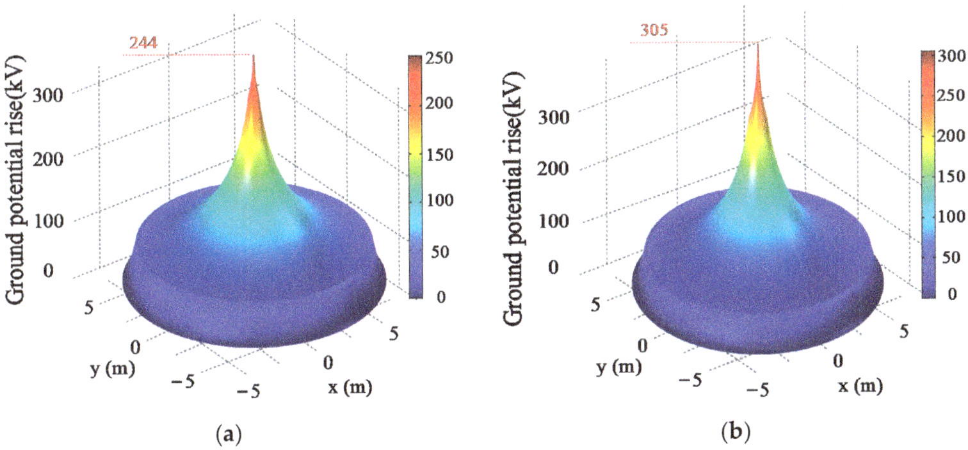

Figure 11. Spatial distribution of surface potential at peak time, (**a**) considering the structure of the discharge channel and its growth process, or (**b**) ignoring the structure of the discharge channel and its growth process.

4.4. Discharge Channel Structure

The structure and growth range of the discharge channel directly affected the safe distance of the pipeline in the ground near the grounding electrode and the insulation distance between the independent grounding electrode and the main grid. Figure 12 shows the growth process and current density distribution of the discharge channel in the above case. At the end of the first impulse current, the discharge length of the channel reached 1.5 m. Considering the electrode length, the distance between the end of the discharge channel and the surface was 2.5 m. Moreover, the soil structure at this time was taken as the initial structure under the subsequent impulse current.

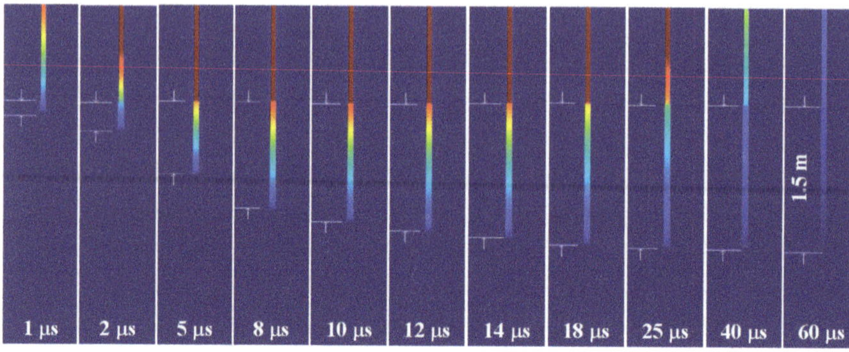

Figure 12. Growth process of the first impulse discharge channel.

In order to avoid the influence of waveform parameters on the growth distance of the discharge channel and the recovery characteristics of soil resistivity, the peak value of the subsequent impulse current was consistent with that of the first impulse current, both of which were 30.8 kA.

The time interval between the two impulse currents directly affected the recovery degree of soil resistivity. The literature [20,21] shows that the smaller the time interval, the smaller the recovery degree of the ionized state of the first impulse discharge channel and the lower the resistivity, which results in a more favorable environment for subsequent impulse current discharges. At this time, a single discharge channel is more likely to occur; that is, the subsequent discharge channel overlaps with the first discharge channel and continues to develop on this basis, so as to further expand the length of the discharge channel. Therefore, we chose this case as the modeling object. Based on the results of the literature [20,21], we chose the time interval to be 0.2 ms, the residual resistivity of soil secondary breakdown was ρ_{res_2} =0.2 $\times$ ρ_{res_1}, and the subsequent discharge channel completely covered the first discharge channel and developed in the same direction on the basis of it.

Figure 13 shows the growth process of the discharge channel under the subsequent impulse current; the discharge channel developed to 2.6 m, and the distance from the ground surface was 3.6 m. This was in the same order of magnitude as the discharge channel length by lightning penetration in the literature [17,18]. The difference was mainly caused by lightning current amplitude, the number of return impulses and soil parameters.

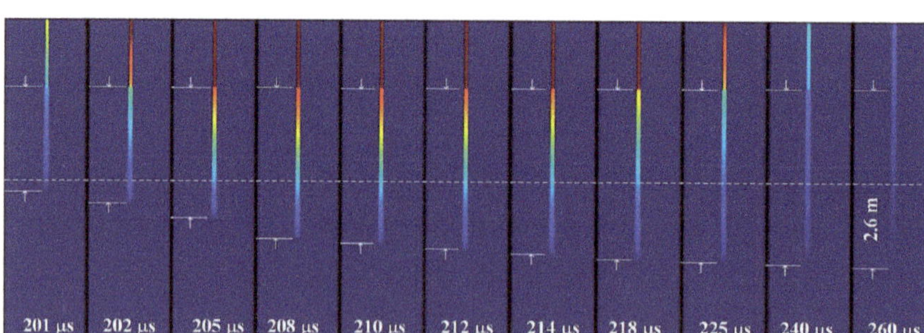

Figure 13. Growth process of second impulse discharge channel.

There were also differences in the growth degree of the discharge channel between the second impulse and the first impulse. The subsequent discharge channel was increased by 1.1 m, which was shorter than the first one. This was because more current could discharge along the vertical direction of the discharge channel, as the length of the discharge channel increased. The energy at the end of the channel decreased and the electric field intensity decreased as well.

Based on the simulation test results, the soil discharge characteristic parameters were obtained, and the analysis model was established. Finally, the model calculation results were compared with the experimental results in the literature for verification. However, most of the existing true-type tests were single impulse discharge tests, mainly due to the lack of successive impulse discharge test platforms that can be used in the field. In future research, we can develop a set of large successive impulse current test platforms that can be used for field tests, carry out successive impulse discharge true-type tests, and further calibrate the models. Our team is carrying out this work.

5. Conclusions

In this paper, the structural characteristics of soil discharge channels under successive impulse currents were studied by experiment and simulation. The differences between the discharge channel of successive impulses and that of the single impulse as well as its variation with time intervals were analyzed.

An X-ray transmission imaging system was used to obtain the images of the discharge channel in soil; the volume of discharge channel was extracted based on the gray analysis method, and its changing law with time intervals was analyzed. The results show that a smaller time interval results in a higher likelihood of a subsequent impulse current discharged by the first impulse discharge channel with a highly ionized state. The lower the energy used to generate a new discharge channel, the smaller the volume of the discharge channel in the soil. On the contrary, the larger the time interval, the higher the probability of multiple discharge channels and the larger the total volume of discharge channels.

A higher number of discharge channels causes the discharge area to be concentrated mainly near the grounding electrode. However, when there is only one discharge channel, the subsequent impulse current continues to discharge further on the basis of the first impulse discharge channel. Although the total volume of the discharge channel is smaller than that of a multi-channel system, the length of the discharge channel, i.e., the maximum influence distance, is larger than that of multiple channels. This increases the impact on underground pipelines.

According to the structural characteristics and development theory of discharge channels, a lightning analysis method of grounding devices considering the structure of discharge channels in soil is proposed, and the development length of discharge channels is analyzed. In the case of this paper, the length of the arc channel can reach several meters, and the maximum length increases with the increase of the number of subsequent impulses and the amplitude of lightning.

Author Contributions: Investigation, D.L., Y.Z., S.X. and C.Z.; methodology, Y.C.; project administration, S.C. All authors have read and agreed to the published version of the manuscript.

Funding: This work is supported by the Science and Technology Project of State Grid Corporation of China ("Study on the Transient Characteristics of Grounding System and the Test and Evaluation Method of Current Dispersion Performance under the Successive Impulse Current", 5500-202026088A-0-0-00).

Institutional Review Board Statement: Not applicable.

Informed Consent Statement: Not applicable.

Data Availability Statement: The study did not report any data.

Acknowledgments: Thank Zhou Mu and Jialun Li for their advice and help in the process of paper compilation.

Conflicts of Interest: The authors declare no conflict of interest.

Abbreviations

Symbol/Abbreviation	Definition	Unit
SGS	standard gray sandbox	/
DS	discharge sandbox	/
GRGD	growth rate of gray difference	/
GPR	Ground potential rise	/
$\mathbf{G}_n$	the matrix of gray image before the discharge channel is generated	/
$\mathbf{G}_d$	the matrix of gray image after the discharge channel is generated	/
$\mathbf{G}_c{}'$	Grayscale difference matrix caused by discharge channel and background error	/
$\mathbf{G}_c$	Grayscale difference matrix caused by discharge channel	/
$\mathbf{G}_0$	Background error constant matrix	/
$h_{c(ij)}$	Thickness of discharge channel at each pixel	cm
a	GRGD (growth rate of gray difference)	/
$s_{c(ij)}$	The area corresponding to each pixel	cm^2
V_c	Discharge channel volume	cm^3
E_c	Critical breakdown field strength	kV/m
ρ_0	Initial soil resistivity	Ωm
ρ_{res_1}	Residual resistivity of soil after first breakdown	Ωm
ρ_{res_2}	Residual resistivity of soil after second breakdown	Ωm
φ	Potential	V
E	Electric field intensity	kV/m
D	Electric flux density	C/m^2
J	Total current density	A/m^2
J_c	Conduction current density	A/m^2
J_d	Displacement current density.	A/m^2

References

1. Sima, W.X.; Liu, S.W.; Yuan, T.; Luo, D.H.; Wu, P.; Zhu, B. Experimental study of the discharge area of soil breakdown under surge current using X-ray imaging technology. *IEEE Trans. Ind. Appl.* **2015**, *51*, 5343–5351. [CrossRef]
2. Liu, S.W.; Sima, W.X.; Yuan, T.; Luo, D.H.; Bai, Y.; Yang, M. Study on X-ray imaging of soil discharge and calculation method of the ionization parameters. *IEEE Trans. Power Deliv.* **2017**, *32*, 2013–2021. [CrossRef]
3. Androvitsaneas, V.P.; Damianaki, K.D.; Christodoulou, C.A.; Gonos, I.F. Effect of soil resistivity measurement on the safe design of grounding systems. *Energies* **2020**, *13*, 3170. [CrossRef]
4. Ala, G.; Favuzza, S.; Francomano, E.; Giglia, G.; Zizzo, G. On the distribution of lightning current among interconnected grounding systems in medium voltage grids. *Energies* **2018**, *11*, 771. [CrossRef]
5. Steinsland, V.; Sivertsen, L.H.; Cimpan, E.; Zhang, S.J. A new approach to include complex grounding system in lightning transient studies and EMI evaluations. *Energies* **2019**, *12*, 3142. [CrossRef]
6. Zhang, B.; Li, Z.Z.; Wang, S. Onset electric field of soil ionisation around grounding electrode under lightning. *High Volt.* **2020**, *5*, 614–619. [CrossRef]
7. Luo, D.H.; Cao, Y.X.; Sima, W.X.; Yuan, T.; Xie, S.J.; Zhang, Y. Study on changing law of 3D structure of successive impulse discharge channel in the ground. In Proceedings of the 2020 IEEE International Conference on High-Voltage Engineering (ICHVE 2020), Beijing, China, 6–10 September 2020.
8. Reffin, M.S.; Nor, N.M.; Ahmad, N.N.; Abdullah, S. Performance of practical grounding systems under high impulse conditions. *Energies* **2018**, *11*, 3187. [CrossRef]
9. Fakhraei, M.; Mahmoudian, M.; Rodrigues, E.M.G. Grounding system modeling and evaluation using integrated circuit based fast relaxed vector fitting approach considering soil ionization. *Appl. Sci.* **2020**, *10*, 5632. [CrossRef]
10. Moreno, J.; Simon, P.; Faleiro, E.; Asensio, G.; Fernandez, J.A. Estimation of an upper bound to the value of the step potentials in two-layered soils from grounding resistance measurements. *Materials* **2020**, *13*, 290. [CrossRef] [PubMed]
11. Ali, A.W.A.; Ahmad, N.N.; Nor, N.M.; Reffifin, M.S.; Abdullah, S.A.S. Investigations on the performance of a new grounding device with spike rods under high magnitude current conditions. *Energies* **2019**, *12*, 1138.
12. Zhang, B.P.; He, J.L.; Zeng, R. Spatially discontinuous ionization phenomenon in inhomogeneous soil. *Sci. China Technol. Sci.* **2010**, *53*, 918–921. [CrossRef]
13. Wen, X.S.; Feng, Z.Q.; Lu, H.L.; Tong, X.F.; Lan, L.; Chen, W.G.; Tan, B. Sparkover observation and analysis of the soil under the impulse current. *IET Sci. Meas. Technol.* **2016**, *10*, 228–233. [CrossRef]
14. Cabrera, V.M.; Cooray, V. On the mechanism of space charge generation and neutralization in a coaxial cylindrical configuration in air. *J. Electrost.* **1992**, *28*, 187–196. [CrossRef]

15. Victor, M.; Cabrera, M.; Lundquist, S.; Cooray, V. On the physical properties of discharges in sand under lightning impulses. *J. Electrost.* **1993**, *30*, 17–28. [CrossRef]
16. Wang, J.; Liew, A.C.; Darveniza, M. Extension of dynamic model of impulse behavior of concentrated grounds at high currents. *IEEE Trans. Power Deliv.* **2005**, *20*, 2160–2165. [CrossRef]
17. Elzowawi, A.; Haddad, A.; Griffiths, H. Visualization of electric discharge in porous materials. In Proceedings of the International Conference on High Voltage Engineering and Application (ICHVE), Poznan, Poland, 8–11 September 2014.
18. Rakov, V.A. *Triggered Lightning*; Springer: Berlin, Germany, 2009.
19. Rakov, V.A.; Uman, M.A. *Lightning: Physics and Effects*; Cambridge University Press: New York, NY, USA, 2003.
20. Luo, D.H.; Sima, W.X.; Yuan, T.; Sun, P.T.; Chen, W.; Wang, J. Influence-factor analysis and parameter calculation of soil discharge and recovery characteristics under successive impulse currents. *IEEE Trans. Power Deliv.* **2019**, *34*, 514–523. [CrossRef]
21. Sima, W.X.; Luo, D.H.; Yuan, T.; Liu, S.W.; Sun, P.T.; Li, T.X. Study on the image observation method and characteristics of soil discharge process under multiple impulse currents. *IEEE Trans. Power Deliv.* **2018**, *33*, 2125–2134. [CrossRef]
22. Petropoulos, G.M. The high-voltage characteristics of earth resistances. *J. Inst. Electr. Eng. Part II Power Eng.* **1948**, *95*, 59–70.
23. Bellaschi, P.L.; Armington, R.E. Impulse and 60-cycle characteristics of driven grounds—III effect of lead in ground installation. *Electr. Eng.* **1943**, *62*, 334–345. [CrossRef]
24. Bellaschi, P.L. Impulse and 60-cycle characteristics of driven grounds. *Electr. Eng.* **1941**, *60*, 123–128. [CrossRef]
25. Velazquez, R.; Mukhedkar, D. Analytical modelling of grounding electrodes transient behavior. *IEEE Trans. Power Appar. Syst.* **1984**, *103*, 1314–1322. [CrossRef]
26. Li, J.L.; Yuan, T.; Yang, Q.; Sima, W.X.; Sun, C.X.; Zahn, M. Numerical and experimental investigation of grounding electrode impulse-current dispersal regularity considering the transient ionization phenomenon. *IEEE Trans. Power Deliv.* **2011**, *26*, 2647–2658. [CrossRef]
27. Luo, D.H.; Yuan, T.; Sima, W.X.; Cao, Y.X.; Ye, S.Y.; Yao, X.Y. Study on the process of soil recovery and method of volume parameter extraction of branch discharge channel under successive impulse currents. *High Volt. Tech.* **2020**, *46*, 1791–1799.
28. Liu, Y.Q.; Theethayi, N.; Gonzalez, R.M.; Thottappillil, R. The residual resistivity in soil ionization region around grounding system for different experimental results. In Proceedings of the IEEE Symposium on Electromagnetic Compatibility, Boston, MA, USA, 18–22 August 2003.
29. Bellaschi, P.L.; Arminggton, R.E.; Snowden, A.E. Impulse and 60-cycle characteristics of driven grounds II. *AIEE Trans.* **1942**, *61*, 349–363.
30. Liew, A.C.; Darveniza, M. Dynamic model of impulse characteristics of concentrated carths. *Proc. IEEE* **1974**, *121*, 123–135.
31. Zhu, B. Study on the Variation Characteristics of Soil Electrical Parameters and the Influence on Grounding Device's Performance. Ph.D. Thesis, Chongqing University, Chongqing, China, 2015.
32. Geri, A. Behaviour of grounding systems excited by high impulse currents: The model and its validation. *IEEE Trans. Power Deliv.* **1999**, *14*, 1008–1017. [CrossRef]

energies

MDPI

Article

Pattern Recognition of Development Stage of Creepage Discharge of Oil–Paper Insulation under AC–DC Combined Voltage Based on OS-ELM

Fubao Jin [1,*], Shanjun Zhang [2] and Yuanxiang Zhou [3]

1 School of Water Resources and Electric Power, Qinghai University, Xining 811600, Qinghai, China
2 Department of Information Science, Kanagawa University, Yokohama 222-0033, Japan; zhang@info.kanagawa-u.ac.jp
3 State Key Laboratory of Control and Simulation of Power Systems and Generation Equipment, Department of Electrical Engineering, Tsinghua University, Beijing 10084, China; zhouyx@tsinghua.edu.cn
* Correspondence: jinfubao@qhu.edu.cn; Tel.: +86-189-9718-8893

Abstract: The recognition of the creepage discharge development process of oil–paper insulation under AC–DC combined voltage is the basis for fault monitoring and diagnosis of converter transformers; however, only a few related studies are available. In this study, the AC–DC combined voltage with a ratio of 1:1 was used to develop a recognition method for the creepage discharge development process of an oil–paper insulation under a cylinder–plate electrode structure. First, the pulse current method was used to collect the discharge signals in the creepage discharge development process. Then, 24 characteristic parameters were extracted from four types of creepage discharge characteristic spectra after dimensionality reduction. Finally, based on the online sequential extreme learning machine (OS-ELM) algorithm, these characteristic parameters were used to recognize the development stage of the creepage discharge of the oil–paper insulation. The results showed that when the size of the sample training set used in the OS-ELM algorithm is close to the number of hidden layer neurons, a high recognition accuracy can be obtained, and the type of activation function has little influence on the recognition accuracy. Four stages of the creepage discharge development process were recognized using the OS-ELM algorithm; the trend was the same as that of the characteristic parameters of the entire creepage discharge development process. The recognition accuracy was 91.4%. The algorithm has a high computing speed and high accuracy and can train data in batches. Therefore, it can be widely used in the field of online monitoring and evaluation of electrical equipment status.

Keywords: AC–DC combined voltage; oil–paper insulation; creepage discharge; OS-ELM; pattern recognition

Citation: Jin, F.; Zhang, S.; Zhou, Y. Pattern Recognition of Development Stage of Creepage Discharge of Oil–Paper Insulation under AC–DC Combined Voltage Based on OS-ELM. *Energies* **2021**, *14*, 552. https://doi.org/10.3390/en14030552

Academic Editor: Issouf Fofana
Received: 18 December 2020
Accepted: 18 January 2021
Published: 21 January 2021

1. Introduction

The insulating structure and operation conditions of converter transformers differ from those of conventional AC transformers. The valve-side winding of a converter transformer can withstand AC–DC combined voltage (ADCV) and harmonic voltage components. The international council on large electric systems (CIGRE) statistics report that the failure rate of converter transformers is about twice that of AC transformers and that considerable cases of insulation failure are caused by creepage discharge at the oil–paper interface [1–3]. Thus, the oil–paper interface remains the vulnerable part of a converter transformer insulation and has therefore been studied widely. Studies have been conducted on the creepage discharge characteristics of oil–paper insulation under ADCV using partial discharge measurements and other methods, and beneficial results have been obtained [1–3]. The partial discharge phenomenon can reflect the insulation failure process. Therefore, it can (a) be used as an indicator of the performance degradation of a transformer oil–paper insulation, (b) determine the severity of discharge development by extracting and analyzing partial discharge signals, and (c) be used for effective condition assessment of oil–paper insulation. Therefore, in-depth research on the development characteristics of the creepage discharge of oil–paper insulation under ADCV can effectively reveal the condition of oil–paper insulation. Furthermore, it is an exigent problem to be resolved in the operation and maintenance of ultrahigh voltage DC converter transformers; in addition, it has great theoretical and engineering significance.

Existing studies on the creepage discharge characteristics of oil–paper insulation under ADCV focus mainly on two aspects: (1) they investigate the factors affecting creepage discharge, including different voltage types and electric field components [4], discharge characteristics [5–8], aging [9,10], and other factors. It has been revealed that the temperature rise, aging, and DC electric field components accelerate the creepage discharge development and (2) they intend to mainly find the characteristic parameters and recognition methods that can effectively characterize the creepage discharge development under ADCV. The currently available techniques for the recognition of partial discharge development process mainly include fingerprint recognition of discharge gray image [11], support vector machines (SVM) [12,13], back-propagation neural networks (BPNN) [14], and fuzzy clustering analysis of radar spectra [15]. Nevertheless, only limited research has been conducted on the pattern recognition of the creepage discharge development process of oil–paper insulation under ADCV. For the recognition of the creepage discharge development process, the fundamental requisite is to divide the development stage of the creepage discharge. However, in this method, some subjective components are available only based on the changing trend of the discharge characteristic parameters with time. Moreover, the aforementioned recognition methods can perform pattern recognition only after collecting all data samples. Thus, if new data are added, all the data must be reevaluated, which is time consuming and inefficient. Therefore, to resolve this problem, this study builds a creepage discharge test platform for oil–paper insulation under ADCV. It comprehensively analyzes four two-dimensional characteristic spectra during the development of the creepage discharge. These characteristic spectra reflect the relationship between discharge interval time, discharge amplitude, and number of discharges during the development. Twenty-four characteristic parameters of the creepage discharge development process are extracted from the statistics of the discharge signal during creepage discharge. Based on the variation law of the characteristic parameters of creepage discharge, a method of recognizing the creepage discharge development process is established. Then, based on the online sequential extreme learning machine (OS-ELM) algorithm, the discharge development stage is identified in the creepage discharge development process of the oil–paper insulation.

2. Test System and Method

2.1. Sample and Electrode Model

Karamay 25# transformer oil was used in the test. Prior to the test, the gas volume fraction of the transformer oil was set to be less than 2% and the microwater volume fraction to be less than 10^{-5}, through the process of degassing, drying, and slag removal, which meets the requirements of GB/T7595. The insulating paper was cut into square specimens of dimensions 90 mm × 90 mm × 1 mm during the test and dried at high temperature (100 °C) for 72 h. Then, the insulating paper was impregnated with a pretreated transformer oil for 72 h under a vacuum of approximately 10 Pa and temperature of 60 °C.

Creepage discharge occurs easily at the spacer and barrier positions, where the longitudinal electric field is stronger than the radial electric field. Therefore, in the test, the cylinder-plate electrode model specified by CIGRE Method II and IEC60243 was used, and an oil-immersed insulating paper was placed between the high-voltage electrode and the ground electrode. The electrode structure and specific parameters are illustrated in Figure 1 [1,16–19].

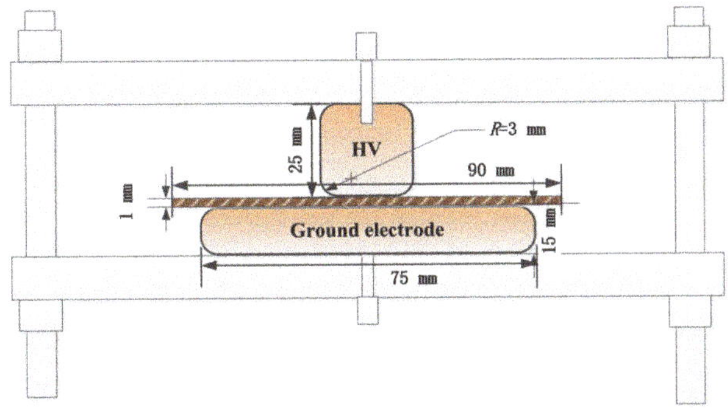

Figure 1. Cylinder-plate electrodes configuration.

2.2. Experimental Platform

The structure of the test system is shown in Figure 2. The AC and DC power used for the test were produced by nonpartial discharge (PD) equipment. The test circuit design was the same as that in Refs. [8,9]. The PD measuring instrument used the pulse current method for measurement; further, the sampling frequency of the system was 5 MHz, and the PD bandwidth was 0.04–1 MHz. An ADCV was applied to the test sample via a protective resistor. When the entire test system was loaded to 70 kV (peak value) without the test sample, no partial discharge signal was present, and the sensitivity of the test system was 0.1 pC.

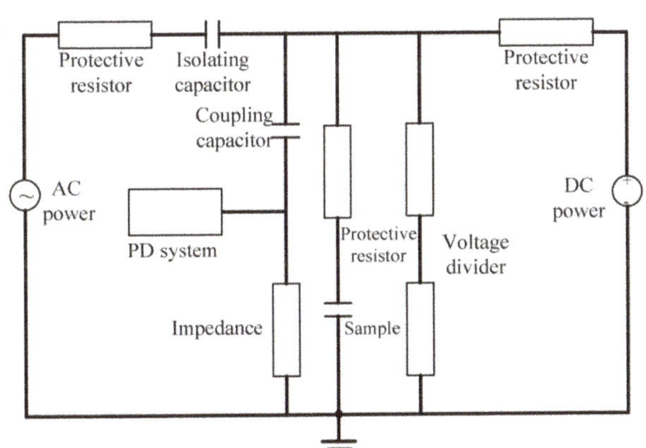

Figure 2. Schematic of the experimental platform.

2.3. Experimental Method

The ratio of the ADCV during the test was selected according to the general criteria and test requirements proposed by IEEE and CIGRE for oil-immersed converter transformer tests. IEEE has specified that the ratio of the ADCV must be 1:1, 1:3, 1:5, or 1:7, where AC takes a valid value and DC takes the average [20]. To obtain a relatively stable and sustainable development of the creepage discharge signal, the ratio of ADCV was selected as 1:1 in this work to study the creepage discharge development process of the oil–paper insulation.

Two methods are currently available for applying voltage, namely, constant voltage method and the step-up method, which have the same effect on the insulation breakdown process. Herein, the constant voltage method was used because it could simulate the actual operation condition of converter transformers better. The voltage ratio was maintained unchanged at 1:1 during the application of voltage, and the boost step was set to 1 kV for both AC and DC voltages. The constant voltage was applied after boosting the voltage to 1.1 times the AC discharge inception voltage, where both the AC and DC voltages were 37 kV.

To accurately record the creepage discharge development process of the oil–paper insulation under ADCV, a set of data was recorded at an interval of 1 min from the beginning of application of the constant voltage, and each recording was performed continuously for 1 min until breakdown occurred. This test procedure was repeated 12 times, and a total of 1210 sets of data were obtained, mainly using which the characteristic parameters of creepage discharge would be subsequently extracted.

3. Pattern Recognition Based on OS-ELM

The conventional extreme learning machine (ELM) transmits all data to the system and then trains and calculates them. However, if new sample data are added, it is required to retrain all sample data, resulting in prolonged learning time and reduced learning rate. To resolve this problem, Liang et al. proposed the aforementioned OS-ELM algorithm, which is an improved version of the ELM approach. In the OS-ELM algorithm, the used data are discarded and are not reused. Further, this algorithm can add new data to the model in individual succession or in batches for training to avoid repeated training of the used sample data; therefore, this algorithm possesses a higher learning rate and higher accuracy [21–25].

As a single-hidden layer feedforward neural network, the OS-ELM algorithm is composed of three layers: input layer, single hidden layer, and the output layer. As shown in Figure 3, the connection weight between the input layer and hidden layer, and the threshold value of the hidden layer neurons are randomly generated, which need not be repeatedly iterated and adjusted in the training process. Compared with the pattern recognition effects of the conventional ELM, SVM, and BPNN methods, the OS-ELM algorithm has the advantages of higher speed, learning rate, generalization performance, and accuracy [26,27]. Therefore, the OS-ELM algorithm was selected in this study for pattern recognition of the creepage discharge development process of the oil–paper insulation under ADCV.

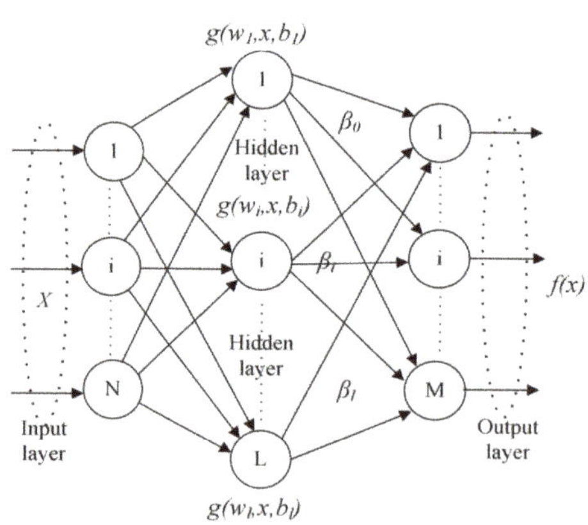

Figure 3. Structure of online sequential extreme learning machine (OS-ELM).

The key part of the OS-ELM algorithm is the learning process of the output weight of the single hidden layer network, and it comprises two stages: (1) the initial stage, wherein the output weight β of the single hidden layer is obtained by training a small number of sample data, and (2) online learning stage, wherein the updated sample data are used to update the output weight β obtained in stage (1).

In stage (1), N sample data (X_i) are present, where $X_i = [x_1, x_2, \ldots, x_n]^T \in R^m$, m denotes the dimension of the sample dataset, $T_i = [t_1, t_2, \ldots, t_n]^T \in R^n$, and n denotes the dimension of the output vector. First, the number of hidden layer neurons L is determined; then, the connection weight w of the input layer and the threshold b of the hidden layer neurons are randomly set; subsequently, the appropriate activation function $g\,(w, x, b)$ is selected; finally, the initial output matrix H_0 of the hidden layer is calculated. According to the ELM theory, the minimum β_0 that can satisfy $||H_0\beta - T_0||$ was calculated, where

$$H_0 = \begin{bmatrix} g(w_1, x_1, b_1) & \cdots & g(w_L, x_1, b_L) \\ \vdots & \ddots & \vdots \\ g(w_1, x_N, b_1) & \cdots & g(w_L, x_N, b_L) \end{bmatrix}_{N \times L} \tag{1}$$

$$T_0 = \begin{bmatrix} t_1 & \cdots & t_L \end{bmatrix}_{M \times L} \tag{2}$$

According to the generalized inverse matrix algorithm, the initial output weight β_0 was calculated, following which the initial learning stage ends, as represented below:

$$\beta_0 = P_0^{-1} H_0^T T_0, P_0 = H_0^T H_0 \tag{3}$$

Subsequently, stage (2) begins. Assuming that N_1 new sample data enter the model, it is desirable to obtain β_1 such that Equation (4) can be the minimum.

$$\left\| \begin{bmatrix} H_0 \\ H_1 \end{bmatrix} \beta - \begin{bmatrix} T_0 \\ T_1 \end{bmatrix} \right\| \tag{4}$$

$$H_0 = \begin{bmatrix} g(w_1, x_{N+1}, b_1) & \cdots & g(w_L, x_{N+1}, b_L) \\ \vdots & \ddots & \vdots \\ g(w_1, x_{N+N_1}, b_1) & \cdots & g(w_L, x_{N+N_1}, b_L) \end{bmatrix}_{N_1 \times L} \tag{5}$$

According to the generalized inverse matrix algorithm, the output weight β_1 is calculated as follows:

$$\beta_1 = P_1^{-1} \begin{bmatrix} H_0 \\ H_1 \end{bmatrix}^T \begin{bmatrix} T_0 \\ T_1 \end{bmatrix}, P_1 = \begin{bmatrix} H_0 \\ H_1 \end{bmatrix}^T \begin{bmatrix} T_0 \\ T_1 \end{bmatrix} \tag{6}$$

The output weight β_{k+1} can be obtained using the Woodbury formula and the recursive relationship as follows:

$$\beta_{k+1} = \beta_k + P_{k+1}^{-1} H_{k+1}^T (T_{k+1} - H_{k+1}\beta_k) P_{k+1}^{-1} = P_k^{-1} - P_k^{-1} H_{k+1}^T \left(I + H_{k+1} P_k^{-1} H_{k+1}^T \right)^{-1} \times H_{k+1} P_k^{-1} \tag{7}$$

Equation (7) provides the basic principle of the OS-ELM algorithm, and weight β of the single-hidden layer neural network can be adjusted through two stages. In the OS-ELM training process, if the data obtained in stage (1) are the only data, i.e., no updated sample data are present, OS-ELM is the same as the standard ELM algorithm. Therefore, the OS-ELM algorithm can be understood as a special ELM algorithm in which all training data arrive simultaneously.

4. Experimental Results and Discussion

4.1. Creepage Discharge Development Process of Oil–Paper Insulation under ADCV

Because the partial discharge of the oil–paper insulation under ADCV is more random than that under AC voltage, the signal representation method for the former significantly differs from that for the latter. In this study, the interval time Δt of two adjacent discharges is used instead of the phase ϕ in the partial discharge under AC voltage. Then, the statistical characteristic spectra are formed by combining the interval time Δt, partial discharge magnitude Q, and number of discharges n. Time-resolved pulse sequence analysis (TPRSA) was used to study the creepage discharge development process of oil–paper insulation under ADCV [8,9].

In this study, four sets of data in the initial stage, two middle stages, and the adjacent end stage of one test were selected for statistical analysis, and the Q-Δt_{pre}-n three-dimensional characteristic spectra are formed—here, Δt_{pre} denotes the time interval between the current discharge and the previous discharge. Figure 4 shows the variation law of the Q-Δt_{pre}-n characteristic spectra of the abovementioned four sets of data during the creepage discharge development process.

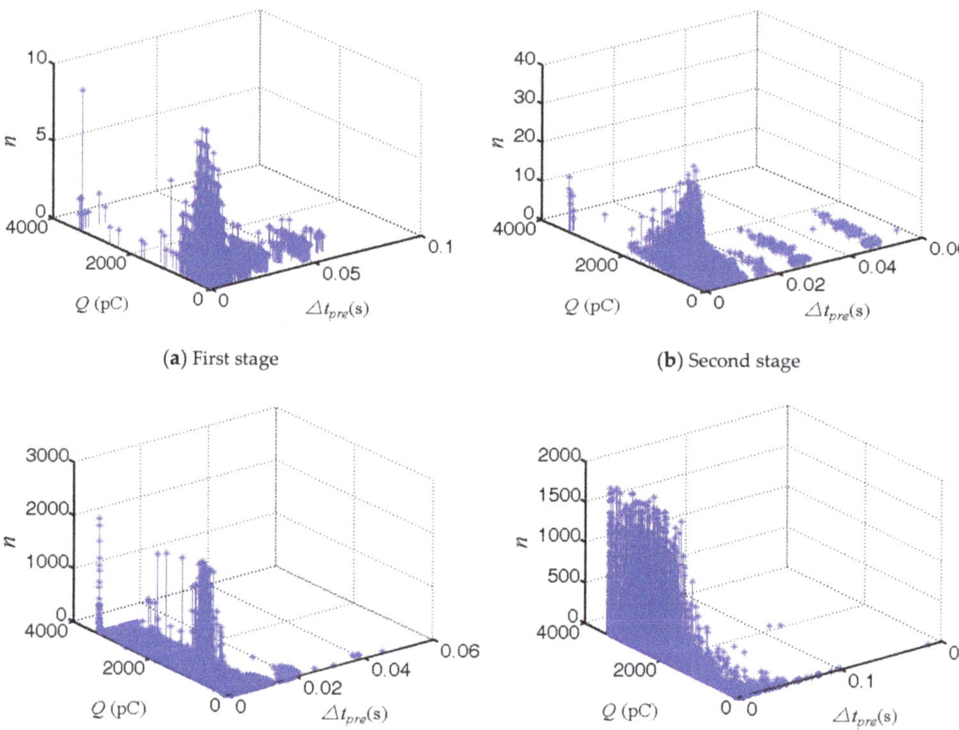

(**a**) First stage (**b**) Second stage

(**c**) Third stage (**d**) Fourth stage

Figure 4. Q-Δt_{pre}-n characteristic spectra of four sets of data during the creepage discharge development process.

Figure 4 clearly indicates the existence of significant differences between the characteristic spectra at each stage. In the first stage (Figure 4a), n is small, Δt_{pre} is large, and the distribution range is 0.01–0.06 s. Further, Q is mainly concentrated at approximately 400 pC, and the distribution of Q is highly scattered. In the second stage (Figure 4b), n increases gradually, and Q is mainly distributed in the region less than 1300 pC; however, Q greater than 2000 pC appears gradually. Further, Δt_{pre} is gradually shortened and is distributed within the range of 0.04 s. In the third stage (Figure 4c), n increases significantly compared with that in the second stage. Further, Q is mainly distributed around 1700 pC but gradually enters the range of 2000–3000 pC, and Δt_{pre} is mainly concentrated in the range of 0.02 s. In the fourth stage (Figure 4d), n continues to increase gradually; further, Q is mainly concentrated in the region greater than 2000 PC, and Δt_{pre} is mainly concentrated in the range of 0.01 s.

Although the TPRSA spectra can reflect the time domain distribution of the creepage discharge development process better, directly using these spectra for pattern recognition is difficult because they contain large amounts of information. Therefore, to alleviate the difficulties in the recognition of the creepage discharge development process, the original data must be transformed, the dimension must be reduced, and the characteristic parameters reflecting the creepage discharge development process must then be extracted.

The characteristic spectra, such as n-Q, n-Δt, Δt_{pre}-Q, and Q_{mean}-Δt_{pre}, were obtained by dimensionality reduction. The Δt_{pre}-Q two-dimensional (2D) characteristic spectra after dimensionality reduction corresponding to each stage of Figure 4 are shown in Figure 5. From the characteristic spectra of each stage shown in Figure 5, the variation law between Q and Δt_{pre} in each stage of the creepage discharge development process can be obtained more clearly. Δt_{pre} is relatively dispersed from the initial stage and the development stage, the critical breakdown is relatively concentrated, and Δt_{pre} is gradually reduced. Further, Q transforms from being a small value initially to a large value at the critical breakdown—its distribution transforms from being a dispersed distribution initially to a concentrated distribution at the critical breakdown. Moreover, it can be used to analyze the n-Δt and Q_{mean}-Δt_{pre} characteristic spectra in a reduced dimension, which is not discussed in detail in this paper. A comprehensive analysis of the variation law between the discharge characteristic parameters in the abovementioned characteristic spectra can effectively reflect the creepage discharge development process. Accordingly, the characteristic parameters were extracted from the abovementioned four characteristic spectra and were used to analyze the creepage discharge development process.

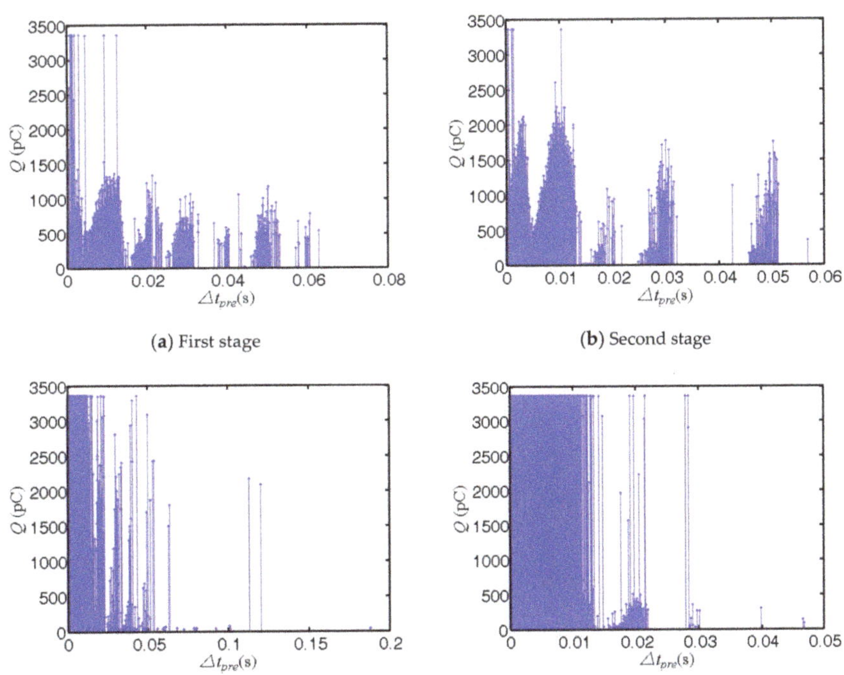

(**a**) First stage (**b**) Second stage

(**c**) Third stage (**d**) Fourth stage

Figure 5. Δt_{pre}-Q two-dimensional characteristic spectra after dimensionality reduction corresponding to each stage of Figure 4.

4.2. Pattern Recognition of Creepage Discharge Development Process Based on OS-ELM

For the n-Q, n-Δt, Δt_{pre}-Q, and Q_{mean}-Δt_{pre} characteristic spectra, their skewness (S_k), kurtosis (K_u), asymmetry (A_{sy}), cross-correlation coefficient (C_c), and the Weibull distribution parameters α and β were extracted. Twenty-four characteristic parameters were extracted from each set of data from the aforementioned four 2D characteristic spectra (Table 1).

Table 1. Statistical characteristic parameters of creepage discharge.

Characteristic Spectra	Characteristic Parameters					
	S_k	K_u	A_{sy}	C_c	α	β
n-Q	√	√	√	√	√	√
n-Δt	√	√	√	√	√	√
Q-Δt_{pre}	√	√	√	√	√	√
Q_{mean}-Δt_{pre}	√	√	√	√	√	√

Figure 6 shows the variation law of each characteristic parameter in the Δt_{pre}-Q characteristic spectra during the entire creepage discharge development process. According to the variation law of the characteristic parameters, such as S_k, K_u, and β, shown in the figure, the creepage discharge development process can be divided into four stages.

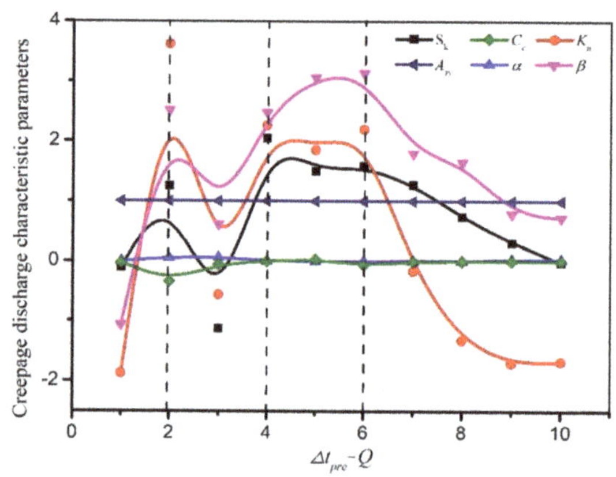

Figure 6. Variation law of characteristic parameters in the Δt_{pre}-Q characteristic spectra during the entire creepage discharge development process.

In the first stage, the characteristic parameters rise rapidly. The changes in the characteristic parameters indicate that Q increases gradually at this stage, Δt_{pre} shortens gradually, and the creepage discharge enters the initial stage. In the second stage, the parameters initially exhibit a decreasing trend and then rise slowly. This indicates that Q initially rises and then stabilizes, Δt_{pre} is gradually shortened, and the creepage discharge enters the development stage. The third stage maintains the current state, and creepage discharge continues to develop. In this stage, Q increases and tends to stabilize, Δt_{pre} is stable, and the creepage discharge enters a stable state. The fourth stage exhibits a trend of gradual decline, and the creepage discharge enters the critical breakdown stage. The variation law of the characteristic parameters of the Δt_{pre}-Q discharge characteristic spectra can aid in better analyzing the relationship between Q and Δt_{pre} during the creepage discharge development process and also in better understanding the physical development process of creepage discharge.

In this study, the OS-ELM algorithm code was written using MATLAB. The characteristic parameters extracted during the entire creepage discharge process were used as sample sets for the recognition of the creepage discharge development process.

First, the influence of the number of hidden layer neurons, type of activation function, and size of the sample data on the training accuracy of the initial stage were analyzed. The corresponding results can be used to rationally select the parameters of the OS-ELM algorithm. Figure 7 shows the relationships among the training accuracy of the initial stage, activation function type, and number of hidden layer neurons. The number of training samples is 750. The horizontal axis is the number of hidden layer neurons, and the vertical axis is the training accuracy under different activation functions. The figure shows that the difference between the accuracies of the same neuron number is small under different activation functions. This indicates that the activation function has less influence on the accuracy under creepage discharge sample. The training accuracy improves gradually as the number of hidden layer neurons increases and tends to stabilize when the number of hidden layer neurons becomes close to the number of training samples. Therefore, the relationship between the sample size, number of hidden layer neurons, and activation function type should be collectively considered to select the algorithm parameters for realizing a better pattern recognition.

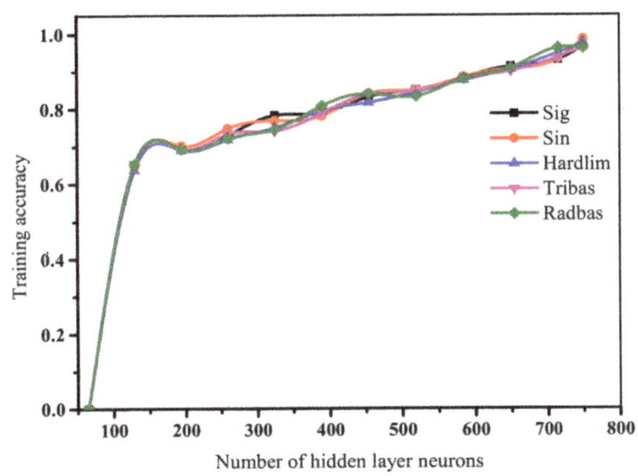

Figure 7. Relationship between activation function, number of hidden layer neurons, and training accuracy.

In this study, the sigmoid function was selected as the activation function, and the number of hidden layer neurons was 650. There were 12 groups of experimental data, of which seven were randomly selected as sample training sets to train the OS-ELM, ELM, SVM, and BPNN algorithms. The remaining five groups were used as sample test sets for recognizing the creepage discharge development process, and a confusion matrix was created based on the recognition results. The confusion matrix is shown in Figure 8. The figure displays that the entire creepage discharge development process can be recognized by four stages, namely, C1, C2, C3, and C4, by the OS-ELM algorithm. In the confusion matrix (Figure 8), the horizontal and vertical directions represent four states: creepage discharge initiation (C1), creepage discharge development (C2), creepage discharge acceleration (C3), and creepage discharge critical flashover (C4).

In this matrix, each row represents the actual state data, and each column represents the identified state data. The elements on the main diagonal represent the correct number of the creepage discharge development stage identified based on the test data, and the nonmain diagonal elements represent the misidentified number. In total, 500 sets of data were used as test samples. In the first row, the total number of states is 120, and the four states are 113, 5, 2, and 0. These data indicate that the total number of real states should be 120. After the diagnosis of the algorithm, the number of diagnosed C1 is 113, and the number of misdiagnosed C2, C3, and C4 is 5, 2, and 0, respectively. The last column represents the recall rate, which reflects the ratio of the correct number of samples to the actual number of samples. The last row of elements represents the precision rate, which reflects the ratio of the predicted number of correct samples to the number of predicted samples. The location at which the recall rate and accuracy rate intersect represents the recognition accuracy, which reflects the ratio of the correct number to the total number of samples. This shows that both the recall rate and the precision of the creepage discharge development process are significantly high (>88%) based on the OS-ELM algorithm, and the sample recognition accuracy is 91.4%. The recognition accuracy of ELM is lower than that of OS-ELM, and the recognition accuracy of SVM is 85.6%, which is lower than ELM.

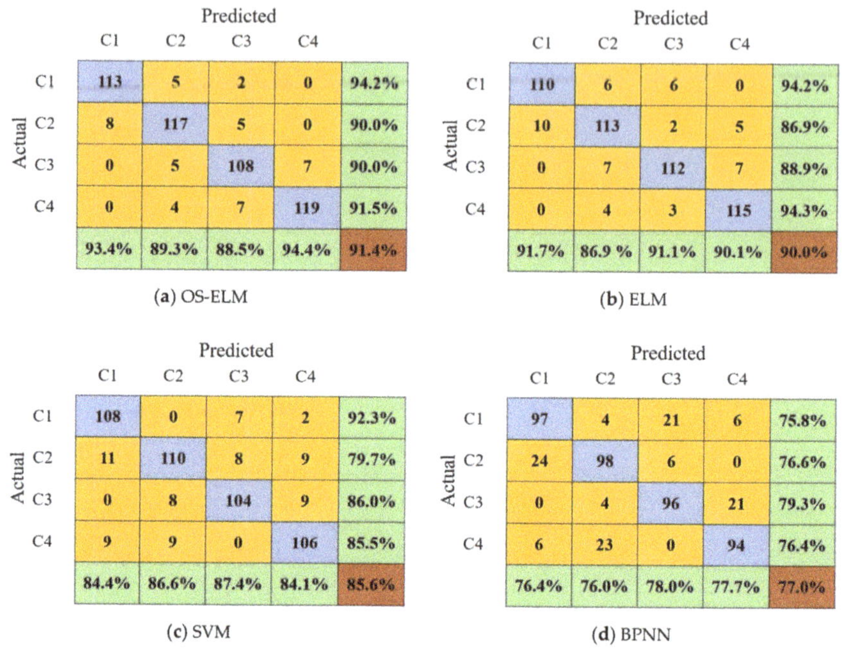

Figure 8. Pattern recognition result of four algorithms.

In contrast, the recognition accuracy of BPNN is 77%, which is much lower than that of OS-ELM. The recognition result of OS-ELM is the same as the result of dividing the creepage discharge development process according to the variation law of the characteristic parameters of the creepage discharge development process depicted in Figure 6. Based on this method, the OS-ELM algorithm can theoretically support the pattern recognition of the creepage discharge development process of oil–paper insulation.

In the ELM algorithm, the weight matrix and threshold from the input layer to the hidden layer are generated randomly; this avoids the iterative process of repeated training and threshold modification in SVM and BPNN. By contrast, the weight matrix from the hidden layer to the output layer uses the least squares method to find the optimal solution to minimize the loss function; thus, it does not require an iterative process. Therefore, the OS-ELM algorithm can shorten the training adjustment time and has a very fast training speed. In the training time comparison part, 200 out of 1210 sets of experimental data were randomly selected as sample training sets to train four algorithms. OS-ELM and ELM algorithms selected the same number of samples and hidden layers; SVM algorithm cross-validation obtains the best parameter penalty factor and RBF kernel function variance. The training time of the four algorithms is shown in Table 2. Both OS-ELM and ELM have short training times, while BPNN requires a longer time. SVM adopts a cross-validation method; therefore, its training time is the longest. It can be observed that, combined with the experimental results, the OS-ELM algorithm is better than the other three methods in terms of recognition rate and training speed, and the OS-ELM algorithm has a significantly short training time.

Table 2. Comparison of the training time of the four algorithms.

Algorithm	Training Time/s
OS-ELM	0.0532
ELM	0.0532
SVM	574.1774
BPNN	264.7327

The recognition accuracy of OS-ELM has improved, compared with that of the other methods, such as SVM, BPNN, and ELM. The OS-ELM algorithm can separately train the data arriving in batches, avoiding repeated training of all data, thereby greatly reducing the time required for network training adjustment and effectively improving the training speed. Owing to the abovementioned advantages, the OS-ELM algorithm is applicable in the field of electrical equipment condition monitoring and evaluation engineering.

5. Conclusions

A creepage discharge test platform for oil–paper insulation under ADCV was developed in this study for the pattern recognition of the creepage discharge development process. According to the variation law of the characteristic parameters of the creepage discharge development process of the oil–paper insulation, the process was divided into four stages. In the first stage, the characteristic parameters increased rapidly; in the second stage, they decreased first and then gradually increased; in the third stage, they stabilized, and then gradually decreased in the fourth stage.

For the creepage discharge data sample type, the hidden layer activation function type had only a slight effect on the training accuracy. In the OS-ELM algorithm parameter selection process, when the sample set size was close to the number of hidden layer neurons, the training rate improved and a higher training accuracy rate was obtained.

The sample recognition accuracy was 91.4% in this study. Furthermore, the recall rate and the precision of the creepage discharge development process were both very high (>88%). The OS-ELM algorithm has an advantage that it can train samples in batches and, therefore, has good engineering application prospects in the field of electrical equipment condition monitoring and evaluation.

Author Contributions: F.J. conceived and designed the experiments, S.Z. optimized the algorithm; Y.Z. grasped the research direction. F.J. wrote—review and edit the paper. All authors have read and agreed to the published version of the manuscript.

Funding: This research was funded by the Ministry of Education Chunhui Plan (QDCH2018004), National Natural Science Foundation of China (51865049), and National Basic Research Program of China (973 Program) (2011CB209400). The authors gratefully acknowledge these supports.

Institutional Review Board Statement: Not applicable.

Informed Consent Statement: Not applicable.

Data Availability Statement: Not applicable.

Conflicts of Interest: The authors declare no conflict of interest.

References

1. Jin, F.B.; Zhang, S.J.; Zhou, Y.X.; Ma, H.; Tapan, K.S. Pattern Recognition of Development stage of Creepage Discharge of Oil–Paper Insulation Under AC–DC Combined Voltage Based on OS-ELM. In Proceedings of the 2020 IEEE International Conference on High Voltage Engineering and Application (ICHVE 2020), Beijing, China, 6–10 September 2020.
2. China South Power Grid. *Typical Fault Analysis of HVDC Transmission System Equipment*; China Electric Power Press: Beijing, China, 2009; pp. 15–173. (In Chinese)
3. Zhou, Y.X.; Sha, Y.C.; Li, J.Z.; Wang, J.Y. Partial Discharge Characteristics in Oil-paper Insulation Under Combined AC-DC Voltage. *IEEE Trans. Dielectr. Electr. Insul.* **2014**, *21*, 1529–1539.
4. Li, J.H.; Zhang, L.; Han, X.T.; Yao, X.; Li, Y.M. PD Detection and Analysis of Oil-Pressboard Insulation Under Pulsed DC Voltage. *IEEE Trans. Dielectr. Electr. Insul.* **2017**, *24*, 324–330. [CrossRef]

5. Wei, Z.; Li, C.R. Discharge of Oil-pressboard Insulation in AC-DC Composite Field: Phenomenon and Characteristics. *IEEE Trans. Dielectr. Electr. Insul.* **2016**, *23*, 237–245.
6. Wu, H.; Li, C.R.; Qi, B.; Zhao, X.L.; Lv, J.Z.; Zhao, L.J. The Electric Field Distribution in Oil-Paper Insulation Under Combined AC-DC Voltage. In Proceedings of the IEEE International Conference on Condition Monitoring and Diagnosis (CMD 2012), Bali, Indonesia, 23–27 September 2012; pp. 1097–1101.
7. Zhou, Y.X.; Huang, M.; Chen, W.J.; Lu, L.C.; Jin, F.B.; Huang, J.W. Space Charge Behavior Evolution with Thermal Aging of Oil-paper Insulation. *IEEE Trans. Dielectr. Electr. Insul.* **2015**, *22*, 1381–1388. [CrossRef]
8. Zhou, Y.X.; Jin, F.B.; Sha, Y.C.; Huang, M.; Huang, J.W.; Liu, Z.H.; Lu, L.C. Effects of DC Prestressing on Partial Discharge in Oil-impregnated Pressboard Insulation. *IEEE Trans. Dielectr. Electr. Insul.* **2016**, *23*, 4668. [CrossRef]
9. Zhou, Y.X.; Jin, F.B.; Huang, M.; Le, T.H.; Huang, J.W.; Liu, Z.H.; Lu, L.C. Effects of Thermal Aging on Creepage Discharge in Oil-Impregnated Pressboard Under Combined AC-DC Voltage. *IEEE Trans. Dielectr. Electr. Insul.* **2015**, *22*, 2737–2746. [CrossRef]
10. Li, J.; He, Z.M.; Grzybowski, S. Electrical Aging Lifetime Model of Oil-Impregnated Paper Inder Pulsating DC Voltage Influenced by Temperature Insulation. *IEEE Trans. Dielectr. Electr. Insul.* **2013**, *20*, 1992–1997. [CrossRef]
11. Li, S.M.; Si, W.; Li, Q.Q. Partition and Recognition of Partial Discharge Development Stages in Oil-Pressboard Insulation with Needle-Plate Electrodes Under Combined AC-DC Voltage Stress. *IEEE Trans. Dielectr. Electr. Insul.* **2017**, *24*, 1781–1793. [CrossRef]
12. Wang, K.; Liao, R.J.; Yang, L.; Li, J. Optimal Features Selected by NSGA-II for Partial Discharge Pulses Separation Based on Time-Frequency Representation and Matrix Decomposition. *IEEE Trans. Dielectr. Electr. Insul.* **2013**, *20*, 825–838. [CrossRef]
13. Ma, Z.Q.; Yang, X.; Lin, C.Y.; Zhou, D.; Chen, J.G. Partial Discharge Developing Stages Identification Based on Cluster-Hierarchical Decision SVM in Oil-Paper Insulation. In Proceedings of the 2016 IEEE International Conference on High Voltage Engineering and Application (ICHVE 2016), Chengdu, China, 19–22 September 2016.
14. Wang, K.; Li, J.Z.; Zhang, S.Q.; Gao, F.; Grzybowski, S. A New Image-Oriented Feature Extraction Method for Partial Discharges. *IEEE Trans. Dielectr. Electr. Insul.* **2015**, *25*, 1015–1024. [CrossRef]
15. Li, S.M.; Li, Q.Q.; Si, W.; Yao, J.Y. An Image-Oriented Recognition Method for PD Development Stage in Oil-Pressboard Insulation with Needle-Plate Model Under AC-DC Mixed Voltage. In Proceedings of the International Conference on Electrical Materials and Power Equipment (ICEMPE 2017), Xi'an, China, 14–17 May 2017; pp. 275–282.
16. International Electrotechnical Commission. *Specification for Pressboard and Press Paper for Electrical Purposes*; IEC 60641; IEC: Geneva, Switzerland, 2008.
17. International Electrotechnical Commission. *Electrical Strength of Insulating Materials Test Methods*; IEC 60243; IEC: Geneva, Switzerland, 1998.
18. International Electrotechnical Commission. *High Voltage Test Technique-Partial Discharge Measurement*; IEC 60270; IEC: Geneva, Switzerland, 2000.
19. Kennedy, W.M. Recommended Dielectric Tests and Test Procedures for Converter Transformers and Smoothing. *IEEE Trans. Power. Deliv.* **1986**, *3*, 161–166. [CrossRef]
20. IEEE Std C57. *129TM IEEE Standard for General Requirements and Test Code for Oil-Immersed HVDC Converter Transformers*; IEEE: Piscataway, NJ, USA, 2008.
21. Huang, G.B.; Zhu, Q.Y.; Siew, C.K. Extreme Learning Machine: A New Learning Scheme of Feedforward Neural Networks. In Proceedings of the International Joint Conference on Neural Networks (IJCNN 2004), Budapest, Hungary, 25–29 July 2004; pp. 985–990.
22. Huang, G.B.; Zhu, Q.Y.; Siew, C.K. Extreme Learning Marching: Theory and Application. *Neurocomputing* **2006**, *70*, 489–501. [CrossRef]
23. Zong, W.E.; Huang, G.B.; Chen, Y.Q. Weighted Extreme Learning Machine for Imbalance Learning. *Neurocomputing* **2013**, *101*, 229–242. [CrossRef]
24. Liang, N.Y.; Huang, G.B.; Saratchandran, P.; Sundararajan, N. A Fast and Accurate Online Sequential Learning Algorithm for Feedforward Networks. *IEEE Trans. Neural Netw.* **2006**, *17*, 1411–1423. [CrossRef] [PubMed]
25. Huang, G.B.; Chen, L.; Siew, C.K. Universal Approximation Using Incremental Constructive Feedforward Networks with Random Hidden Nodes. *IEEE Trans. Neural Netw.* **2006**, *17*, 879–892. [CrossRef] [PubMed]
26. Zhang, Q.Q.; Song, H.; Yong, J.; Sheng, G.H. Online Sequential Extreme Learning Machine for Partial Discharge Pattern Recognition of Transformer. In Proceedings of the IEEE Power Engineering Society Transmission and Distribution Conference, Denver, CO, USA, 16–19 April 2018.
27. He, J.; Liu, C.H.; Tang, Q.H.; Zhang, C.H.; Cao, M.; Xu, X. Partial Discharge Pattern Recognition Algorithm Based on Sparse Self-coding and Extreme Learning Machine. In Proceedings of the 2018 2nd IEEE Conference on Energy Internet and Energy System Integration (EI2 2018), Beijing, China, 20–22 October 2018.

energies

Review

Recent Development of Two Alternative Gases to SF_6 for High Voltage Electrical Power Applications †

John Owens [1,*], Ang Xiao [1], Jason Bonk [1], Michael DeLorme [1] and Agnes Zhang [2]

[1] 3M Company, Saint Paul, MN 55144-1000, USA; axiao3@mmm.com (A.X.); jbonk@mmm.com (J.B.); mdelorme@mmm.com (M.D.)
[2] 3M China Ltd., Shanghai 200336, China; azhang11@mmm.com
* Correspondence: jgowens@mmm.com

† This paper is an extended version of our paper published in 2020 IEEE International Conference on High Voltage Engineering and Application (ICHVE), Beijing, China, 6–10 September 2020; pp. 1–4.

Abstract: For many years, SF_6 has been the preferred dielectric medium in electrical power applications, particularly in high voltage gas-insulated equipment. However, with the recognition that SF_6 has an extremely long atmospheric lifetime and very high global warming potential, governments have pursued emission reductions from gas-filled equipment. The electrical power industry has responded to this environmental challenge applying SF_6-free technologies to an expanding range of applications which have traditionally used SF_6, including gas-insulated switchgear, gas-insulated circuit breakers and gas-insulated lines or bus bars. Some of these SF_6-free solutions include gas mixtures containing fluorinated compounds that have low climate impact, among them, a fluoronitrile and a fluoroketone developed as 3M™ Novec™ 4710 Insulating Gas and 3M™ Novec™ 5110 Insulating Gas, respectively. Both fluoronitrile and fluoroketone mixtures are successfully used in gas-insulated equipment currently operating on the grid where they reduce greenhouse gas emissions by more than 99% versus SF_6. This paper reviews these leading components of alternative-gas mixtures with updates on the performance, safety and environmental profiles in electrical power applications.

Keywords: sulfur hexafluoride; SF_6; insulation; dielectric medium; SF_6-free; SF_6-alternative; fluoroketone; fluoronitrile

Citation: Owens, J.; Xiao, A.; Bonk, J.; DeLorme, M.; Zhang, A. Recent Development of Two Alternative Gases to SF_6 for High Voltage Electrical Power Applications. *Energies* **2021**, *14*, 5051. https://doi.org/10.3390/en14165051

Academic Editors: Pawel Rozga and Issouf Fofana

Received: 24 June 2021
Accepted: 12 August 2021
Published: 17 August 2021

1. Introduction

Sulfur hexafluoride, SF_6, has been a critical component in high voltage applications for several decades with the installed base of gas-filled equipment continuing to grow. Its combination of chemical, electrical, and physical properties has made SF_6 the preferred dielectric medium in gas-insulated switchgear (GIS), gas circuit breakers (GCB) and gas insulated lines (GIL). While much of the equipment operating on the electrical grid today depends upon the use of SF_6, the industry has been searching for an alternative due to environmental concerns over the properties of this highly stable insulating gas. A long atmospheric lifetime of 3200 years results in a global warming potential (GWP) for SF_6 of 23,500, making it the most potent greenhouse gas identified to date.

Identification of viable alternatives to SF_6 is complicated by the unique combination of properties required in dielectric applications. Unfortunately, the very properties that make SF_6 an ideal insulating gas, namely chemical inertness, are the same properties that make it exceptionally long lived in the atmosphere. Therefore, any replacement of SF_6 as an insulating gas must implicitly have some form of reactivity to facilitate degradation in the atmosphere and overcome the environmental concerns. The materials also need to be nonflammable and low enough in toxicity to allow for safe handling using practices similar to those currently used within the industry. Alternatives certainly need to have very high dielectric strength, providing performance as close to SF_6 as possible. Since the gas-filled equipment will be used in a variety of conditions, the materials must remain gaseous over the expected operating temperatures of these systems. The dielectric medium must also be stable over the working life of this equipment without contributing to corrosion or other adverse effects on the device. Most importantly, to be sustainable alternatives, new compounds need to have acceptable combinations of environmental properties, including no ozone depletion potential and significantly reduce the greenhouse gas emissions from these applications compared to SF_6, since this is the principal reason for transitioning to new technology.

Two compounds, a fluoronitrile and a fluoroketone, were found to combine the requisite properties for electric power applications. They both have been shown to function as a key dielectric component in insulating gas mixtures while providing significantly lower climate impact. As a result, the electric power industry has begun implementing SF_6-alternative gas mixtures based upon these compounds over the last several years [1–4]. The fluoronitrile and fluoroketone are recognized within the electric power industry as 3M™ Novec™ 4710 Insulating Gas and 3M™ Novec™ 5110 Insulating Gas, respectively [5]. In some publications they are referred to as C4-FN or C5-FK or even simply C4 or C5. For the duration of this paper, these compounds will be identified as Novec 5110 gas and Novec 4710 gas. This paper is a review of these components in SF_6-alternative gas mixtures covering material properties and performance in dielectric applications as well as safety and environmental considerations.

2. Performance in Dielectric Applications

2.1. Properties of Pure Novec Insulating Gases

The Novec Insulating Gases exhibit several physical properties that are similar to SF_6. They are highly fluorinated, nonflammable, high density gases with extremely low freezing points and excellent dielectric properties. At any given pressure, the pure Novec gases display dielectric breakdown voltages that are superior to that of SF_6 as shown in Figure 1. Table 1 provides a summary of these key physical properties. It also lists the environmental attributes of each gas. Like SF_6, the Novec Insulating Gases are non-ozone depleting since they do not affect stratospheric ozone leading to an ozone depletion potential (ODP) of zero. However, their measurably shorter atmospheric lifetimes lead to significantly lower GWPs. As will be shown below, the shorter atmospheric lifetimes are also the key attribute that enables substantial reductions in the overall greenhouse gas (GHG) emissions resulting from gas-insulated equipment using these alternatives.

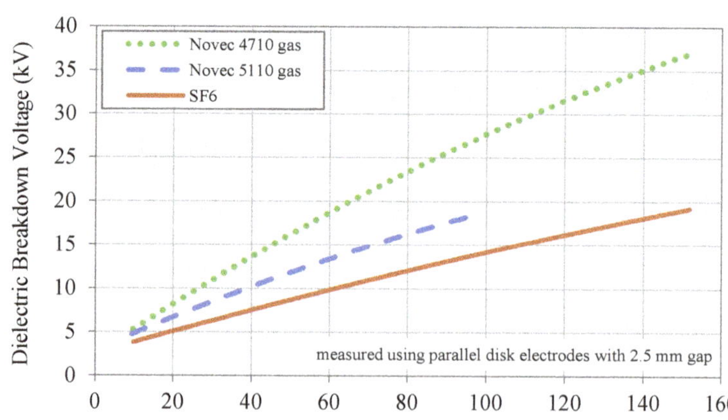

Figure 1. Dielectric breakdown voltage of pure gases [5].

Table 1. Alternative gas properties compared to SF_6 [5].

Property at 1 Bar, 25 °C	Sulfur Hexafluoride	Novec 4710	Novec 5110
Chemical Formula	SF_6	$(CF_3)_2CFCN$	$(CF_3)_2CFC(O)CF_3$
Molecular Weight	146	195	266
Boiling Point (°C)	−63.9 [a]	−5	27
Vapor Pressure (kPa)	2372	297	94
Freezing Point (°C)	−50.8	−118	−110
Flash Point (°C)	none	none	none
Gas Density (kg/m^3)	5.9	7.9	10.7
Thermal Conductivity (W/m·K)	0.013	0.025	0.004
Breakdown Voltage (kV) 2.5 mm gap with parallel electrodes	14.0	27.5	18.4 [b]
Atmospheric Lifetime (year)	3200	30	0.04 (15 days)
Ozone Depletion Potential	zero	zero	zero
GWP (100-year ITH)	23,500	2100	<1

[a] Sublimation Point, [b] at saturation.

2.2. Properties of Gas Mixtures

Due to their higher boiling points and corresponding lower vapor pressures, the Novec gases are used in gaseous mixtures rather than as pure materials. Dilution in gaseous mixtures allows the equipment to operate at temperatures well below the boiling points of these materials without condensation. Once gases form a homogeneous mixture, they do not physically separate unless liquefied by cooling below the condensation temperature or compressed to very high pressures. Similarly, although gas density will vary with height in a vertical column, the mixture does not separate over time with the higher molecular weight components concentrating at lower elevations. Figure 2 shows the change in gas density as a function of height in a column of gas for several gases. The pressure exerted by the column of gas above any point creates a greater density compared to higher elevations. Thus, the density of a gas decreases at higher elevations. Larger variations occur as the molecular weight of the gas increases since the greater mass produces higher pressures at the lower elevations. However, the concentrations of individual components in a gas mixture do not change with height. The pressure exerted by the column of gas mixture above a molecule of any component is the same, resulting from the density of the gas mixture above it rather than any individual pure gas. As a result, all components of a mixture are exposed to the same gravitational force and pressure. Therefore, no driving force is created to cause a separation. A similar conclusion was reached in the 1982 EPRI Report EL-2620 [6]: "In the absence of condensation, a gas mixture will not separate into its component gases over a short or long period of time even when the molecular weights of the component gases are markedly different." Accordingly, gas separation has not been observed experimentally [5–7]. For example, a gas mixture containing Novec 4710 gas and CO_2 was stored in a 2-m vertical tube at −15 °C for 6 months with no change in composition detected over the height of the tube [7].

Table 2 shows a comparison of representative gas mixtures that are used in high voltage systems relative to pure SF_6. The dielectric breakdown voltage of a gas mixture varies with the concentration of Novec gas as well as the total pressure of the mixture. As shown in Figure 3, it is possible to compensate for the lower dielectric strength of a dilute gas mixture by increasing the total gas pressure used within the system. In fact, that is the strategy often employed by manufacturers of gas-insulated equipment. Numerous systems using Novec gas mixtures are currently operating on the grid, including installations of GIS, GCB and GIL. These systems have been designed to deliver performance comparable to similarly rated SF_6 equipment, [8,9].

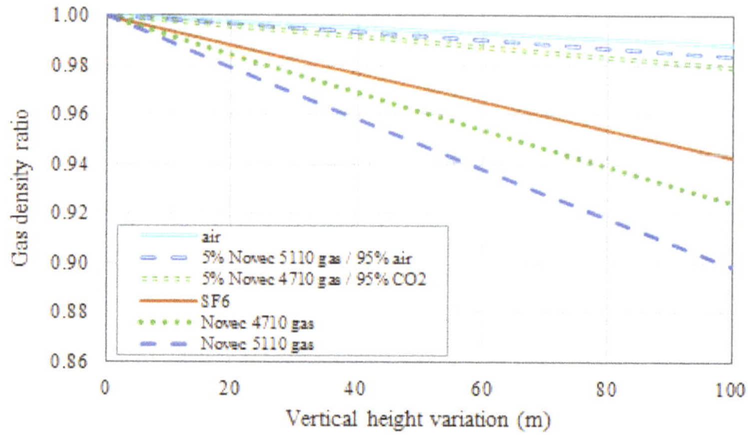

Figure 2. Variation in gas density as a function of gas column height. Gas mixtures are described in Table 2.

Table 2. Gas mixture properties compared to SF_6.

Gas Formulation (mole%)	100% SF_6	5% Novec 4710/95% CO_2	5% Novec 5110/95% Air
Typical GIS Pressure (bar)	4	6	6.5
Gas Density @ 25 °C (kg/m^3)	24.75	12.48	10.67
Condensation temperature (°C)	−38	−27	0
Dielectric breakdown voltage relative to SF_6	—	~1	~1

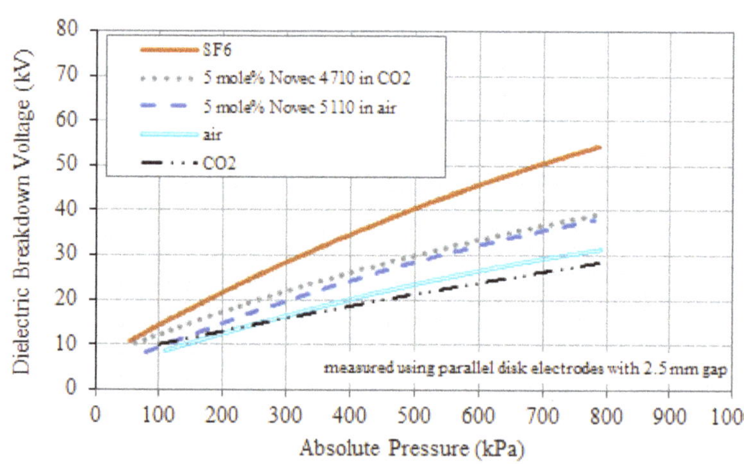

Figure 3. Dielectric breakdown voltage of gas mixtures compared to pure SF_6.

3. Safety Considerations

A key aspect for use of any SF_6-alternative technology is the ability to use it safely within gas-filled equipment. Personnel may come into contact with an insulating gas through handling during initial filling and maintenance of the equipment, leakage during normal operation and when decommissioning the system. The safety of Novec Insulating Gases has been evaluated through a series of toxicological studies [10,11]. These 3M-sponsored studies were approved by the laboratories' Institutional Animal Care and

Use Committees and animal care complied with all applicable national and local regulations. All toxicological studies that followed OECD guidelines (Organization for Economic Co-operation and Development) were performed under GLP conditions (Good Laboratory Practice). Both gases demonstrated low acute toxicity hazard as reflected in their Globally Harmonized System (GHS) classification of Category 4 or higher. Both Novec gases also presented a low hazard profile in repeat-dose inhalation toxicity studies where irritant-associated effects were noted in tissues at the portal of entry (nose and mouth), the respiratory and gastrointestinal tracts, at the highest exposure concentrations. In addition, both gases have demonstrated no genotoxicity potential where Novec 4710 gas was found to be not mutagenic in both in vitro and in vivo assays and Novec 5110 gas was shown to be not mutagenic in an in vitro genotoxicity assay. While Novec 5110 gas has not yet been evaluated in an in vivo study the next nearest homologue (an analogous fluoroketone with chain length one carbon longer) has been shown to be not mutagenic through in vivo tests. Thus, based on all available data the weight of evidence indicates that the both Novec Insulating Gases would not be classified as CMR hazards (carcinogenicity, mutagenicity, reproductive toxicity).

As an additional step, the assessment of the available data and associated hazard classification recommendation for Novec 4710 gas was confirmed and validated in an independent, third-party assessment [12]. This technical assessment confirmed that "Based on the available data, no self-classification for the CMR hazard categories is currently warranted or anticipated in the future." A summary of the key results for both Novec gases is shown in Table 3.

Table 3. Key toxicological results on pure Novec Insulating Gases.

Novec 4710 Gas	Novec 5110 Gas
Low acute inhalation toxicity (4-h LC_{50} > 10,000, <15,000 ppmv)	Low acute inhalation toxicity (4-h LC_{50} > 148, <213 mg/L) [1]
Low repeated-dose inhalation toxicity (based upon 28-day study)	Low repeated-dose inhalation toxicity (based upon 28-day study)
Negative for in vivo genotoxicity using both micronucleus and Comet assays	Not mutagenic in bacterial reverse mutation assays
Negative for reproductive and developmental toxicity	Expected to be negative for reproductive and developmental toxicity based upon read across from next nearest homologue

[1] Defined as a liquid under Globally Harmonized System based upon vapor pressure.

Considering the results from the full range of studies, the 3M Medical Department established occupational exposure limits (OEL) of 65 ppm and 225 ppmv (8-h time weighted averages) for Novec 4710 gas and Novec 5110 gas, respectively. Small releases of insulating gases can occur during filling, maintenance, and decommissioning operations when gas-tight connections are sealed and unsealed. However, airborne concentrations measured during gas transfer operations are normally less than 10 ppmv [5]. Workplace airborne SF_6 concentrations observed in indoor gas-insulated switchgear applications are typically below 1 ppmv [13]. As a result, the OELs stated above provide a sufficient margin of safety in these applications and the observed airborne concentrations of Novec Insulating Gases described above are well below the action level of $\frac{1}{2}$ the OEL as defined by US Occupational Safety and Health Administration (OSHA). On this basis, risk analyses have established that gas mixtures containing the Novec Insulating Gases are safe to handle in gas-filled equipment under all expected operational conditions [1,2].

Independent groups have also conducted toxicological tests [14–16] with Novec 4710 gas using non-OECD test protocols. Variation in test parameters such as the animal species, exposure time and the condition of the gas will provide significantly different results. As a result, OECD and international standards such as GHS have standardized hazard testing criteria, requiring test methods that are scientifically sound and validated according to international procedures in order to provide information relevant to a human health assessment while minimizing the need for animal testing. The results from tests conducted using non-standard protocols have led to some confusion over the toxicological profile for the Novec gases.

The data reported by Li and colleagues [14] for acute inhalation tests conducted in the rat over a 4-h time interval found an LC_{50} value of 15,000–20,000 ppmv, which is consistent with the LC_{50} value discussed above. Additional tests were conducted at high concentrations over a time interval of 24-h. Such an exceptionally long test period is far beyond the 4-h exposure required for acute inhalation testing that is used for GHS classification of a chemical and does not aid in performing a human health risk assessment. The alleged effects on various organs systems observed in the 24-h exposure were actually a result of pulmonary edema-induced hypoxia (insufficient oxygen reaching the internal organs) and not a direct response of the test material on these organs. The 3M-sponsored, 28-day inhalation toxicity study referenced above found the respiratory tract to be the target organ, exhibiting signs of an irritant-like effect, but no histopathological changes were noted in other organ systems. Overall, the results in the paper are consistent with the LC_{50} values published to date and do not contradict the recommended 65 ppmv occupational exposure limit.

Preve and colleagues [15] have repeatedly cited toxicological data developed outside of the recommended and validated testing protocols. The acute inhalation LC_{50} data used in their publications were derived using different animal models (mouse). The OECD protocols for acute inhalation toxicity (OECD 403, 433 and 436) all state that the preferred test species is the rat as it has been previously been demonstrated that mice are often more sensitive in acute inhalation studies than other mammals, a factor which complicates the use of data generated in mice for risk assessment purposes [17]. Similarly, the discussions in these papers regarding mutagenicity aspects appear to overlook both the available data on the Novec gases as well as the recommendations for the use of read-across techniques encouraged by regulatory bodies such as the European Chemicals Agency (ECHA). As a result, the data generated in those studies do not augment the information for a human health risk assessment.

Zhang and colleagues published the results from a series of inhalation toxicity studies conducted in the mouse [16]. As expected, the results demonstrated the higher sensitivity of the mouse in acute inhalation studies compared to the rat but again did not demonstrate any additional relevance for a human health risk assessment. While the authors stated that there is still much work to be conducted on the toxicity of C4 nitrile and a need for an occupational exposure level, this assessment clearly does not reflect the significant amount of data readily available on this material which includes GLP-conducted acute, sub-chronic, developmental and reproductive, and genetic toxicity studies. Based upon these studies, 3M has developed an occupational exposure limit of 65 ppm which is published on the 3M safety data sheets and product literature.

Additional considerations apply when handling any insulting gas after arcing events. In the case of electrical arcing in equipment containing SF_6, high-toxicity decomposition byproducts such as HF, S_2F_{10} and SO_2 can be generated. These byproducts are highly hazardous and pose a potential toxicity risk to those exposed. Depending on the nature of the arcing event, the Novec gas mixtures may also undergo some degree of decomposition. Even though testing demonstrated that arced Novec gas mixtures can be less hazardous than arced SF_6 mixtures [1,2], similar precautions should be taken when handling such gas mixtures. Employees performing maintenance procedures on electrical switches containing arced SF_6 are required to use proper handling procedures and wear personal protective equipment. Similar precautions should be taken with arced Novec gas mixtures.

4. Environmental Considerations

4.1. Global Warming Potentials

One metric for analyzing the potential environmental impact of SF_6 alternatives is a comparison of the global warming potential (GWP) for the gases used within the different technologies. The GWP is an index that provides a relative measure of the possible climate impact of a compound which acts as a greenhouse gas in the atmosphere. It effectively calculates the amount of energy absorbed by a compound over a period of time relative to that of a reference compound, CO_2. The GWP as defined by the Intergovernmental Panel on Climate Change (IPCC) [18] is calculated as the integrated radiative forcing due to the release of 1 kg of that compound relative to the warming due to 1 kg of CO_2 over the same time interval (the integration time horizon (*ITH*)), as shown in Equation (1):

$$\mathrm{GWP}_i = \frac{\int_0^{ITH} R_i C_{i_0} \exp\left(\frac{-t}{\tau_i}\right) dt}{\int_0^{ITH} R_{CO_2} C_{CO_2}(t) dt} \tag{1}$$

where R is the radiative forcing per unit mass of a compound (the change in the flux of radiation through the atmosphere due to the infrared (IR) absorbance of the compound), C is the atmospheric concentration of a compound, τ is the atmospheric lifetime of a compound, t is time and i is the compound of interest. The commonly accepted *ITH* is 100 years.

Only two variables in the GWP calculation are affected by the physical characteristics of the compound—the radiative forcing due to IR absorbance and the atmospheric lifetime. All fluorinated compounds absorb IR energy in the "window" at 8 to 12 μm which is largely transparent in the natural atmosphere. This IR absorbance, coupled with a long atmospheric lifetime, results in a high GWP for many perfluorinated compounds such as SF_6.

The most effective approach to producing a lower GWP alternative is to develop a compound with a significantly shorter atmospheric lifetime. For highly fluorinated compounds this means synthesizing a molecule containing functionality or structural features that allow it to decompose more readily in the natural atmosphere. This is precisely the approach that was taken with the Novec Insulating Gases. Novec 5110 gas incorporates a carbonyl group that undergoes direct photolysis when exposed to sunlight in the lower atmosphere leading to a GWP value of less than 1 [19]. Novec 4710 gas contains a nitrile group that reacts with hydroxyl radicals in a process similar to the degradation mechanism for most organic compounds that enter the lower atmosphere. Multiple studies have reported an atmospheric lifetime and GWP value for Novec 4710 gas. At first glance, these values may appear to vary considerably. However, as the review below demonstrates, the results are consistent within recognized experimental uncertainty.

The initial studies were performed in the 3M Environmental Laboratory to investigate the atmospheric lifetime of Novec 4710 gas. A series of experiments measured the rate of degradation for Novec 4710 gas due to reaction with hydroxyl radicals relative to methane or pentafluoroethane as a reference compound. Hydroxyl radicals were generated via photolysis of ozone in the presence of water vapor. Concentrations of the reactants were measured continuously by Fourier transform infrared spectroscopy (FTIR) using a 10-m pathlength within a 5.7 L gas cell maintained at 300 K. Additionally, gas samples were analyzed by gas chromatography with mass spectrometry during one of the experiments to confirm the concentrations of Novec 4710 gas. The average atmospheric lifetime calculated from four separate experiments was 30 years for Novec 4710 gas [20].

The radiative efficiency for Novec 4710 gas was calculated at 0.225 $Wm^{-2}ppbv^{-1}$ using the method of Pinnock et al. [21] with an IR cross-section measured using 0.5 cm^{-1} resolution. This radiative efficiency value takes into account the necessary stratospheric temperature adjustments and atmospheric lifetime corrections. The radiative efficiency combined with a 30-year lifetime results in a GWP of 2100 using the IPCC calculation method [18].

A study published by Sulbaek Andersen and colleagues conducted smog chamber experiments to investigate the atmospheric fate of Novec 4710 gas [22]. Experiments were performed within a 101 L photoreactor maintained at 296 K. Hydroxyl radicals were generated by photolysis of ozone in the presence of hydrogen gas. The atmospheric lifetime was determined from these experiments to be approximately 22 years. Combining this lifetime with the radiative efficiency they measured at 0.217 $Wm^{-2}ppbv^{-1}$ using an FTIR resolution of 0.25 cm^{-1} resulted in a GWP value reported as 1490. The lifetime reported in this study was calculated using the measured reaction rate constant and an average hydroxyl radical concentration in the atmosphere. For compounds considered to be well-mixed in the atmosphere (i.e., lifetimes more than a few months), it is more common to calculate the lifetime relative to a reference compound such as methyl chloroform since there is a comprehensive analysis of its abundance in the atmosphere as well as its rate of emission and removal. The atmospheric lifetime calculated from this method is 32 years, resulting in a GWP of 2090.

Another series of experiments were conducted by Blázquez and colleagues in which they examined the temperature dependence of the reaction of hydroxyl radical with Novec 4710 gas [23]. Hydroxyl radicals were produced by photolysis of HNO_3. Measurements were made from 278 to 358 K. A linear equation (in the form of the Arrhenius equation) was fit to these kinetic data. The atmospheric lifetime was reported as 47 years using kinetics extrapolated to 272 K. The radiative efficiency was measured in this study to be 0.279 $Wm^{-2}ppbv^{-1}$ using a 1 cm^{-1} spectral resolution. These data combined to report a GWP value of 3646. While the lower temperature for the kinetic calculations is more representative of the average tropospheric temperature, a comparison of values across all studies requires data to be compared from equivalent conditions. The kinetic data measured at 298 K in this study results in an atmospheric lifetime of 31 years. Calculation of the GWP using this lifetime and the above radiative efficiency produces a value of 2620.

While there is variability in the GWP values resulting from these independent studies, the values are well within the uncertainty reported by IPCC of ±35% [18] as shown in Figure 4. The average lifetime and GWP values from the 3 studies are 31 years and 2260, respectively, which agree well with the original values report by 3M. On this basis, 3M continues to report the lifetime and GWP values derived from their internal studies of 30 years and 2100, respectively.

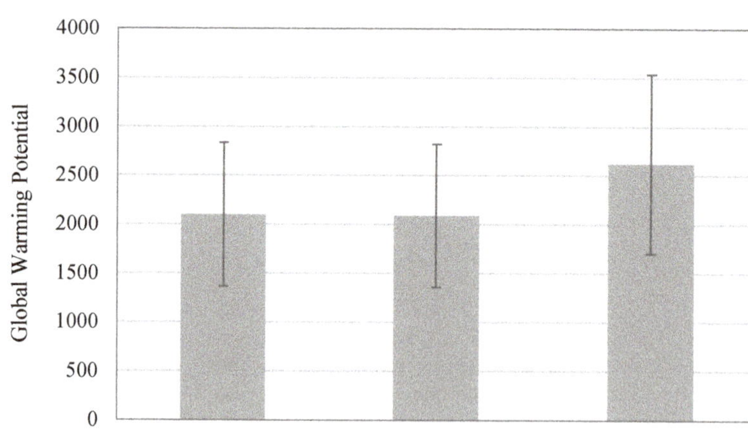

Figure 4. GWP values for Novec 4710 gas with uncertainty cited by IPCC.

The GWP for a gas mixture is calculated using the GWP value for each individual component multiplied by its weight fraction in the mixture according to Equation (2):

$$GWP_{mixture} = \sum_i x_i \, GWP_i \quad (2)$$

where x_i and GWP_i are the weight fraction and GWP of component i, respectively.

4.2. Greenhouse Gas Emissions

A comparison of GWP values for representative gas mixtures used as alternatives to SF_6 is shown in Table 4. However, this type of comparison only provides a partial assessment of the environmental impact from insulating gas technologies. The mass of gas released, even from the same volumetric leakage rate, can be significantly different due to the considerably different gas densities. Table 4 also shows the GHG emission reductions achieved by the alternative-gas mixtures are even lower than would have been apparent through a simple comparison of GWPs.

Table 4. Initial climate performance of alternative-gas mixtures compared to SF_6.

Gas Formulation (mole%)	100% SF_6	5% Novec 4710/95% CO_2	5% Novec 5110/95% Air
Pressure (bar)	4	6	6.5
GWP of gas mixture	23,500	398	<1
GWP reduction vs SF_6	—	98.3%	>99.9%
GHG content (kg CO_2e/m^3)	553,929	4969	3.5
GHG emission reduction from discrete emission relative to SF_6	—	99.1%	>99.9%

Another disadvantage to assessing the climate impact of gases solely through comparison of GWP values is the inherent limitations within the GWP calculation itself. It is important to note that the commonly recognized GWP for a substance is calculated over a 100-year ITH. This ITH is a compromise between shorter-term and longer-term effects [18]. However, this means that the full climate impact of a very long-lived gas, such as SF_6, is not fully accounted for in the GWP calculation. Figure 5 displays a plot of the quantity of gas remaining in the atmosphere following a 1 kg release. A compound such as Novec 4710 gas with an atmospheric lifetime of 30 years is expected to be essentially fully degraded within the GWP calculation timeframe. Contrast that with SF_6 which, due to its atmospheric lifetime of 3200 years, remains in the atmosphere far longer than the 100-year ITH. As a result, only a fraction of its potential impact on climate change is included in the GWP calculation.

Installations of gas-filled electric power equipment are expected to remain in use for decades with low level emissions occurring throughout this time due to leakage. Many regions require reporting of these GHG emissions on an annual basis, even though, as shown in Figure 5, a portion of the gas leaked in any year can remain in the atmosphere for far longer. An assessment of the cumulative GHG emissions would account for not only the mass of gas emitted annually but also the amount of the gas that accumulates in the environment during its use. Both factors can have a measurable influence on the overall climate impact of a technology.

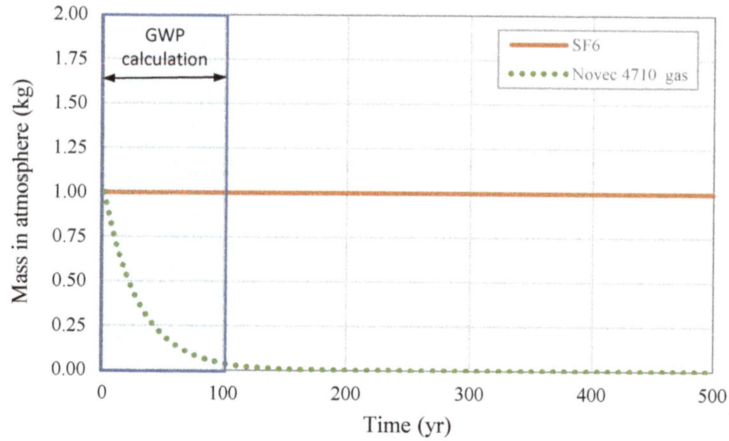

Figure 5. Residence time of insulating gas in the atmosphere, assuming 1 kg release of each compound at time zero.

Figure 6 compares the cumulative GHG emissions that would occur due to leakage of insulating gas over a 40-year lifetime of an installed base of gas-filled equipment. The comparison assumes volumetric emissions from the equipment equivalent to 1 T/year of SF_6 over that lifetime. The calculations are carried out for 100 years corresponding to the timeframe used in GWP assessments in order to illustrate the limitation of relying on the GWP parameter alone. Results for alternative-gas mixtures with GWPs of 398 and 1 are plotted along with SF_6. Comparison of the GWPs for these mixtures to SF_6 suggests that these alternatives represent a 98.3% and >99.9% improvement, respectively. Additionally, if the different gas densities are factored into the calculation, the reduction in GHG emissions improves to 99.1% and >99.9%, respectively, as shown in Table 4. However, the shorter atmospheric lifetimes of the alternative gases mean that both materials degrade much more rapidly over time compared to SF_6, preventing measurable accumulation of these alternatives in the environment. This limits the cumulative GHG emissions from the alternative-gas insulation technologies. When calculated over a 100-year timeframe both gas mixtures reduce GHG emissions by more than 99.9%, regardless of the GWP of the alternative-gas mixture.

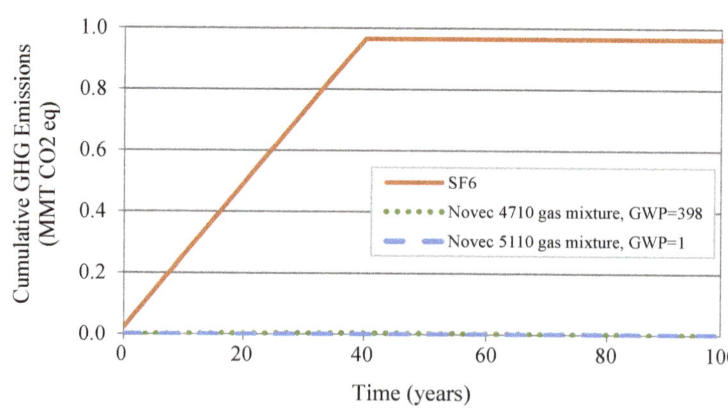

Figure 6. Cumulative greenhouse gas emissions, assuming emission equivalent to 1T/yr of SF_6 for 40 years of operation.

A lifecycle assessment (LCA) comparing the climate impacts of these alternative-gas technologies came to similar conclusions [24]. The analysis compared 145 kV GIS bays operating with the alternative-gas mixtures to identical equipment designed for SF_6 throughout the gas-use phases of the equipment lifecycle (filling, operation, decommissioning). The LCA demonstrated that the alternative-gas technologies result in large reductions of the carbon footprint of these applications with a climate impact that is negligible compared to SF_6, confirming the results of the GHG calculations shown above.

5. Conclusions

Gas mixtures containing a fluoroketone or a fluoronitrile, Novec™ 5110 Insulating Gas and Novec™ 4710 Insulating Gas, respectively, are being implemented as low climate-impact alternatives to SF_6. When used at higher pressure, these gas mixtures can deliver dielectric performance comparable to SF_6 in high voltage systems. The safety of Novec gases has been evaluated through a series of toxicological studies which demonstrate that the hazard profiles of the gas mixtures containing these materials are safe to handle in gas-filled equipment. Both alternative gases have significantly lower GWPs than SF_6. Moreover, their shorter atmospheric lifetimes prevent measurable accumulation of these gases in the atmosphere. This results in substantial reduction (>99.9%) in GHG emissions over the expected working life of equipment using these alternatives, irrespective of the GWP for the individual gas mixture components. As a result, these advanced materials enable insulation technologies that can make a meaningful contribution to reducing the environmental impact of high voltage applications. Therefore, limiting alternative-gas technologies based on GWP alone could be counterproductive to the goal of reducing the climate impact from electric power applications. In fact, the European Commission report in 2020 stated "In specific sites where the voltage rate must be maintained and space is restricted, such as substations at power plants or in urban areas, currently designs based on fluoronitriles may be the only viable alternative to SF_6 based switchgear" [25]. Gas insulated equipment containing Novec 4710 gas mixtures first started operating on the grid in 2017, while equipment containing Novec 5110 gas mixtures first started operating on the grid in 2015. More than 100 equipment bays containing alternative gas mixtures have now been installed by multiple utilities located primarily in Europe with recent installations in Asia and North America.

Author Contributions: Performance in dielectric applications, A.X., A.Z. and J.O.; safety considerations, J.B. and M.D.; environmental considerations, J.O. All authors have read and agreed to the published version of the manuscript.

Funding: This research was funded by 3M company and received no external funding.

Institutional Review Board Statement: The 3M-sponsored toxicological studies were approved by the laboratories' Institutional Animal Care and Use Committees and animal care complied with all applicable national and local regulations. All toxicological studies that followed OECD guidelines were performed under good laboratory practice conditions.

Informed Consent Statement: Not applicable.

Data Availability Statement: The data presented in this review are available from 3M or in the European Chemicals Agency database at https://echa.europa.eu/information-on-chemicals (accessed on 14 August 2021) under EC reference numbers 690-995-3 and 806-451-7.

Conflicts of Interest: The authors declare no conflict of interest.

References

1. Hyrenbach, M.; Paul, T.; Owens, J. Environmental and safety aspects of AirPlus insulated GIS. *CIRED Open Access Proc. J.* **2017**, *1*, 132–135. [CrossRef]
2. Kieffel, Y.; Biquez, F.; Vigouroux, D.; Ponchon, P.; Schlernitzauer, A.; Magous, R.; Cros, G.; Owens, J. Characteristics of g^3—An alternative to SF_6. *CIRED Open Access Proc. J.* **2017**, *1*, 54–57. [CrossRef]

3. Kieffel, Y.; Irwin, T.; Ponchon, P.; Owens, J. Green Gas to Replace SF_6 in Electrical Grids. *IEEE Power Energy Mag.* **2016**, *14*, 32–39. [CrossRef]
4. Cigré. *Technical Brochure 802—Application of Non-SF6 Gases or Gas-Mixtures in Medium and High Voltage Gas-Insulated Switchgear*; Cigré: Paris, France, 2020.
5. Xiao, A.; Owens, J.; Bonk, J.; Zhang, A.; Wang, C.; Tu, Y. Environmentally Friendly Insulating Gases as SF_6 Alternatives for Power Utilities. In Proceedings of the ICEMPE 2019—2nd International Conference on Electrical Materials and Power Equipment, Guangzhou, China, 7–10 April 2019; pp. 42–48.
6. Wooton, R.E.; Kegelman, M.R. *Gases Superior to SF_6 for Insulation and Interruption*; Report EL-2620; Electric Power Research Institute (EPRI): Washington, DC, USA, 1982.
7. Pohlink, K.; Meyer, F.; Kieffel, Y.; Biquez, F.; Ponchon, P.; Owens, J.; Van San, R. *Characteristics of Fluoronitrile/CO_2 Mixture—An Alternative to SF_6*; Cigré: Paris, France, 2016; Paper D1-204.
8. Stoller, P.; Hengstler, J.; Doiron, C.; Scheel, S.; Simka, P.; Müller, P. *Environmental Aspects of High Voltage Gas Insulated Switchgear That Uses Alternatives to SFf_6 and Monitoring and Long-Term Performance of a Pilot Installation*; Cigré: Paris, France, 2018; Paper D1-202.
9. Laruelle, E.; Maksoud, L.; Kieffel, Y.; Lüscher, R.; Ficheux, A. *SF_6 Alternative—What to Learn from the High Voltage Experience*; Cigré: Madrid, Spain, 2019; Paper 0028.
10. *Material Toxicity Summary Sheet, 3M™ Novec™ 4710 Insulating Gas*; 3M Company: St. Paul, MN, USA, 2019.
11. *Material Toxicity Summary Sheet, 3M™ Novec™ 5110 Insulating Gas*; 3M Company: St. Paul, MN, USA, 2019.
12. Ramboll US Consulting, Inc. Global Product Safety and Stewardship Practice, "CMR Self Classification". Available online: https://bit.ly/2LQHkEO (accessed on 14 August 2021).
13. Castonguay, J. In-situ measurements of SF_6 leak rates in indoor gas-insulated switchgears (GIS). In *Gaseous Dielectrics IX*; Christophorou, L., Olthoff, J., Eds.; Kluwer Academic/Plenum Publishers: New York, NY, USA, 2001; pp. 549–554.
14. Li, Y.; Zhang, X.; Zhang, J.; Xiao, S.; Xie, B.; Chen, D.; Gao, Y.; Tang, J. Assessment on the toxicity and application risk of C4F7N: A new SF_6 alternative gas. *J. Hazard. Mater.* **2019**, *368*, 653–660. [CrossRef] [PubMed]
15. Preve, C.; Maladen, R.; Piccoz, D. *Innovative SF_6 Free Load Break Switch with Shunt Vacuum Interruption (SVI) Technology*; Cigré: Paris, France, 2020; Paper A3-116.
16. Zhang, X.; Fanchao, Y.; Li, Y.; Tian, S.; Xie, B.; Gao, Y.; Xiao, S. Acute toxicity and health effect of perfluoroisobutyronitrile on mice: A promising substitute gas-insulating medium to SF_6. *J. Environ. Sci. Health Part A* **2020**, *14*, 1646–1658. [CrossRef]
17. Ten Berge, W.F.; Zwart, A.; Appelman, L.M. Concentration-time mortality response relationship of irritant and systemically acting vapors and gases. *J. Hazard. Mater.* **1986**, *13*, 301–309. [CrossRef]
18. IPCC. *Climate Change 2013: The Physical Science Basis. Contribution of Working Group I to the Fifth Assessment Report of the Intergovernmental Panel on Climate Change*; Stocker, T.F., Qin, D., Plattner, G.-K., Tignor, M., Allen, S.K., Boschung, J., Nauels, A., Xia, Y., Bex, V., Midgley, P.M., Eds.; Cambridge University Press: Cambridge, UK; New York, NY, USA, 2013.
19. Ren, Y.; Bernard, F.; Daële, V.; Mellouki, A. Atmospheric fate and impact of perfluorinated butanone and pentanone. *Environ. Sci. Technol.* **2019**, *53*, 8862–8871. [CrossRef]
20. 3M Company. *Internal Report No. E11-0512*; 3M Environmental Laboratory: St. Paul, MN, USA, 2015.
21. Pinnock, S.; Hurley, M.D.; Shine, K.P.; Wallington, T.J.; Smyth, T.J. Radiative forcing of climate by hydrochlorofluorocarbons and hydrofluorocarbons. *J. Geophys. Res.* **1995**, *100*, 23227–23238. [CrossRef]
22. Sulbaek Andsersen, M.; Kyte, M.; Andersen, S.T.; Neilsen, C.J.; Nielsen, O.J. Atmospheric chemistry of $(CF_3)_2CFCN$: A replacement compound for the most potent industrial greenhouse gas, SF_6. *Environ. Sci. Technol.* **2017**, *51*, 1321–1329. [CrossRef] [PubMed]
23. Blázquez, S.; Antiñolo, M.; Nielsen, O.J.; Albaladejo, J.; Jiménez, E. Reaction kinetics of $(CF_3)_2CFCN$ with OH radicals as a function of temperature (278–358 K): A good replacement for greenhouse SF_6. *Chem. Phys. Lett.* **2017**, *687*, 297–302. [CrossRef]
24. Billen, P.; Maes, B.; Larraín, M.; Braet, J. Replacing SF_6 in Electrical Gas-Insulated Switchgear: Technological Alternatives and Potential Life Cycle Greenhouse Gas Savings in an EU-28 perspective. *Energies* **2020**, *13*, 1807. [CrossRef]
25. European Commission Report C(2020) 6635. REPORT FROM THE COMMISSION: Assessing the Availability of Alternatives to Fluorinated Greenhouse Gases in Switchgear and Related Equipment, Including Medium-Voltage Secondary Switchgear. Available online: https://ec.europa.eu/clima/sites/clima/files/news/docs/c_2020_6635_en.pdf (accessed on 14 August 2021).

MDPI
St. Alban-Anlage 66
4052 Basel
Switzerland
Tel. +41 61 683 77 34
Fax +41 61 302 89 18
www.mdpi.com

Energies Editorial Office
E-mail: energies@mdpi.com
www.mdpi.com/journal/energies